Met. O. 895

METEOROLOGICAL OFFICE

METEOROLOGY FOR MARINERS

THIRD EDITION

LONDON: HMSO

First published 1957
Third edition 1978
Sixth impression 1992

ISBN 0 11 400311 4

CONTENTS

PART VI. ICE AND EXCHANGE OF ENERGY BETWEEN SEA AND ATMOSPHERE

LIST OF ILLUSTRATIONS

FOREWORD

THE aim of this book is to present the elementary theory of meteorology in a way that is suitable for Merchant Navy Officers and to explain its practical application to safe and economic ship operations. Much of the information in this Third Edition was contained in the Second Edition but to keep abreast of the progress in the developing science of meteorology it has now been brought up to date. The chapters on ocean surface currents and sea ice are included because of their bearing on safety of navigation and the influence warm and cold currents have on weather and climate. A full description of ships' weather-reporting codes and the use of the International Analysis Code has not been included as a code change may be introduced during the next decade, prior to the publication of the Fourth Edition, and the relevant information is published in a Meteorological Office booklet *Ships' Code and Decode Book*, 9th Edition 1977.

We are advised by oceanographers that oceanography has progressed to such a state that an introduction to or an outline of the subject can no longer be adequately described in a few chapters, and Part VII of the Second Edition is now too out of date to be reprinted. It is, therefore, with some reluctance that we have been obliged to accept that Part VII can no longer be included in this book but hope that mariners in search of further knowledge will not neglect this interesting subject.

When the previous edition of this book was published reference was made to ship routeing. Experience has subsequently shown that with the improvement in weather forecasting and the very high service speeds of ships operating on the North Atlantic and North Pacific very considerable economies can be found if navigators pay greater regard to the weather conditions likely to prevail whilst they are on passage. It is frequently possible in ocean navigation for ships to avoid deep depressions, areas of rough seas and heavy swells, thereby making better passage time, to save fuel and reduce both ship and cargo damage. Chapter 13 now contains more detail on the subject of ship routeing in order to assist those who endeavour to navigate with financial economy in mind.

It will be noted that little reference is made in this book to meteorological instruments or to the making of meteorological observations at sea. The companion book, the *Marine Observer's Handbook*, deals in more detail with these aspects of marine meteorology, and also with such subjects as ocean waves, meteorological, astronomical and optical phenomena and ice nomenclature, besides including a comprehensive collection of cloud photographs. This book is on sale at Government Bookshops.

Many of the diagrams contained in this book are based upon observations made by observers aboard British Voluntary Observing Ships. It is thus in some measure a tribute to the work of many thousands of Voluntary Marine Observers.

Although there is no substitute for meteorological surface observations received from Voluntary Observing Ships, meteorological satellites now make a valuable contribution to the global observing systems and mariners benefit from data obtained from these satellites in that tropical cyclones can no longer remain undetected. To predict the movement of tropical storms still remains a major problem but a knowledge of their position and history can assist the shipmaster in deciding on his future course of action and advice on this difficult and important matter has been revised and is contained in Chapter 11.

The contents of this book should satisfactorily cover the meteorological syllabuses for Department of Trade examinations for certificates of competency for Deck Officers in the Merchant Navy. It may also serve as a useful book of reference for those who have completed their courses of study and wish to maintain a good working knowledge of meteorology.

Mr B. F. Bulmer, B.Sc., who served for many years in the Marine Division and Marine Climatological Section of the Meteorological Office, accepted the task of revising this book and we wish to acknowledge gratefully his efforts in so ably completing the work.

Marine Division,
Meteorological Office.

PART I. THE METEOROLOGICAL ELEMENTS

CHAPTER I

THE ATMOSPHERE, ITS CONSTITUTION AND PHYSICAL PROPERTIES

Composition of the Atmosphere

The atmosphere surrounds the whole surface of the earth, both land and sea. Just as life in the depth of the ocean is subjected to a pressure owing to the weight of the sea above, so all of us, whether on land or sea, are living at the bottom of an ocean of air and are subjected to a pressure exerted by the weight of air above us. At sea level this pressure is about 1 kilogram per square centimetre. If we ascend a mountain or go up in an aeroplane, the pressure we experience at the new level is reduced, since the weight of air below our new level no longer contributes to the pressure exerted on us. Our bodies are adapted to the sea-level pressure of the atmosphere so that we are not conscious of it, but when we make a rapid ascent as in an aeroplane the reduced pressure affects the eardrums and renders breathing difficult owing to the reduction in the quantity of oxygen available in a given volume of air. An increase of elevation from sea level to 1 kilometre will decrease the pressure by about 100 grams per square centimetre, but the rate of decrease becomes less at higher levels. Roughly half the mass of atmosphere lies below the level of the summit of Mt Blanc (4820 metres) and about two-thirds below the summit of Mt Everest (8840 metres). Because the rate of decrease of density with height becomes progressively less the atmosphere becomes more and more tenuous at great heights and has no definite upper limit. Even at 130–160 kilometres above the earth the air is still sufficiently dense for meteors to become white-hot and so visible when entering the atmosphere during darkness. The meteor is raised to white heat by the very rapid compression of the air ahead of it. This sudden compression is called an adiabatic compression; that is, it takes place without loss or gain of heat from outside sources. Adiabatic processes are more fully explained in a later paragraph of this chapter.

The chemical composition of dry air is remarkably constant everywhere over the earth's surface and up to a height of at least 19 kilometres. Chemical analysis shows that the amount of each gas expressed as a percentage by volume of the total is as follows:

Gas	*Volume %*
NITROGEN	78·09
OXYGEN	20·95
ARGON	0·93
CARBON DIOXIDE	0·03

There are also very small amounts of NEON, HELIUM, KRYPTON, OZONE and other gases present.

While some of these gases play a vital part in the maintenance of plant and animal life, the exact composition of dry air is unimportant to the meteorologist, who is chiefly interested in the amount of water vapour with which it is associated. This is because such variations as normally occur in the chemical constitution of dry air have little effect upon its meteorological properties, whereas changes in the water-vapour content of air play a vital role in meteorological processes. The amounts of water vapour in the air vary from very small amounts up to a maximum value which is determined by the temperature of the air. This maximum (or saturation) value increases with increasing temperature. At temperatures below freezing the weight of water vapour which can be held is very small, while at high temperatures it can amount to 4% by weight of the air and water-vapour mixture. Water vapour is continuously being added to the atmosphere by evaporation from the earth's surface, particularly from oceans, lakes and rivers, and is continuously being removed from the atmosphere by condensation, resulting in precipitation in various forms, mainly rain and snow.

Vertical Structure of the Atmosphere

Figure 1.1 illustrates the vertical structure of the atmosphere but is drawn to scale only up to a height of 40 kilometres. If a uniform scale was used throughout, the important lower levels would be too overcrowded for the necessary detail to be shown clearly. The lowest region is called the TROPOSPHERE. Besides containing the greater mass of air and almost all the water vapour, the most important feature of this region is the decrease of temperature with height which persists through nearly its whole thickness. The average rate of decrease is about 0·6°C per 100 metres but considerable variations from this value are common. Sometimes shallow layers are found through which temperature increases with height. Apart from a few exceptions which are dealt with later, the ordinary features of weather, including most forms of cloud and storms of all kinds, develop and expend their energies within the troposphere, the thickness of which can vary between about 9 kilometres and 16 kilometres. At a level which is taken to be the upper limit of the region this rapid fall of temperature ceases quite abruptly. For a considerable height above this level the rate of change of temperature with height is small, and the region is described as the STRATOSPHERE. The layer of transition, which may often be quite a sharp boundary between the troposphere and the stratosphere, is known as the TROPOPAUSE. The height of the tropopause varies from about 9 kilometres at the poles to 16 kilometres at the equator, and a curious result of this variation in height is that the tropopause is colder above the tropics than at the poles. Over the equator the temperature at the tropopause is about −76°C; in temperate latitudes and polar regions it is on the average about −54°C. An increasing amount of flying is now being done in the lower stratosphere, partly because jet engines work more efficiently in the thin air at those levels, and also because this region is largely free from the hazards of convection clouds and the icing and turbulence associated with them. (*See* page 11, Table 1.1.)

Higher in the stratosphere a small quantity of ozone is found, mainly in a layer between about 20 and 40 kilometres above the earth's surface. This ozone layer is important because ozone strongly absorbs radiation of certain wavelengths emitted by the sun, chiefly those in the ultra-violet region of the spectrum. If this ozone layer was not present an excessive amount of ultra-violet radiation would reach the earth with harmful effects upon many forms of life.

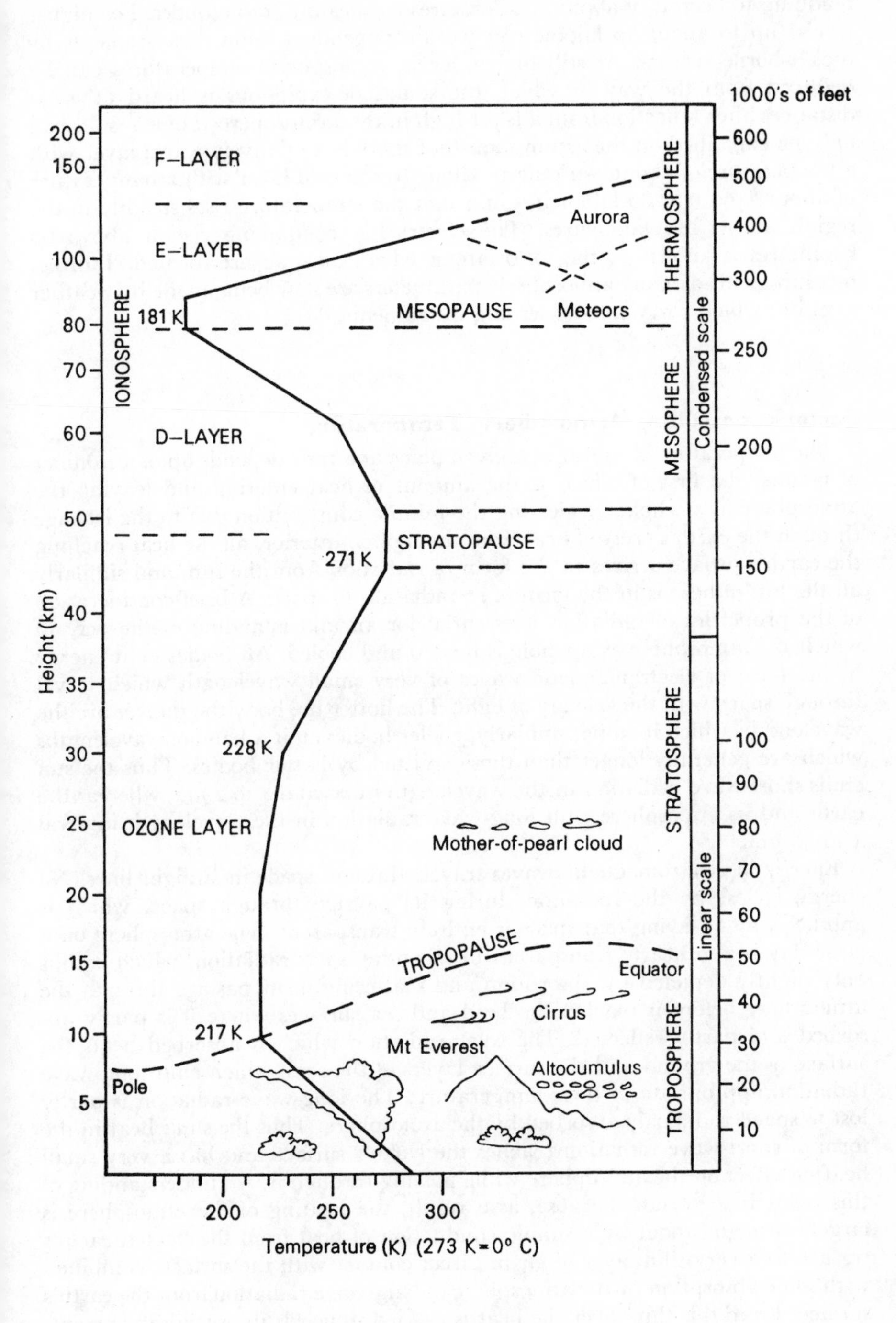

Figure 1.1. The vertical structure of the atmosphere

Direct measurements of temperature, humidity and winds can at present be made up to heights of about 40 kilometres by means of radiosondes. For higher levels (up to about 70 kilometres) we are dependent upon measurements by rocket-borne sensors. At still higher levels, atmospheric temperatures can be deduced from the way in which the sound of explosions is heard at great distances after reflection from a layer high in the stratosphere. These results can only be explained on the assumption that there is a relatively warm layer with a temperature of about −2°C at 48 kilometres, a cold layer with a temperature of about −90°C at 80 kilometres and that the temperature rises steadily in the region above 100 kilometres. The electrically conducting region above 60 kilometres is known as the IONOSPHERE where aurorae are located. Indirect measurements of many atmospheric parameters are now being made by weather satellites which carry remote-sensing instruments.

Factors controlling Atmospheric Temperature

The temperature of the air at a given place and time depends upon a number of factors, the first of which is the amount of heat entering and leaving the atmosphere as a whole. Neglecting the minute contribution due to the leakage through the earth's crust of heat from the molten interior, all the heat reaching the earth's surface arrives in the form of radiation from the sun, and similarly all the loss of heat is in the form of re-radiation to space. A brief consideration of the properties of radiation is essential for an understanding of the way in which the atmosphere as a whole is heated and cooled. All bodies emit energy in the form of electromagnetic waves of very small wavelength which travel through space with the velocity of light. The hotter the body the shorter are the wavelengths which it emits; similarly, cooler bodies emit a band of wavelengths which are generally longer than those emitted by hotter bodies. Thus the sun emits short-wave radiation in the wavelength interval 0·4 to 2 μm, whereas the earth and its atmosphere emit long-wave radiation in the wavelength interval 4 to 50 μm.

Energy in electromagnetic waves travels through space in straight lines. No energy is lost by the radiation during its passage through space, which is another way of saying that space is entirely transparent. The atmosphere on a clear day is also nearly transparent to incoming solar radiation, which is thus only slightly depleted by absorption and scattering in its passage through the atmosphere before it reaches the land and sea surfaces where it is partly absorbed and partly reflected. The solar radiation which is absorbed heats the surface of the ground and the surface layers of the seas, which emit long-wave radiation appropriate to their temperature. The long-wave radiation is partly lost to space and partly absorbed by the atmosphere. Thus the sun's heat in the form of short-wave radiation reaches the earth's surface and has a very small heating effect on the atmosphere while passing through it. An understanding of this point is important because, as a result, the heating of the atmosphere is largely brought about by a simple conduction of heat from the heated earth's surface to a very thin layer of air in direct contact with the surface, combined with some absorption by the atmosphere of long-wave radiation from the earth's surface. From this thin layer the heat is carried upwards in ascending currents and eddies even to the top of the troposphere by the processes of convection and turbulence which will be described later.

Water present in the atmosphere as vapour or as cloud modifies the above process in an important way, since, while it absorbs only a small fraction of the incoming short-wave solar radiation, it strongly absorbs the outgoing long-wave radiation.

On cloudy nights this outgoing radiation is absorbed by cloud sheets which also emit radiation in all directions. The portion emitted downwards represents a gain of radiation to the ground and the atmosphere below the clouds, and consequently the fall of air temperature between the clouds and the surface is less than the fall on a clear night. Even on a clear night some of the long-wave radiation is absorbed by water vapour in the atmosphere and similarly re-emitted in all directions. The downward travelling portion helps to reduce the fall of air temperature in the same way. In both these cases the greatest reduction in the fall of temperature occurs close to the ground. This absorption of radiation by water vapour explains why the fall of temperature on clear nights in deserts and continental interiors in winter is so much greater than in localities having a maritime climate (e.g. western Europe), where the amount of water vapour held in air in which no cloud is present is often quite large. On cloudy days the temperature does not rise as high as on clear days, mainly because the upper surfaces of the clouds reflect back into space a large proportion (amounting to between 56 and 81% in the case of dense clouds) of the sun's radiation which reaches them.

Since the mean temperature of the atmosphere as a whole does not vary appreciably from year to year, it can be concluded that the total energy of solar radiation absorbed by the earth and its atmosphere is balanced by the total energy of radiation emitted to space from the atmosphere.

Figure 1.2 shows diagrammatically, for a summer midday, some of the processes involved in the exchange of heat in the form of short-wave radiation between the sun and the earth and long-wave radiation between the earth and space. The width of the beams also gives an approximate indication of the relative amounts of energy involved in the processes of transmission, scattering, reflection and absorption of solar and terrestrial radiation under the various conditions shown. The beams showing long-wave radiation refer to the net amount lost to space, that is, to the difference between the long-wave radiation from the earth's surface and the back radiation from the atmosphere. Scattering, as its name suggests, is the term used to describe the effect of throwing out of the energy of solar radiation in all directions, produced by the molecules of the various gases in the atmosphere and by other suspended particles which are also very small compared with the wavelength of the short-wave radiation. Absorption of solar energy produces a corresponding rise of temperature in the absorbing substance; this is equally true for a gas as well as for liquids and solids.

Temperature Variation over the Earth's Surface

The primary cause of temperature differences over the earth's surface is the variation in the sun's altitude. If this was the only cause we might expect the highest temperatures at the equator and the lowest at the poles. In fact the distribution of temperature everywhere over the earth is controlled by several factors, such as the distribution of land and sea, the amount of cloud cover throughout the year and the temperature and extent of ocean currents. Dynamical factors such as the rotation of the earth on its axis, which has an important effect on the pressure distribution and the resulting system of winds,

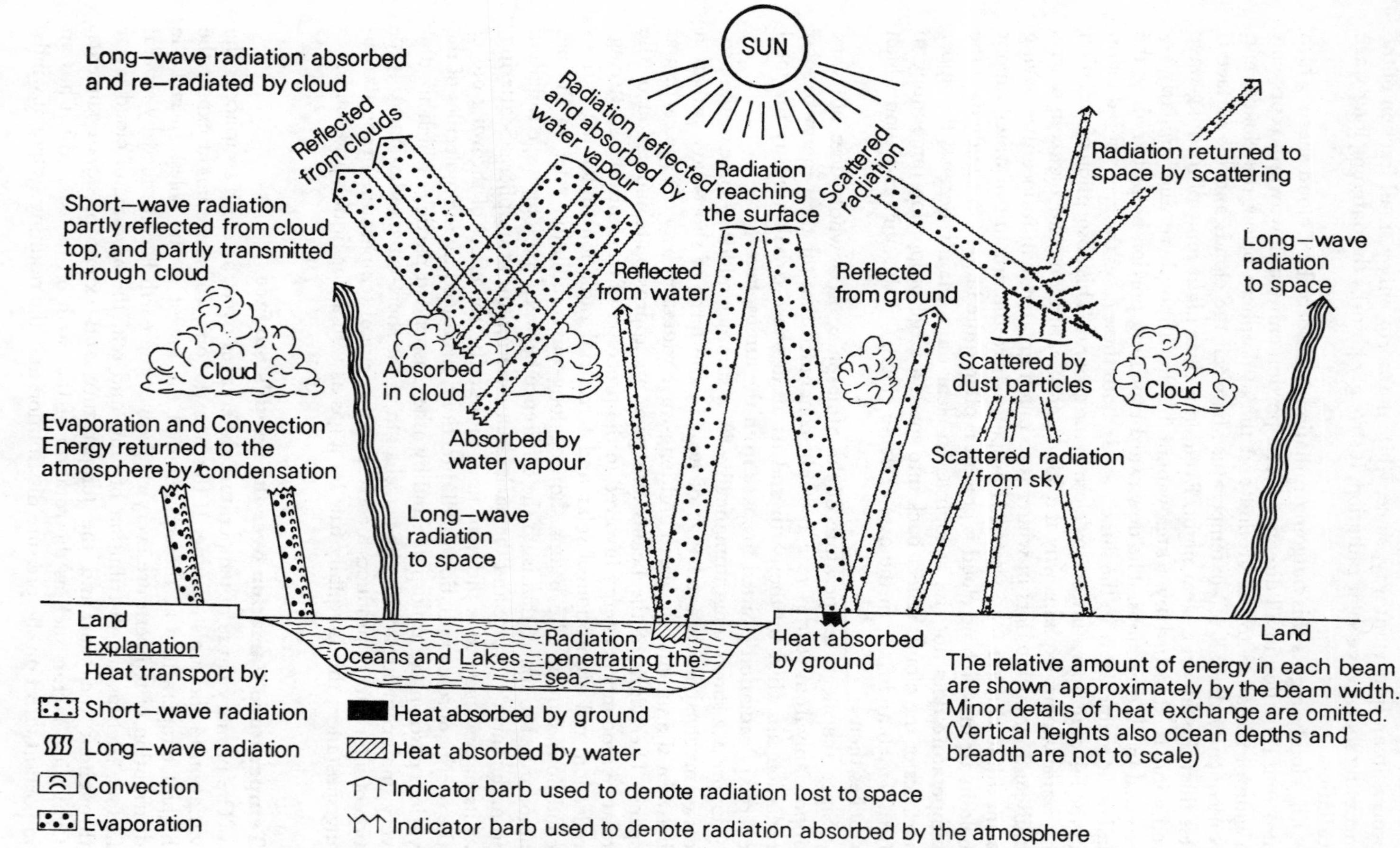

Figure 1.2. Exchange of heat through the atmosphere and at the earth's surface

also play their part, while the variations of solar elevation and length of daylight through the seasons are everywhere responsible for the seasonal changes of temperature. The result is that the highest temperatures are found not at the equator, but over land which is covered by the sub-tropical high-pressure belts in summer. The lowest temperatures occur in winter in the interior of the great land masses. Where, as in Antarctica, a land mass occupies the polar latitudes, the lowest temperatures are located comparatively near to the pole. In the northern hemisphere, in contrast, because the land masses are remote from the pole, the lowest temperatures occur in the interior of the northernmost land masses, namely in Siberia and North Canada. Near the North Pole, as far as is known, the temperatures are less extreme. The available data suggest that the interior of Antarctica in winter is the coldest place on earth (*see* page 97).

Isotherms (Greek *isos*, equal; *therme*, heat)

Isotherms are lines joining places having the same temperature. The distribution of mean sea temperatures, for January and July, over the surface of the oceans is shown by isotherms in Figures 8.1 and 8.2, while the corresponding isotherms for mean air temperature are given in Figures 8.3 and 8.4.

Water Vapour in the Atmosphere

Evaporation, or the escape of water vapour from the surface of water, snow or ice into the air above, goes on continuously all over the earth, so that the lower levels of the atmosphere always contain appreciable amounts of water vapour, especially over the warmer parts of the oceans. The amount of water vapour which the air can hold is limited solely by the temperature of the air. At each temperature there is a definite maximum value to the amount of water vapour which the air can contain as vapour.

The same facts can be described by saying that for every temperature there is a definite maximum value to the pressure which can be exerted by the water vapour, and that water vapour can only be retained as long as it does not exert a pressure in excess of this value. Any excess is condensed, i.e. returned to the liquid or solid form, usually first as cloud, but ultimately as precipitation which returns to the earth's surface in the form of rain, snow, hail, etc.

Just as, over a long period of years, temperatures over the world as a whole appear to have altered little, so the evidence available suggests that there has been little, if any, sustained change in the annual amounts of cloud averaged over the whole earth through a long period of years. Therefore total evaporation in, say, a year is roughly equal to the total precipitation over the world in the same year.

The amount of water vapour in the air is important in many meteorological questions, besides directly affecting human comfort. One way of describing it would be as the weight of water vapour contained in a specified quantity of air. Figure 1.3 shows how the maximum vapour concentration (expressed in grams of vapour per cubic metre) varies with temperature. It is important to note that the maximum amount of moisture that the air can hold increases more rapidly with temperature as the temperature increases. Between 0°C and 30°C the capacity of air to hold water is roughly doubled with each 10°C rise in temperature.

The saturation vapour pressure which is another measure of the maximum water content of air shows a similar variation with temperature, characteristic values being: at 0°C 6·1 mb, at 10°C 12·3 mb, at 20°C 23·4 mb, while at 100°C the pressure becomes equal to the normal atmospheric pressure and boiling occurs. (When atmospheric pressure is less than normal, boiling will occur at a lower temperature.)

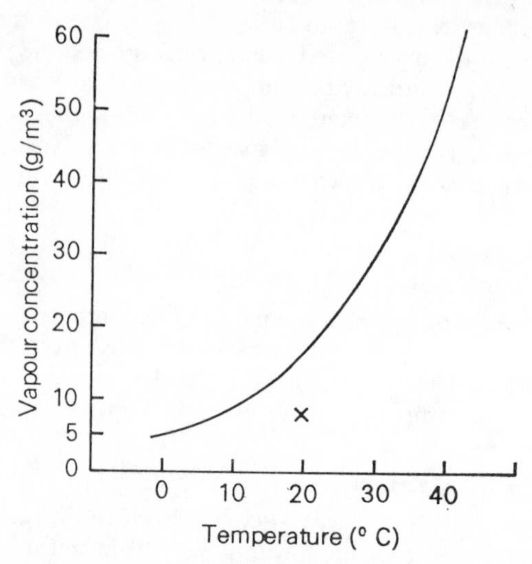

Figure 1.3. Values of vapour concentration at saturation

Relative Humidity, Dew-point and Frost-point

A mass of air at a certain temperature can thus contain only a given maximum amount of water vapour. When this maximum figure is reached the air is said to be saturated. More precisely it is the space occupied by the air which is saturated when the water vapour contained in it exerts the maximum pressure possible at that temperature without condensation taking place. In general the air at a given locality contains rather less water vapour than is needed for saturation. When a meteorological observer finds that the air is holding half the quantity of water vapour needed to make it saturated at that temperature, he describes this fact by saying that the relative humidity of this air is 50%. In the same way, when his observations show him that the air is saturated he can express the fact by saying that the relative humidity is 100%, i.e. the ratio (expressed as a percentage) of the actual amount of vapour held to that which the air could hold at saturation in this case is 100. Point X in Figure 1.3 represents air at a temperature of 20°C and relative humidity 50%.

Apart from a feeling of dampness there is sometimes little to distinguish air of 100% relative humidity from air having a slightly lower relative humidity. However, this is only true if the temperature of this air is kept unchanged. If the air is saturated, a very small fall of temperature is sufficient to cause condensation. If the air is cooled by coming into contact with a colder ground surface the result will be the formation of dew. If the air near the ground is being cooled, mist or fog will result, while if the air which is being cooled is at a higher level cloud will result. The temperature at which condensation to water droplets occurs is called the DEW-POINT (*see* the dew-point Table on page 18). If the dew-point is above freezing, condensation will be in the form of water;

if it is below freezing then moisture will be deposited on cold ground directly in the form of ice crystals (hoar frost).

The vapour pressure over a plane ice surface is less than that over a plane water surface at the same temperature; this is because the molecules are more tightly bound in the solid state. It follows that if the vapour pressure, *e*, is below 6·11 mb (the saturation vapour pressure at 0°C) it is possible to find two temperatures, one the dew-point, at which the saturation vapour pressure over *water* is equal to *e*, and the other, called the FROST-POINT, at which the saturation vapour pressure over ice is equal to *e*. When air is just saturated at freezing point the dew-point and the frost-point have the same value, namely 0°C; but at lower temperatures the dew-point falls below the frost-point by an amount which increases as the temperature falls. For example, when the vapour pressure is 2·00 mb the dew-point is −14·4°C and the frost-point nearly −12·8°C.

Formation of water droplets in the free atmosphere cannot take place in the absence of minute particles, known as nuclei. Such nuclei are numerous in the atmosphere though they vary greatly in numbers and origin. Salt particles, usually derived from sea spray, and minute particles of dust or sand, originating from deserts and from the products of urban areas, are the principal sources of nuclei. At levels in the atmosphere, where the temperature is below freezing, condensation occurs upon such nuclei in the form of supercooled water droplets. Whether it is also possible for ice crystals to form in the free atmosphere by direct condensation of water vapour on suitable nuclei or whether they always result from the freezing of liquid drops is still a controversial matter. It is a fact, however, that at temperatures down to about −10°C nearly all cloud particles are liquid droplets and that below about −40°C they usually consist of ice crystals. Between these two temperatures many clouds are mixed, i.e. they contain both liquid water and ice, but as the temperature falls ice crystals tend to predominate.

These facts are of great importance to the meteorologist concerned with cloud physics and the mechanism of precipitation. They may also be of significance to an airman because knowledge of the vertical distribution of temperature and of dew-point or frost-point is essential for forecasting cloud or (in wartime) the likelihood of condensation trails, while the composition of clouds has an important bearing on aircraft icing. Knowledge about frost-point and about supercooled water droplets in the upper air is admittedly of little practical importance to the mariner; reference is made to them here in order to 'complete the picture' of the role of water vapour in the atmosphere.

Relative humidity varies widely in the lower levels of the atmosphere. In very dry air over desert regions values as low as 5 to 10% commonly occur, but over the oceans, where evaporation into the surface layers continuously occurs, such low values are uncommon and the air is often not far from saturation. At two positions in the Western Approaches to the British Isles (namely lat. 59°N, long. 19°W and lat. 52½°N, long. 20°W) the mean annual value of the relative humidity is 89%; the corresponding figure for Kew Observatory in south-east England is 79%.

Hygrometers* (Greek *hugros*, moist)

Instruments for measuring humidity are called hygrometers. Relative humidity can be measured in several ways, but only two methods are in

* *See* also *Marine Observer's Handbook,* Chapter 2.

common use. The simpler and more reliable method employs a thermometer, the bulb of which is wrapped around by a piece of muslin which is kept moist by some simple device, usually by cotton threads which are tied round the muslin and run to a container holding distilled water. This thermometer is known as the WET-BULB THERMOMETER. Simultaneous readings of an ordinary thermometer (known as the DRY-BULB THERMOMETER), together with the wet-bulb thermometer, can be used to determine the relative humidity by means of tables. This is because there is a real physical relationship between the difference of the readings, dry-bulb minus wet-bulb, and relative humidity. In dry air the evaporation is large and results in a large cooling of the surface of the wet bulb, so that the difference between the dry- and wet-bulb readings is large. Correspondingly in damp air near saturation the difference between the dry- and wet-bulb readings is small, and decreases to nothing when the air is saturated and no evaporation is occurring at the wet bulb. The other variety of these instruments in common use depends upon a property of human and animal hairs, by which they are able to absorb moisture and vary in length with the dryness or dampness of the atmosphere. The change in length of a bundle of hairs can be magnified through a system of levers and communicated to a pen and arm, which make a record on a chart and thus enables a continuous record of the changes of relative humidity to be kept. In practice the readings of these instruments frequently have to be checked against those of dry- and wet-bulb thermometers, to guard against errors caused by changes in the properties of the hair and friction in the bearings.

Diurnal Variation of Temperatures

From sunrise until the sun is on the meridian, the rate at which any place on the earth's surface receives heat continuously increases with the elevation of the sun, in the absence of any change in the transparency of the atmosphere. Some increase is also due to the progressively shorter path of atmosphere traversed. These increases are liable to local and temporary interruption due to changing cloud conditions or to drifting dust or other pollution. There is a corresponding decrease in radiation received from the sun from noon to sunset. All this time the earth is giving out heat by radiation, and this loss goes on increasing as long as the surface temperature is rising. The increase in temperature by day results from an excess of incoming solar radiation over outgoing terrestrial radiation. Under normal conditions the maximum temperature must be reached at sometime after local noon, when a balance is achieved between the incoming and outgoing streams of radiation. In the British Isles, the maximum temperature usually occurs about 1400 (local time) in winter and 1500 in summer.

After sunset the earth loses heat by radiation less rapidly than by day because the temperature is lower. Both by day and night the earth loses more energy (about 50% more) in the form of long-wave radiation than it gains by back radiation from the atmosphere. In the early part of the day the incoming short-wave solar radiation more than compensates for this loss. After midday the short-wave solar radiation decreases and ceases at nightfall. There is nothing to counteract the loss of heat by long-wave radiation from the earth and so the earth's surface temperature falls continuously until a minimum is reached when the incoming radiation equals the outgoing radiation just after sunrise.*

---

* *See* page 5 about the effect of cloud on outgoing radiation.

The short-wave radiation from the sun which strikes the surface of the land is absorbed in a very small depth of earth (*see* page 4), and consequently produces a large diurnal range of temperature, which is partly shared by the lowest layers of air near the surface. The sun's radiation which reaches the surface of the sea penetrates it to a considerable depth, with the result that diurnal range of sea temperature and of the air in contact with it is very small even in the tropics.

Diurnal Variation of Relative Humidity

As we have already explained, if we increase the temperature of a mass of air in which the moisture content remains constant, the relative humidity of that air will decrease. Thus, even when the quantity of water vapour in the air at a locality does not change through the 24 hours, if there is a diurnal variation of air temperature there must be a diurnal variation of relative humidity, with the lowest relative humidity in this case coinciding with the time of maximum temperature. In general the amount of moisture in the air also varies through the day, being largest in the afternoon, so that the diurnal variation of relative humidity depends upon the daily variations of both temperature and moisture content, which are, however, both much less pronounced over the sea than over the land. On the average the effect of temperature change on relative humidity may be expected to outweigh the change due to changes in the quantity of moisture so that on the average the relative humidity falls as the temperature rises in its normal diurnal progress. On individual occasions this is by no means always the case. Humidity is discussed further in connection with Cargo Care (*see* page 17).

Temperature Variation with Height

As the lower atmosphere, or troposphere, derives its heat indirectly from the sun, through heating of the earth's surface and of the layer of the air in contact with it, it is not surprising that on the whole the temperature of the troposphere is highest at the bottom and decreases with height. The rate at which this decrease takes place is called the LAPSE RATE; on the average its value is 0·6°C per 100 metres increase in altitude. However, the temperature of the troposphere does not always and everywhere decrease with height. The temperature at times, notable just over fog, increases with height. The term INVERSION is used to describe a layer through which temperature increases with height. Sometimes two or three such layers are found at different levels and an inversion is also present on occasions at the tropopause. Table 1.1 shows the average variation of temperature (and pressure) with height.

Table 1.1. Pressures and Temperatures at various heights in the International Standard Atmosphere (International Civil Aviation Organization)

Pressure mb	Height m	Temperature °C	Pressure mb	Height m	Temperature °C
1013·2	0	15	400	7200	−31·7
1000	100	14·3	300	9175	−44·5
900	1000	8·5	200	11 800	−56·5
800	1950	2·3	100	16 200	−56·5
700	3016	−4·6	60	19 450	−56·5
600	4200	−12·3	40	22 075	−54·5
500	5575	−21·2	10	31 200	−45·5

Adiabatic changes (Greek *adiábatos*, impassable)

This is the name for the changes in temperature, pressure and volume which are produced in a substance when no heat is allowed to reach it or leave it while it is being compressed or expanded. Air, like other gases, is subject to certain natural laws. When air is compressed its temperature rises; conversely, when it is allowed to expand its temperature falls. The scientist Tyndall devised an experiment to show this, by placing a piece of dry tinder in a glass cylinder fitted with a piston. When air was rapidly compressed in this cylinder by quick pumping, the tinder was set alight. No heat was given to the air from any external source; the temperature of the mass of compressed air was increased because the work on it by the piston increased the energy of the molecules, making them move more quickly. The rise in temperature of a bicycle pump during vigorous use is a common illustration of the same effect.

In one method of refrigeration a compressed gas is allowed to escape from a steel cylinder and, in the rapid expansion that follows, it cools enough to cause freezing. In expanding against the surrounding pressure the gas uses energy derived from its own molecules, with the result that the average molecular velocity is reduced and the temperature of the gas falls. As long as these changes occur without any heat passing to or reaching the gas from an outside source, these changes are said to be adiabatic and the increase or decrease of temperature thus produced is known as adiabatic (or dynamical) heating, or cooling.

If the lower layers of air in a particular locality become heated more than the surrounding air, the heated air will rise. This might be caused, for example, by the existence of a ploughed field surrounded by grassland. Sunshine on the bare soil would produce a higher temperature than over the grass and the difference would be communicated to the overlying air. Having risen, the air reaches levels where the pressure is lower than it was near the surface. The drop in pressure causes a fall in temperature and because air is a bad conductor of heat the temperature change is substantially adiabatic. Provided that the air is not saturated with respect to water vapour its decrease in temperature with height due to expansion is approximately 1°C per 100 metres. This value is known as the DRY ADIABATIC LAPSE RATE.

If saturated air rises and so expands, it cools at a slower rate. Any cooling leads to condensation and this involves the liberation of latent heat which is absorbed by the air and so reduces its rate of cooling. The amount of this reduction varies with the temperature of the air since warm air can hold a larger quantity of moisture than cold air so that the quantity of latent heat is greater. Thus the SATURATED ADIABATIC LAPSE RATE varies between about 0·4°C per 100 metres in the moist air of some tropical areas to almost 1°C per 100 metres at low temperatures and high levels in the atmosphere. On the average the saturated adiabatic lapse rate at low levels in temperate regions is about one-half of the dry adiabatic lapse rate.

Stability of the Atmosphere

The atmosphere is said to be stable when the distribution of temperature and humidity with height are such that any small displacement of a parcel of air invokes forces which tend to restore the parcel to its former level. Unsaturated air is stable when its lapse rate is less than the value of the dry adiabatic lapse rate; saturated air is stable when its lapse rate is less than the value of the saturated adiabatic lapse rate. Conversely, the atmosphere is said to be unstable when the distribution of temperature and humidity with height is such that the displacement of a portion of it produces forces tending to move it further away from its former level. Unsaturated air is unstable when its lapse rate exceeds the dry adiabatic rate, and saturated air is unstable when its lapse rate exceeds the saturated adiabatic rate. When the lapse rate (whether unsaturated or saturated) is exactly equal to the appropriate adiabatic lapse rate, the displacement causes no force, either to oppose or to increase the displacement, and the air is then described as being in neutral equilibrium.

These facts are shown graphically in Figure 1.4, where the variation of temperature with height (lapse rate) of a layer of air is shown by plotting height as ordinate **Y** and temperature as abscissa **X**. Suppose the actual lapse rate of a portion of the atmosphere is given by **BPB′** (in other words if the temperature at each height is plotted on the diagram, these points when joined form the line **BPB′**), while the dry adiabatic lapse rate is shown by **APA′**. Consider the effect of displacing, under adiabatic conditions, an unsaturated parcel of air from its position at **P**. If the parcel is raised to level **P′** its temperature will change at the dry adiabatic lapse rate to become **A″**. But at this level the temperature of the surrounding air is **B″** which is lower than that of the parcel **A″**, which must therefore continue to rise because, being warmer than its surroundings, it is also less dense. Thus the air at **P** is unstable if displaced upwards. A similar argument will show that the air at **P** is unstable if displaced downwards and that the air at every level along **BPB′** is similarly unstable if displaced either upwards or downwards. The diagram can also be used to demonstrate the instability of saturated air for lapse rates exceeding the saturated adiabatic if we suppose **BPB′** denotes the actual lapse rate of a portion of the atmosphere which is saturated at every level (as often occurs in a fog bank or in thick cloud), and that **CPC′** represents the saturated adiabatic lapse rate. Similarly, a line drawn to represent a lapse rate less than that denoted by **APA′** or **CPC′** would illustrate stable conditions.

By means of radiosonde observations (for a description *see Marine Observer's Handbook*, Chapter 3), supplemented at times by observations from special aircraft, a knowledge of the temperature and humidity of the air at various levels up to the tropopause, and frequently to a higher level, is obtained at fixed hours daily from a large number of stations in the British Isles, Europe, USA and many other parts of the world in temperate and some tropical regions. (*See* Figure 1.5.)

An accurate knowledge of the temperature and humidity of the atmosphere at all levels from the surface to the upper troposphere on any occasion is of great value to the forecaster, because he is then able to make a definite statement on the type and amount of cloud which he expects to develop later on. For example, on a clear morning, if his information shows that the lapse rate is likely to exceed the dry adiabatic at low levels, he knows that volumes (i.e. 'parcels') of air will soon rise from the surface layers to a level where condensation occurs. From the argument in an earlier paragraph he knows that these parcels can continue to ascend freely through their surroundings at any higher level, provided the lapse rate there is greater than the saturated adiabatic lapse rate. So he will first estimate the time when the surface temperature will have risen to a value sufficient to establish a dry adiabatic lapse rate up to a level at which the parcel of air, freely ascending from the surface, can become saturated. Next he looks at the temperatures at higher levels to see whether the lapse rate

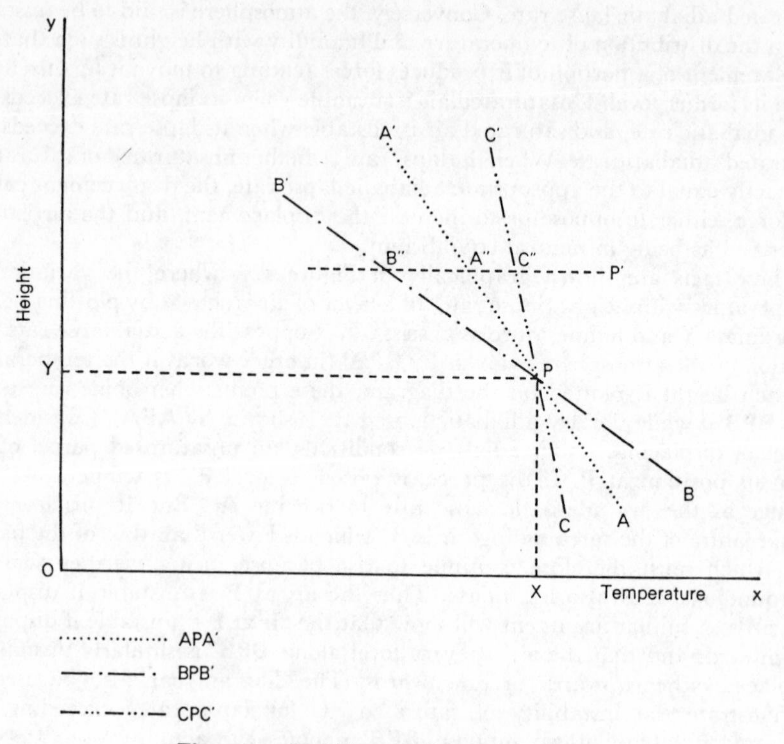

Figure 1.4. Stability conditions of dry air

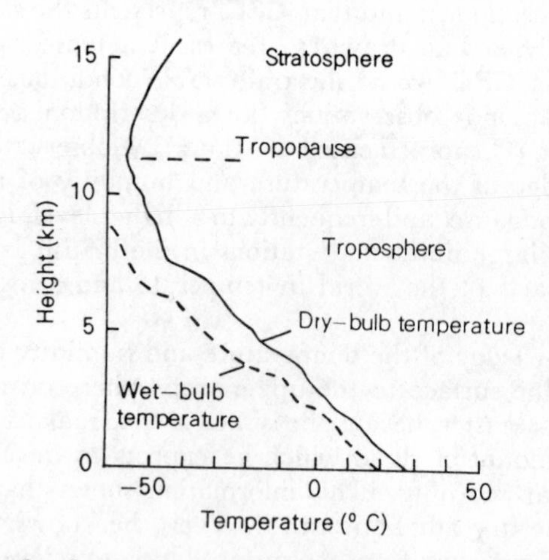

Figure 1.5. Example of a temperature sounding through the lower atmosphere

at these levels is greater or less than the saturated adiabatic lapse rate. Normally it is much simpler to do this by plotting the temperature and humidity upon a TEPHIGRAM (*see* Figure 1.6), which is a printed diagram upon which the curves of dry and saturated adiabatic lapse rates are inscribed for a number of dry-bulb temperatures at intervals of 5°C, in addition to co-ordinates based upon pressure, temperature and other meteorological elements which need not concern us here. From the curve which is obtained by plotting and joining these points upon a tephigram, the thickness of atmosphere through which the lapse rate exceeds the saturated adiabatic is readily determined. When the thickness is considerable, say several kilometres, any cloud developing in the layer will extend upwards to the top of the layer through which the lapse rate exceeds the saturated adiabatic. A lapse rate equal to or exceeding the saturated adiabatic must usually exist throughout a depth of about 3 kilometres to ensure that cumulus clouds can become thick enough to allow showers to develop, similarly the saturated adiabatic layer normally needs to be at least 5 kilometres deep to result in the formation of thunderstorms. If, as often happens, the lapse rate beneath the condensation level is less than the dry adiabatic, and the air there is not saturated, cumulus clouds will not form until a dry adiabatic lapse rate has been established through the whole layer below the condensation level, which will not happen until this layer has received a certain amount of heating from below. This usually occurs after sunshine has had a chance to warm the surface layers for a few hours, as described earlier, or when cold air is moving towards lower latitudes across a relatively warm sea surface. It is rare for the lapse rate in the free atmosphere to exceed the dry adiabatic lapse rate except in a shallow layer; on the other hand the lapse rate frequently exceeds the saturated adiabatic lapse rate through several kilometres in the lower atmosphere.

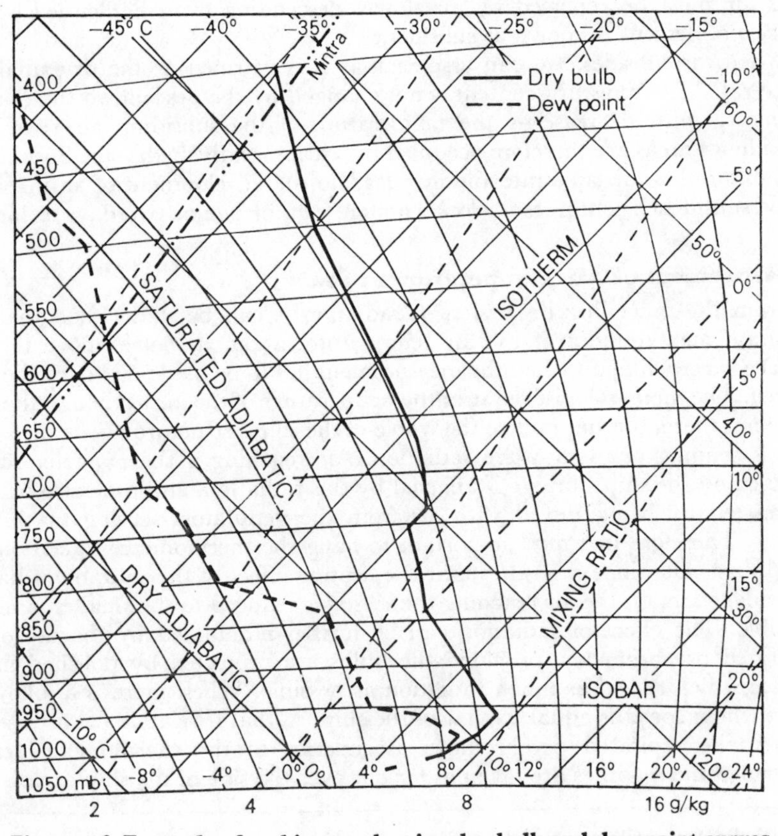

Figure 1.6. Example of tephigram showing dry-bulb and dew-point curves

The process whereby a portion (or portions) of the atmosphere (or other fluid) moves vertically relative to its surroundings in response to the buoyancy forces which arise when the portion acquires a temperature in excess of its surroundings, is known as convection. Bubbles rising to the surface of water heated from below, as in a kettle, are an example of this process. Cumulus clouds in the atmosphere are a further example of the process and are commonly referred to as convection cloud.

Convection currents cannot occur, and the formation of convection cloud is prevented when the lapse rate is less than the saturated adiabatic, especially when there is an inversion of temperature a short distance above the ground. Inversions are common near the ground around sunrise following a clear night and, if the day following is clear, after several hours of sunshine the lowest layers of the atmosphere will often be warmed sufficiently to establish a dry adiabatic lapse rate to a height above the condensation level. When this happens convection cloud will be formed as described earlier.

Subsidence

Stable and unstable air have been discussed above in terms of the behaviour of small parcels of air displaced upwards or downwards from their initial level. However, in some situations, as in the lower layers of a stationary anticyclone, there is considerable movement of air horizontally outwards over a large area. This air must be replaced by air slowly descending from higher levels: the process of descent is known as subsidence.

When air subsides, we can assume that it is warmed at the dry adiabatic lapse rate. The moisture content is not changed by the descent, so the relative humidity must decrease as the temperature of the subsiding air rises, and subsidence areas are therefore occupied by warm, relatively dry air. Subsidence also diminishes the lapse rate and may lead to the development of an inversion if the subsidence continues for long enough, e.g. for one or two days or longer.

Measurement of Air Temperature at Sea*

From the preceding discussion of radiation it can be understood that the accurate measurement of the air temperature about a metre above the sea surface is very difficult when the measurement has to be made on board the ship herself. The meteorologist requires the temperature of the air as though the ship were not there, but in practice the whole of the ship's structure above the water line is sending out long-wave radiation. Air reaching a thermometer screen exposed on the ship's bridge is affected by this radiation, and may also be contaminated by being mixed with air from the ventilators serving machinery spaces. Therefore it is most important to hang the thermometer screen on the windward side, since it ensures that the heating effect of the ship, by radiation and advection, on the air reaching the screen is reduced to the smallest amount possible. The effect of radiation can be further eliminated by the use of an aspirated psychrometer, provided the bulbs are protected by polished metal shields which reflect as much radiation as possible. When there is a following wind whose speed is equal to that of the ship, so that a relative calm exists on board, the aspirated psychrometer is superior to the thermometer screen (provided the readings are taken as far out over the side of the ship as possible,

---

* *See* also *Marine Observer's Handbook*, Chapter 2.

and while the motor in the instrument is still running at full speed), since screen thermometers will be liable to read wrongly owing to lack of ventilation.

Practical Value of Temperature and Humidity Information to the Mariner in Relation to the Care of Cargo

Knowledge about the physical processes involved in variations of temperature and humidity is of considerable value to the ship's officer in connection with the care of cargo. Knowledge about sea temperature changes (*see* pages 95 and 96) is also valuable in this connection.

Appreciable damage, due to moisture effects, sometimes occurs to cargoes carried aboard ship. Some evidence of this is provided by contact with shipping interests and from enquiries received in the Meteorological Office for information about weather conditions at ports and along shipping routes on specified occasions when damage has occurred, but it is difficult to obtain any precise figures regarding the overall amount of damage. When a cargo is loaded into a ship its history prior to loading is important, although the ship's officer frequently has no means of knowing much about this. The temperature and weather conditions under which the cargo is loaded may need to be taken into consideration.

Cargo damage includes such effects as mould formation, germination of grain, corrosion of metals, caking of sugar or chemicals, etc., and may arise either from 'hold sweat', i.e. water which condenses on cold deck-heads, etc., and thence wets the cargo, or from 'cargo sweat', i.e. condensation directly on to the cargo itself. In studying the problem account must be taken of the special properties of cargoes, especially of those which are hygroscopic or which generate heat and release gases. Various sources and sinks of heat within the ship itself must also be considered, particularly when the cargo spaces are not insulated. These include the lower part of the ship's hull plates whose temperature is largely controlled by the sea temperature; the ship's structure above the water line whose temperature varies due to radiation effects, from night to day, and also from shaded to sunny sides; and hot or cold bulkheads adjacent to engine rooms or refrigerated spaces respectively.

Another important factor is the type of ventilation system provided. This may range from the old-fashioned cowl-type natural ventilation to the more modern forced ventilation with all-weather ventilator heads, or even, in a few cases, to full air-conditioning. Except where such expensive air-conditioning equipment is available, however, the ventilation of cargoes to remove unwanted heat or moisture can only be accomplished with the air that happens to surround the ship. If properly used, natural ventilation can give good results on the majority of occasions, but sometimes it is quite inadequate, e.g. when there is no wind relative to the ship. A number of rules of thumb have been developed, based on long experience, which have proved to be fairly sound when applied aboard such ships but they are far less reliable when applied on ships with more sophisticated systems, for with these it is much easier to over-ventilate and to cause a lot of damage in a relatively short time. An officer who believed in ventilating whenever the weather was fine, for example, could do far less damage with natural ventilation than he could with mechanical ventilation. It is therefore essential that officers should understand the relatively simple physical principles involved in correctly using any system of ventilation and make careful measurements of air temperature and dew-point both inside and outside cargo compartments.

As a general guide it is best not to ventilate if the hold temperature is below the dew-point of the outside air. If on the other hand the hold temperature is appreciably higher than the dew-point of the outside air it is generally advisable to ventilate.

Table 1.2. Dew-Point Temperature

Air temperature °C	Depression of wet bulb									
	1	2	3	4	5	6	7	8	9	10
0	−3	−7	−11	−18	−32					
5	2	0	−4	−8	−14	−20				
10	8	6	3	0	−3	−8	−15	−27		
15	13	11	9	7	4	1	−2	−7	−14	−26
20	18	17	15	13	11	9	6	3	0	−5
25	24	22	20	19	17	15	13	11	8	5
30	29	27	26	24	23	21	19	17	16	13

CHAPTER 2

PRESSURE

Atmospheric Pressure, the Barometer

The atmospheric pressure at sea level has an average value of about one kilogram per square centimetre and represents the weight of a vertical column of air of one centimetre cross-section extending to the top of the atmosphere. Because of the low density of air this not inconsiderable weight corresponds with a very long column of air (*see* Chapter 1). With a fluid of higher density the same pressure would be exerted by a column of smaller vertical extent. For example, in the case of water, a column extending to a height of about 10 metres will exert a pressure equal to the normal atmospheric pressure. In other words we can say that the normal atmospheric pressure at the earth's surface can be balanced by a column of water about 10 metres high. By choosing a still more dense fluid, such as mercury, the vertical column becomes much shorter, round about 760 millimetres, and much more convenient to measure.

Basically the simple mercury barometer consists of a glass tube some 90 centimetres (36 inches) long, closed at one end, which is filled with mercury and, while the open end is temporarily sealed, this end is immersed in a cistern filled with mercury. When the seal is removed and the tube is held upright with the closed end uppermost, the column of mercury in the tube sinks until the pressure it exerts is equal to the pressure exerted by the atmosphere on the surface of the mercury in the cistern. The length of this column of mercury then provides a measure of the atmospheric pressure. In the past, atmospheric pressure has been recorded in these terms, i.e. as 29·530 inches (or 750·062 millimetres) of mercury. This mode of expression is, however, not strictly logical since we are not describing a length but a pressure, which in the metric system would normally be expressed in dynes per square centimetre. Since it happened that in the formerly used c.g.s. system of units, normal atmospheric pressure was about 1 000 000 dynes per square centimetre, this value was chosen as a unit of atmospheric pressure and called the 'bar'. However, because in meteorology we are chiefly concerned with the accurate measurement of comparitively small variations from the mean value, a unit one thousandth of this value, the MILLIBAR, was used as the working unit. Although in the now widely adopted SI system of units the unit of pressure is the pascal (= 1 newton per square metre), the millibar is still generally used in meteorological work. One millibar equals 100 pascals.

Barometers are usually graduated in millibars but some instruments still in use are graduated in inches or in millimetres. Tables are available for converting readings in millibars (under specified conditions discussed later) to readings in inches or millimetres of mercury, and vice versa.

An alternative instrument for measuring atmospheric pressure is the aneroid (Greek, without liquid) barometer. This in its simplest form consists of a shallow capsule of thin corrugated metal which is exhausted of air. The atmospheric pressure on the outside of the capsule tends to press the two faces together but this is resisted by the stiffness of the metal. The relative movement of the two faces due to changes of outside pressure can be magnified and indicated as a pressure reading.

These instruments are lighter and more compact than the mercury barometer. Until recently they were less accurate than the mercury instrument which was therefore preferred. In recent years, however, improvements have been introduced and the 'precision aneroid' barometer is now regarded as having similar accuracy to the mercury barometer. Because of its compactness and greater ease of reading, the precision aneroid is now adopted as the standard instrument for issue by the Meteorological Office to Voluntary Observing Ships.

However accurately a barometer may be read, the reading needs to be 'reduced to standard conditions' before it can usefully be compared with other readings which may be made under differing conditions. For example, with a mercury barometer, a given pressure will produce a reading which will differ according to the temperature of the instrument. Accordingly it is necessary to record the temperature and apply a correction to adjust the reading to a standard temperature.

The corrections to be applied differ according as to whether the instrument used is of the mercury type or the aneroid type and will accordingly be discussed separately. Details of both instruments and of the corrections to be applied to them are set out in the *Marine Observer's Handbook*, but a brief summary is given here for the reader's convenience. The tables referred to below are those in the *Marine Observer's Handbook* (10th edition).

Barometer Corrections

Mercurial barometers in use today which were made before 1955 are calibrated so as to read as nearly as possible correctly at 12°C in latitude 45° north at mean sea level. All barometers made or repaired on or after 1 January 1955, however, are made to read as nearly as possible correctly at 0°C and gravity 9·80665 m/s^2 (which is practically the same as gravity at mean sea level in latitude 45°). Thus the corrections to be applied to the readings obtained with a particular instrument will depend on the date of manufacture or repair (before or after 1 January 1955) as shown on its accompanying National Physical Laboratory certificate.

(*a*) CORRECTION FOR TEMPERATURE (Tables 1, 2, 6 or 7 as appropriate)

Mercury expands with increase of temperature, so that the same atmospheric pressure is balanced by a longer column of mercury on a warm day and by a shorter column on a cold day. A correction is applied to obtain the reading of the barometer which it would show at the standard temperature of 12°C or 0°C as the case may be.

(*b*) CORRECTION FOR ALTITUDE (Table 5 or 10 as appropriate)

Atmospheric pressure, a measure of the weight of air above any point, will naturally be less at, say, the top of a lighthouse or on the bridge of a ship than at sea level some distance below. In order to make pressure readings all comparable, the pressure read on the instrument needs to be corrected to obtain the pressure which the instrument would record assuming it could be located at sea level in the same place. The actual difference depends not only upon the height of the instrument but also upon the density of the air between it and sea level. The density of the air in the column between the instrument and sea level varies according to the temperature of the air in that column and this is obtained from the reading of the dry-bulb thermometer in the screen.

(*c*) CORRECTION FOR LATITUDE (Tables 3, 4, 8 or 9 as appropriate)

Owing to the flattening of the earth at the poles and the fact that the vertical component of the centrifugal force due to the earth's rotation (which acts in the opposite direction to gravity) is greatest at the equator, the force of gravity increases steadily from the equator to the poles. The standard value of gravity is now taken as 9·80665 m/s^2 which is practically the same as that at sea level in lat. 45°. Thus a negative correction must be applied in low latitudes, where the barometer always reads high by a small amount because the 'weight' of the mercury is less than at lat. 45° (a positive correction must be applied in latitudes higher than 45°).

(*d*) INDEX CORRECTION

When all these corrections have been made it is found that the readings of individual barometers still differ slightly for a number of reasons including capillarity of the mercury. These corrections are determined for every barometer at the National Physical Laboratory (NPL) and included in the 'index correction', the amount of which is stated on the certificate supplied by the NPL for use with the barometer.

EXAMPLE

In lat. 27°N the barometer reads 1017·3 mb at a height of 16 m above sea level. The attached thermometer reads 25°C, the dry bulb in the screen reads 26°C and the index correction of the barometer is +0·3 mb. Standard conditions 0°C, gravity of 9·80665 m/s^2.

		mb
Uncorrected reading		1017·3
Index correction	+	0·3
		1017·6
Temperature correction for 25°C	−	4·3
		1013·3
Height correction for 16 m at air temperature of 26°C..	+	1·9
		1015·2
Gravity correction in lat. 27°	−	1·6
		1013·6

Marine mercurial barometers supplied by the Meteorological Office are fitted with a correction slide, which enables the corrections to be applied mechanically and dispenses with the use of tables. However, the corrections, as read from this slide, still need to be applied to the pressure reading!

Aneroid barometers need only to be corrected for index error and for altitude. They normally include a device for compensating for changes of temperature, which is ensured either by using a bimetallic link or leaving a calculated small amount of air in the vacuum chamber. Aneroids compensated for temperature are usually so marked.

The readings of aneroid barometers after correction as above should be compared as frequently as possible with corrected readings of a good mercury barometer. The reason for this is that the index error of aneroid barometers is liable to change quite frequently owing to change in the elasticity of the metal of the vacuum chamber. (*See Marine Observer's Handbook*, Chapter 1.)

Pressure Distribution; Isobars

If pressures, corrected as above so as to make them comparable, are measured at a number of positions simultaneously, and the values plotted on a chart, a fair estimate can be made of those at intermediate places and lines can be drawn connecting places where the pressure is the same. The lines of equal pressure so drawn are called ISOBARS and resemble the contours of equal altitude which define hills and valleys on a map. A chart for a particular time which includes data such as wind, weather, visibility and cloud and isobars (usually drawn at intervals of 4 mb, e.g. 1012, 1016 and 1020 mb, etc.) for a number of stations is termed a weather map, or more frequently a SYNOPTIC CHART, because it gives a synopsis or general view of the weather conditions over a large area at a given instant of time. Any synoptic chart will show a distribution of pressure in which there are regions of high pressure and low pressure resembling the hills, ridges and valleys found on a map.

Pressure Gradient

The idea of pressure gradient is best illustrated by another look at a contour map where at any particular place we can estimate the gradient or slope of the ground. The steepest slope is seen to be at right angles to the contour lines and its numerical value is greater where the contour lines are closer together. Similarly, we speak of the pressure gradient as being at right angles to the isobars, its magnitude being measured by the ratio pressure difference/distance, in suitable units. Thus if two neighbouring isobars (4-mb interval) are 97 nautical miles apart and are also straight and parallel, the pressure gradient between them is 4/97, i.e. 0·041 mb/n. mile, and has a direction at right angles to the isobars and is reckoned as positive in the direction going from high to low pressure. A pressure gradient of this magnitude is associated with a wind which in latitude 50°N or S has a value of 30 knots (*see* Chapter 3, page 31).

Pressure and Winds

A study of any synoptic chart of a large area outside the tropics, such as the North Atlantic, at once suggests that there is a close relation between pressure distribution and wind speed and direction; at any point the wind blows nearly parallel to the isobars in the neighbourhood, crossing them at a small angle in the direction from the higher to the lower pressure. The direction of the wind is such that to an observer facing the wind the lower pressure is on his right in the northern hemisphere and on his left in the southern hemisphere. This fact may be expressed in another way by saying that the air circulates clockwise round centres of high pressure and anticlockwise round centres of low pressure in the northern hemisphere (while in the southern hemisphere the air circulation is anticlockwise round centres of high pressure and clockwise round centres of low pressure). Lastly, the wind is strong where the pressure gradient is steep and light where the pressure gradient is small, that is the speed of the wind is closely proportional to the pressure gradient.

The relation between wind direction and barometric pressure was first defined in a law by Buys Ballot in 1857. This is quoted in Chapter 3 which discusses the relation between pressure and wind in greater detail.

The most important features in any pressure distribution are shown in Figures 2.1 to 2.4, which are drawn for north latitudes.

(*a*) Depression or Low (Figure 2.1)

A depression or low-pressure system is shown as a region of relatively low pressure with closed isobars. The isobars form a closed system with the lowest pressure inside that isobar which has the smallest value of pressure, the wind circulation around it being anticlockwise in the northern hemisphere, clockwise in the southern hemisphere. A description of the structure of a typical depression and the weather experienced within the various regions under its influence is given in Chapter 9, on pages 117 to 124.

(*b*) Secondary Depression (Figure 2.2)

The term secondary depression is applied to a small depression within the area covered by a larger primary depression. The example shown in Figure 2.2 is more vigorous than the parent or primary depression. Sometimes, however, no closed centre may be apparent, but only an area of relatively low pressure extending as a somewhat bulbous trough (*see* below for trough) from the parent depression. The weather experienced in a secondary depression is described on page 123.

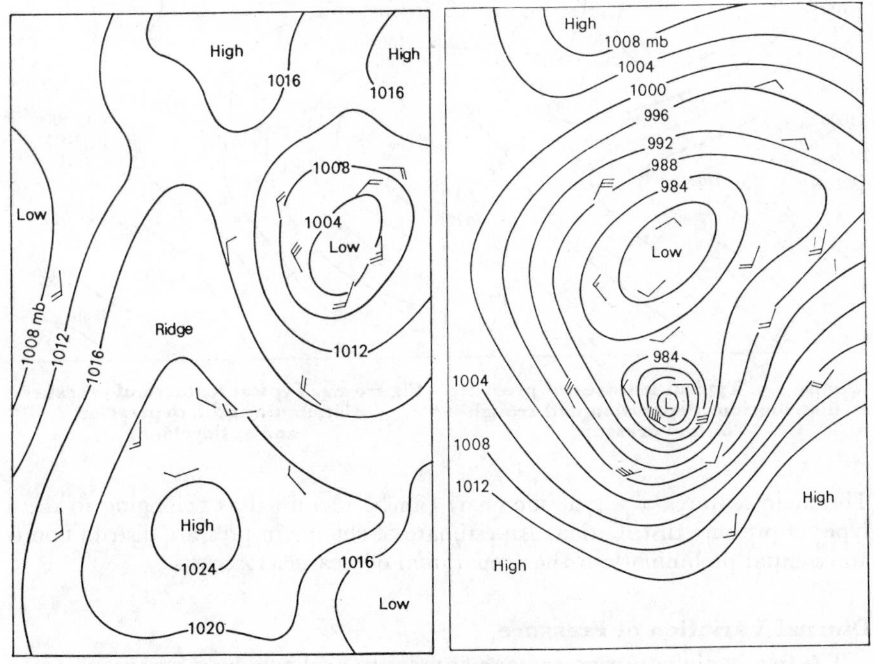

Figure 2.1. Typical features of pressure distribution: anticyclone and depression

Figure 2.2. Typical features of pressure distribution: primary and secondary depression

(*c*) Trough of Low Pressure (Figure 2.3)

A trough is an elongated area of low pressure indicated by isobars extending outwards from a depression. The weather associated with a trough of low pressure is summarized in Chapter 10, page 134.

(*d*) Anticyclone or High (Figure 2.1)

An anticyclone is shown as a region of relatively high pressure with closed isobars. The isobars form a closed system with high pressure on the inside, the wind circulation around it being clockwise in the northern hemisphere and anticlockwise in the southern hemisphere. The weather experienced in an anticyclone is described in Chapter 10, page 132.

(*e*) Ridge of High Pressure (Figures 2.1 and 2.3)

A ridge of high pressure is an elongated area of high pressure indicated by isobars extending outwards from an anticyclone. The weather experienced in a ridge of high pressure is described in Chapter 10, page 136.

(*f*) Col (Figure 2.4)

A col is a saddle-backed region between two lows and two highs. The weather experienced in a col is described on page 136.

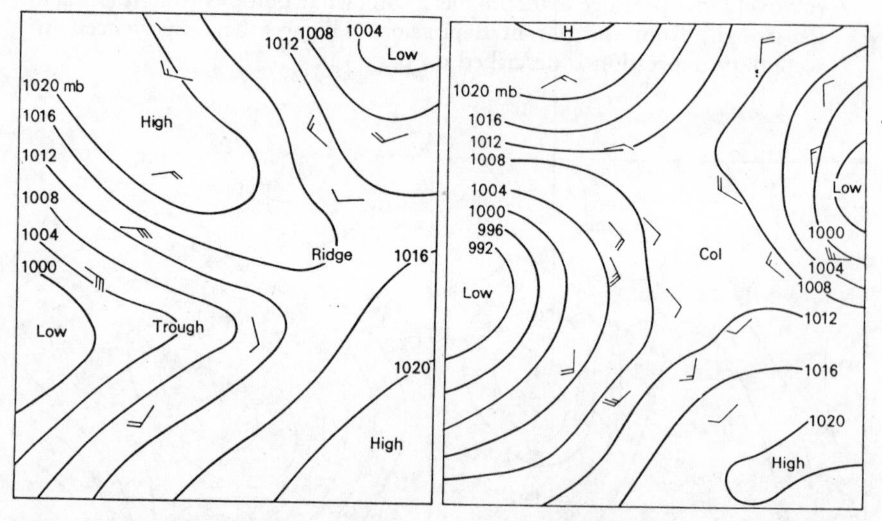

Figure 2.3. Typical features of pressure distribution: depression and trough of low pressure

Figure 2.4. Typical features of pressure distribution: col, depression and anticyclone

The main features of a synoptic chart can be identified as belonging to these types of pressure distribution. An estimate of change in pressure distribution is an essential preliminary to the preparation of weather forecasts.

Diurnal Variation of Pressure

The frequently occurring changes of pressure as shown by a barograph trace may be due to many causes, and in temperate latitudes in winter it is not possible to discern any systematic variation. But in the tropics, and in quiet summer weather of the temperate zones, a diurnal variation of pressure with a definite pattern becomes evident. Over a long period the mean daily range is less than 1 mb in the British Isles but more than 2 mb in subtropical and tropical regions. Figure 2.5 shows a typical curve for the tropics, with maxima at 1000 and 2200 hours and minima at 0400 and 1600 hours local time.

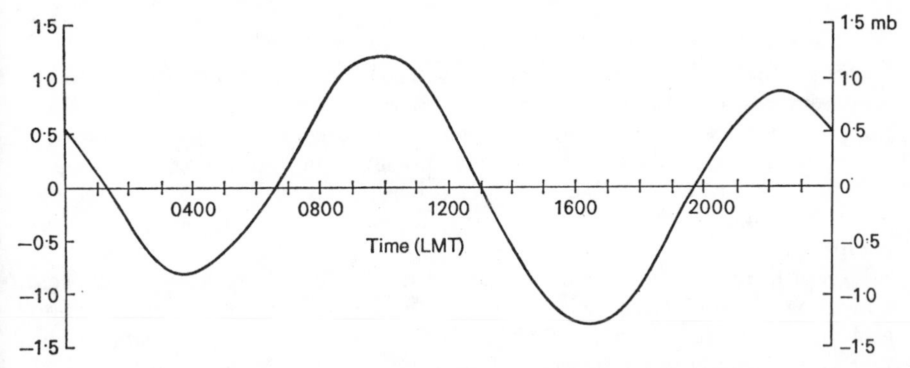

Figure 2.5. Diurnal variation of pressure typical of the tropics

In temperate latitudes irregular pressure changes are usually so much larger than the diurnal variations that the latter need not be taken into account. In tropical regions irregular changes are usually much smaller than the diurnal change from which they cannot be distinguished until the diurnal change has been subtracted. This procedure is important in tropical regions because a fall of pressure 2 or 3 mb below the value appropriate for the locality, after allowing for the regular diurnal change of pressure, may be the first indication of the approach of a tropical revolving storm. The normal value of pressure for a locality and the diurnal change can be obtained from a meteorological atlas or *Admiralty Pilot*.

Pressure Tendency

The name TENDENCY is given to describe the rate of change of pressure with time. In practice it usually refers to the time interval of three hours before an observation. The meteorologist wishes to know the change of pressure at a particular place. The ship's report of barometric tendency does not directly tell him this, because part of that tendency is due to the ship's progressive movement. But by using the figures for the ship's course and speed given in the synoptic message he can eliminate this spurious part of the tendency by making a correction to the reading.

Corrected readings of barometric tendency help to determine the movements of the various pressure systems, since the barometer usually falls in advance of a depression and rises in advance of an anticyclone. If the system is stationary, the tendency can show whether it is filling up or deepening. (*See* Chapter 12.)

Lines drawn on the chart through places having the same tendency are known as ISALLOBARS. The isallobar is of value to the forecaster because it shows the rate of change of pressure with respect to time.

CHAPTER 3

WIND AND WAVES

Introduction

Wind is defined as air in motion. In general the wind motion estimated by an observer or measured by an anemometer is the wind parallel to the ground. Ashore, wind is measured by anemometers which record its velocity in knots. As the wind speed near the surface varies with height above the ground, readings of anemometers are corrected to a standard height of 10 metres so that readings from instruments which may be at different heights can be satisfactorily compared.

Except in naval ships, anemometers are rarely used at sea owing to expense and siting difficulties and because the ship's structure complicates the air flow and makes accurate readings hard to obtain. At sea, observations of wind are therefore normally made by estimation.

Beaufort Scale

The BEAUFORT SCALE forms the basis of wind-force estimation at sea. It was originally introduced in 1808 by Admiral Beaufort, who defined the numbers of the scale in terms of the effect of wind on a man-of-war of his day. Table 3.1 shows how the method has changed through the years. The scale itself has not changed much, however, and it seems likely that conditions which Beaufort would have described by a certain number, judged from the behaviour of a warship of his time, would be described by the same number using the modern criterion whereby the wind force is judged from the appearance of the sea. As will be seen from the Table each Beaufort number has been allotted an equivalent range of wind speeds as well as an equivalent mean wind speed at standard height, so that conversion from Beaufort number to wind speed is simple. At the time of writing (1977) proposals have been made to modify the ranges in knots corresponding with the various Beaufort numbers, but these have not yet been agreed. Wind direction at sea is also normally estimated from the appearance of the sea—the true azimuth from which the wind is blowing being that at right angles to the line of sea waves. For further details about wind observation see *Marine Observer's Handbook*, Chapter 4.

Wind and Air Movements on a Rotating Earth

Wind is movement of air set up in the atmosphere by differences in atmospheric pressure between two localities. These differences in pressure are caused by variations of temperature in columns of air over different places. The atmosphere is always trying to achieve a uniform pressure distribution by transfer of air from one region where an accumulated excess of air has resulted in high pressure, to another region where a deficiency of air has resulted in low pressure. It is a matter of common observation that this adjustment is not carried out by winds blowing direct from high to low pressure but that the wind tends to blow in a circular manner around regions of low pressure or high pressure (*see* page 22), and for the explanation of this we must consider the effect of the rotation of the earth.

Consider first the earth's surface in the proximity of the North Pole. To a close approximation the earth's surface in this region can be regarded as a flat plane surface rotating about the axis **ON** (*see* Figure 3.1). Figure 3.2 represents this plane surface in plan.

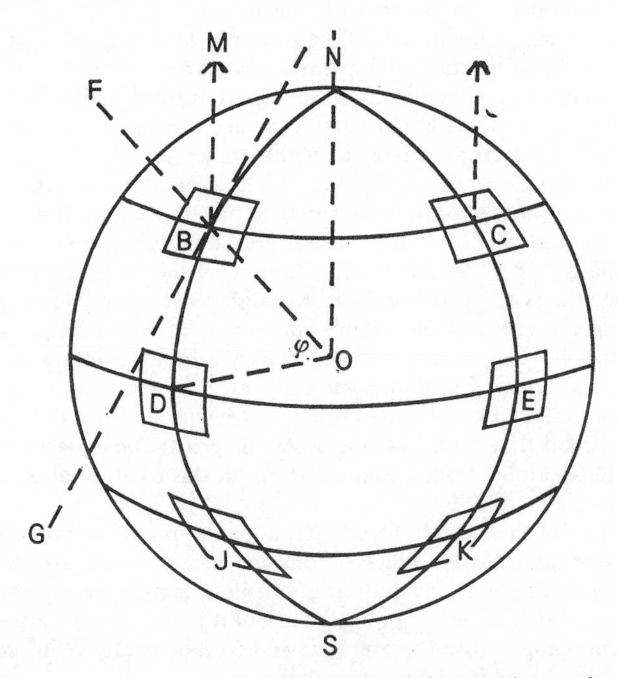

Figure 3.1. Diagram illustrating effect of the earth's rotation

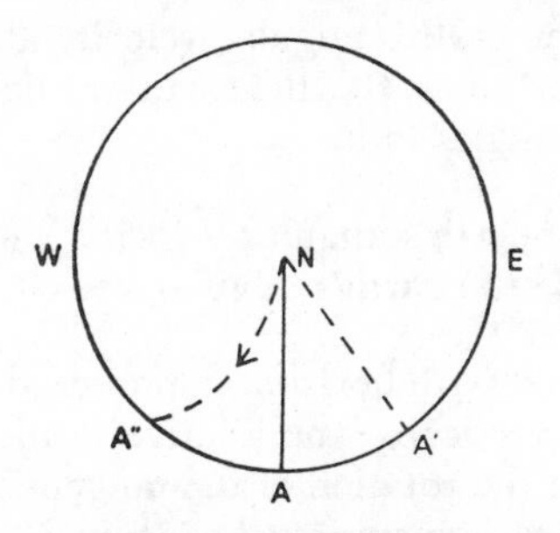

Figure 3.2. The effect of the earth's rotation at the pole

Suppose a parcel of air is displaced from the pole on a meridional path represented by **NA** in Figure 3.2. Because the underlying surface of the earth is rotating, by the time the earth has reached position **A** (moving in a straight line relative to space) the meridian corresponding to the line **NA** at zero hour will have rotated to **NA′**. An observer in space would see the displacement as a straight line and would be aware of the earth's rotation relative to it. But an observer on the earth would be oblivious of the earth's rotation and would see the air's trajectory as a line progressively diverted towards the right from its original meridional displacement as in the curved line **OA″**.

This simplified example of the deflection imposed on moving air on account of the earth's rotation may be extended to a more general proposition which states that any body moving relative to the rotating earth is subject to the Coriolis force (named after its discoverer) which is directed to the right of its direction of motion (in the northern hemisphere).

Reverting to the representation of the rotating earth in Figure 3.1 and having seen how the spin of the underlying earth affects air movement near the pole, let us turn to an intermediate latitude such as that marked **BC** in the diagram. In the vicinity of the point **B** the earth's surface can be considered to approximate to a flat (tangential) plane at right angles to the earth's radius **OB**. Looking at Figure 3.1 and considering how this plane in the vicinity of point **B** behaves as the earth's rotation carries it to position **C**, we can see that it too is subject to a rotation. This can be seen from the differing orientation, in space, of the meridians **NBS** and **NCS**. The amount of the rotation, however, is less than it is near the pole. If we continue, through progressively lower latitudes, to consider what the effect of the earth's rotation is at points on the equator, we can judge this by considering the small area around point **D** as the earth's rotation carries it to **E**. Confining our attention (as before) to rotation relative to the earth's radial axis at points **D** and **E**, we can see from the orientation of the local meridian that such rotation becomes zero at the equator. Accordingly the deflecting (Coriolis) force which depends on this rotation also becomes zero at the equator.

The variation of the Coriolis force from a maximum at the pole to zero at the equator may be quantified as follows: consider once more the area around point **B** in Figure 3.1. The earth's rotation about its axis may be represented by its angular velocity ω. The earth's rotation about a parallel axis through **B** would have the same value. Now the angular velocity about any point of a rotating sphere may be split up into two components around axes at right angles to one another, in the same way as a force **F** can be split up into components along two perpendicular axes. So the earth's angular velocity about **BM** can be resolved into two components—one about **BF**, the earth's radial axis at point **B**, and the other about **BG** at right angles to it.

About **BF** the earth's angular velocity is $\omega \sin \varphi$
and about **BG** the earth's angular velocity is $\omega \cos \varphi$.

As we are only concerned with horizontal motion, the latter expression which represents motion in the vertical plane relative to the earth's surface about **B** can be neglected here. So the rotation of the tangential plane in the vicinity of the point **B** in latitude φ can be expressed as $\omega \sin \varphi$.

Now let us refer to Figure 3.3 which illustrates movement in the tangential plane at point **B**. We have seen that the whole plane is rotating about its vertical axis with an angular velocity $\omega \sin \varphi$. Suppose a parcel of air moves from point **B** to point **H** with velocity V.

In a time t it will travel a distance Vt. In the same time the underlying earth's surface, represented by the tangential plane at **B**, will turn through an angle $\omega \sin \varphi\, t$. Thus, relative to the earth's surface, the parcel of air reaches point **I**. The distance **HI** is $V.t$ times $\omega \sin \varphi\, t$, or $Vt^2\omega \sin \varphi$, provided t is small. In simple dynamics the distance(s) moved by a body subject to an acceleration f acting for a time t is given by the expression $S = \frac{1}{2}ft^2$. So our parcel of air can be regarded as having an acceleration f whose value can be deduced by substituting the value $\frac{1}{2}ft^2$ for the expression derived for the distance **HI**, namely $Vt^2\omega \sin \varphi$. This gives us the value $2V\omega \sin \varphi$ for the acceleration imparted to a body (or air parcel) moving with a velocity V on account of the rotation of the earth. The acceleration is directed at right angles (to the right in the northern hemisphere) to the instantaneous direction of motion.

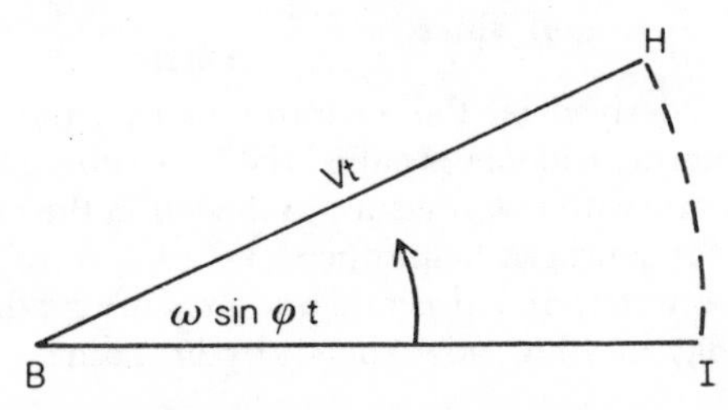

Figure 3.3. Diagram illustrating air movement on a rotating earth

In the southern hemisphere the direction of this geostrophic (or Coriolis) acceleration is reversed. Here it is directed to the left of the instantaneous direction of motion. That this is so will readily be appreciated by referring to Figure 3.1. As the earth spins the tangential plane about the point **J** rotates as it moves to position **K**. The examination of the orientation of the meridian through **J** will show that the meridian rotates in a clockwise direction, in contrast to the anticlockwise rotation of the meridian through point **B** as it moves to position **C**.

Geostrophic Wind

Let us now consider the motion of air assuming it to be moving with uniform speed in a straight line (i.e. in a region where the isobars are straight and parallel as in Figure 3.4).

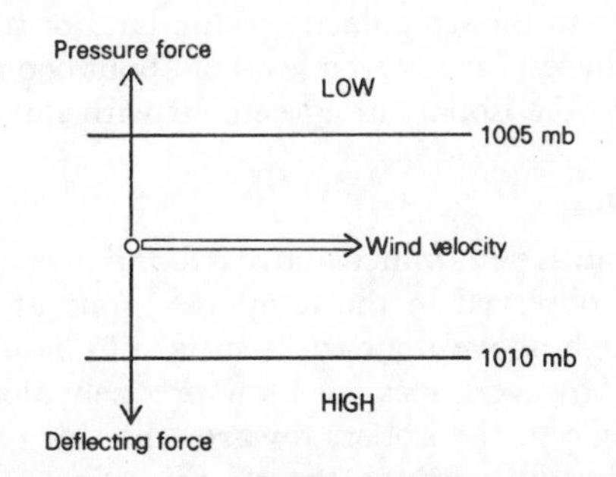

Figure 3.4. Geostrophic wind velocity

Provided we neglect frictional forces, there are only two horizontal forces acting on any parcel of air: one due to the pressure gradient acting in a direction perpendicular to the isobars towards the side of low pressure, the other the deflecting force of the earth's rotation acting perpendicular to the path of the air and to the right of it. Only if the accelerations produced by these two forces balance, i.e. are equal in magnitude and opposite in direction, can the motion remain steady, and V remain constant as was assumed above. Denoting the pressure gradient by G and the density of the air by ρ, the acceleration due to the pressure gradient is given by $\frac{G}{\rho}$. For steady motion this acceleration must balance the acceleration $2\omega V \sin \varphi$ due to the earth's rotation, i.e.

$$\frac{G}{\rho} = 2\omega V \sin \varphi \text{ or } V = \frac{1}{2\,\omega \sin \varphi} \frac{G}{\rho}.$$

The wind velocity V derived on the assumptions that the isobars are straight and parallel, and ignoring friction, is called the GEOSTROPHIC WIND. Its direction is parallel to the isobars, with low pressure to the left in the northern hemisphere and to the right in the southern hemisphere.

The relationship between wind direction and pressure gradient was established by Buys Ballot in 1857 in a law now known by his name.

Buys Ballot's Law

In north latitudes, face the wind and the barometer will be lowest to your right.

In south latitudes, face the wind and the barometer will be lowest to your left.

Wind Circulation around Pressure Systems

The first deduction which can be drawn from this law is that the winds circulate around the centres of low and high pressure; in the northern hemisphere they blow anticlockwise around an area of low pressure and clockwise around an area of high pressure, while in the southern hemisphere the directions are reversed. A look at any synoptic chart will verify that this statement is nearly true except for regions within a few degrees of latitude of the equator, although it will be noticed that over land the wind blows mostly about 30° from the direction of the isobars towards the side of low pressure, and over the sea the inclination of the wind to the direction of the isobars is about 10°, but is usually greater than this in the Trades. In both cases this effect is mainly caused by friction due to the irregularity of the land or sea surface, which also retards the air in the lowest layers. At a level of about 600 metres and higher the wind blows parallel to the isobars in agreement with Buys Ballot's Law.

The Effect of Friction

The geostrophic wind as determined above is found to be a good approximation to the wind actually observed in the temperate zone at a level of about 600 metres, i.e. high enough above the earth's surface to be unaffected by friction.

The surface wind, however, does not blow precisely along the isobars in any latitude but at an angle to the isobars towards the side of low pressure. It can be shown that friction will produce exactly this effect; the actual amount of deviation from the isobars depends not only on the latitude but also on the amount of friction.

Another effect of surface friction is that the wind speed at the surface is reduced below the value of the geostrophic wind. Over the sea, it is found that the surface wind speed usually is about two-thirds that of the geostrophic wind. Over land, the fraction varies more widely and generally lies between two-thirds and one-half of the geostrophic wind speed, according to the degree of exposure of the place where the wind is measured.

Determination of Surface Wind from Isobars

The equation for the geostrophic wind allows the latter to be calculated from the pressure gradient. However, the wind corresponding to a particular gradient varies according to latitude. Synoptic weather maps normally contain a diagram from which geostrophic wind speed can be determined by measuring the distance between successive isobars. If this diagram, on which the various wind speeds are given in knots, is transferred to a celluloid scale, the geostrophic wind at any locality on the synoptic chart can be read off by simply putting the scale across the isobars as shown in Figure 3.5. In the example shown, where the

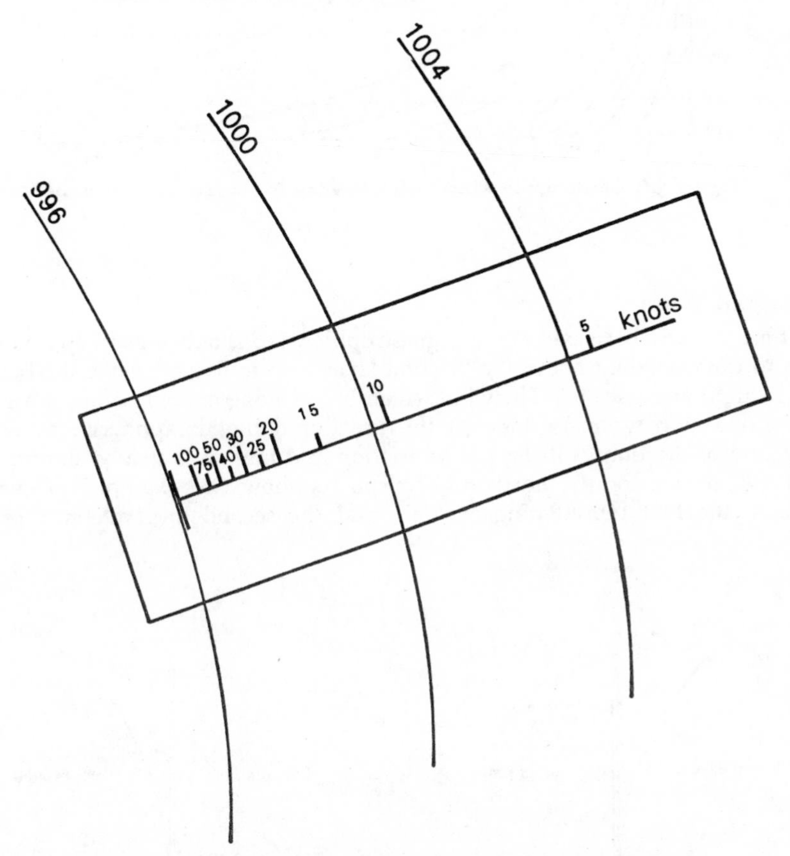

Figure 3.5. The determination of surface wind from isobars

scale is constructed for isobars drawn at intervals of 4 mb, the geostrophic wind between the isobars of 996 and 1000 mb is found to be 11 knots. A correction for latitude can be applied where necessary. Alternatively, celluloid scales are available on which the distance between the isobars can be translated into the value appropriate to any latitude by using a different part of the scale depending on the latitude required (*see* Figure 3.6).

The speed of the surface wind at sea may then be taken as two-thirds of the geostrophic wind value, and the direction as one or two points from the direction of the isobars, towards the side of low pressure.

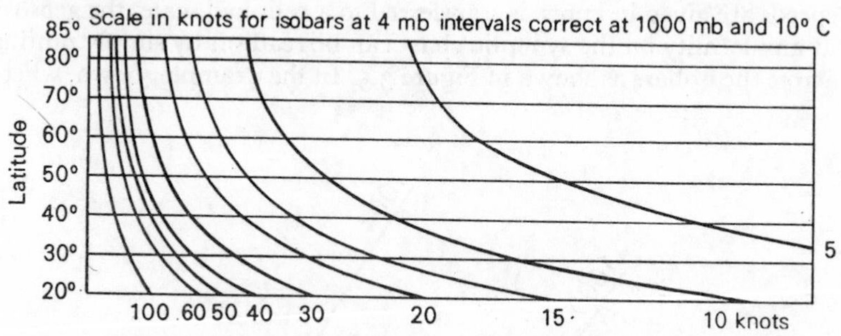

Figure 3.6. Geostrophic wind scale allowing for variation of latitude

Gradient Wind

It has already been said that the geostrophic wind is only a good approximation to the true wind above the 'friction layer' (up to 600 m) when the isobars are straight and parallel. The wind which blows along curved isobars is known as the GRADIENT WIND. In this case, the direction of motion approximates to the direction of the isobars if there is no friction and motion is steady. Figures 3.7 and 3.8, drawn for the northern hemisphere, show two examples of curved isobars, the first representing cyclonic and the second anticyclonic motion.

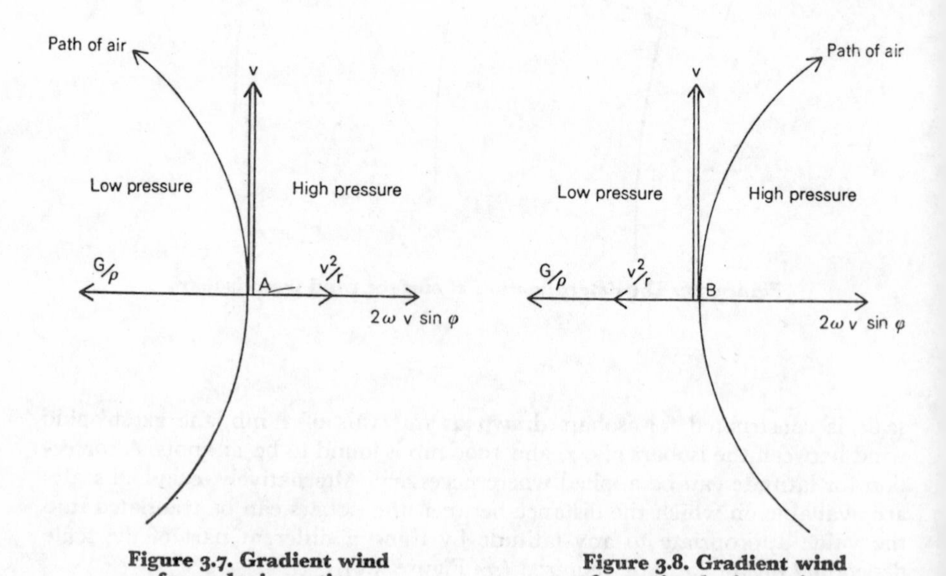

Figure 3.7. Gradient wind for cyclonic motion

Figure 3.8. Gradient wind for anticyclonic motion

Arrows with a double shaft denote velocity V, while single arrows denote the forces which affect the air at **A** and **B**. For steady circular motion, the difference between the accelerations due to the pressure gradient and the deflecting force must be exactly that required to keep the air moving in a circular path. From the formula for centrifugal force, $\frac{mV^2}{r}$, where m is the mass of air and r is the radius of curvature of the isobars, this acceleration can be shown to be $\frac{V^2}{r}$. A common example of centrifugal force is the tendency for a weight to fly off into the air if whirled round on a piece of string. The centrifugal force always acts outwards, at right angles to the tangent to the curved path along which the mass is moving. Thus for a cyclone the centrifugal force is directed against and has the opposite sign to the pressure gradient, while for an anticyclone it acts in the same direction and has the same sign as the pressure gradient. The signs of the term $\frac{V^2}{r}$ are as shown in equations (1) and (2). We know that acceleration due to the pressure gradient $= \frac{G}{\rho} = 2\omega V \sin \varphi$ and we have therefore, for cyclonic motion (Figure 3.7)

$$\frac{G}{\rho} - \frac{V^2}{r} = 2\omega V \sin \varphi \qquad (1)$$

and for anticyclonic motion (Figure 3.8)

$$\frac{G}{\rho} + \frac{V^2}{r} = 2\omega V \sin \varphi. \qquad (2)$$

Each of these quadratic equations can give two values for V but it can be shown that one of these values does not represent real motion.*

If we now transpose equation (1) into the form

$$V = \frac{G}{\rho 2\omega \sin \varphi} - \frac{V^2}{r 2\omega \sin \varphi}$$

it can be seen that, for cyclonic motion, the gradient wind speed is less than geostrophic wind by the quantity

$$\frac{V^2}{r 2\omega \sin \varphi}.$$

Similarly, transposing equation (2) into the form

$$V = \frac{G}{\rho 2\omega \sin \varphi} + \frac{V^2}{r 2\omega \sin \varphi}$$

we see that, for anticyclonic motion, the gradient wind speed is greater than geostrophic wind by the quantity

$$\frac{V^2}{r 2\omega \sin \varphi}.$$

---

* On a very much smaller scale, however, clockwise depressions in the northern hemisphere can occur, e.g. in dust devils, where the initial clockwise rotation might be caused mechanically, by turbulence or otherwise, and where $r 2\omega \sin \varphi$ is very small compared to the other terms so that the pressure gradient and the centrifugal force are in balance.

Table 3.2 below illustrates the difference between geostrophic and gradient wind speed for various radii of curvature of the isobars, in lat. 55°.

Table 3.2. Gradient wind speeds for various values of geostrophic wind and radius of curvature of isobars (Lat. 55°)

Geostrophic wind speed (knots)	Radius of curvature of isobars (nautical miles)											
	25	50	100	150	200	300	400	500	750	1000	1500	2000
	CYCLONIC CURVATURE Gradient wind speed (knots)											
5	4	4	5	5	5	5	5	5	5	5	5	5
10	6	7	8	9	9	9	9	10	10	10	10	10
15	8	10	12	13	13	14	14	14	14	15	15	15
20	10	13	15	16	17	18	18	18	19	19	19	20
30	13	17	20	22	24	25	26	27	28	28	29	29
40	16	20	25	28	30	32	33	34	36	37	38	38
50	18	24	30	33	35	39	40	42	44	45	47	47
60	21	27	34	38	41	45	47	49	52	53	55	56
70	23	30	37	42	46	50	53	56	59	61	64	65
80	24	32	41	46	50	56	59	62	66	69	72	74
90	26	35	44	50	55	61	65	68	73	76	80	82
	ANTICYCLONIC CURVATURE Gradient wind speeds (knots)											
5				5	5	5	5	5	5	5	5	5
10				12	12	11	11	11	10	10	10	10
15				24	19	17	17	16	16	16	15	15
20					32	25	23	22	21	21	21	20
30						47	39	36	33	32	32	31
40							63	53	47	45	43	42
50								79	62	58	55	53
60									80	72	67	65
70									103	88	80	77
80									147	106	94	89
90										128	108	102
	25	50	100	150	200	300	400	500	750	1000	1500	2000

Cyclostrophic Wind

For some types of circulation, notably in the case of tropical revolving storms, the equation for cyclonic motion, namely

$$\frac{G}{\rho} - \frac{V^2}{r} = 2\omega V \sin \varphi,$$

can be further simplified. In such a storm both terms on the left hand side are large and the term on the right-hand side is small since φ is small. If we decide to neglect this small value, we can write

$$\frac{G}{\rho} = \frac{V^2}{r} \text{ or } V = \sqrt{\frac{Gr}{\rho}}.$$

This value of V is known as the CYCLOSTROPHIC WIND and represents the wind which is found to blow nearly parallel with the (more or less) circular and crowded isobars in a tropical revolving storm.

Although the cyclostrophic wind does in fact give a good approximation to the actual wind found in most tropical revolving storms and tornadoes, nevertheless the effect of the earth's rotation does have some influence on the behaviour of these systems. For instance, tropical revolving storms seldom, if ever, form within a few degrees of latitude of the equator (usually about 5°) and, when formed, the winds in them blow around their low-pressure centres in the direction which is appropriate to the hemisphere, that is anticlockwise in the northern hemisphere and clockwise in the southern hemisphere. Although the origin of tropical revolving storms is not yet fully understood, there is now little doubt that they begin their existence at the centre of a small area of low pressure. The fact that they do not seem able to develop really intense wind circulation in the neighbourhood of the equator is probably due to the absence of the effect due to the earth's rotation. As soon as a cyclonic wind circulation can be recognized it is found to resemble the circulation normally found around a low-pressure area in the appropriate hemisphere, throughout the subsequent life of the storm. Thus, although the cyclostrophic wind gives a useful approximation to the winds found in a tropical revolving storm, it is important to bear in mind that the deflecting force due to the earth's rotation has a decisive influence on the behaviour of the storm at all stages in its life history.

Wind Structure at the Lowest Levels

It is a matter of everyday observation that the speed and direction of the wind are constantly varying over short periods. A knowledge of the main features of these variations is very important for the meteorologist, since they have been found to be closely related to the turbulence and lapse rate of the atmosphere as well as to the roughness of the surface, whether land or sea. For the past half-century wind structure near the surface has been thoroughly examined with the aid of daily records made by recording anemographs at numerous inland, coastal and island reporting stations in the British Isles and other countries. An example of a typical anemograph record in Figure 3.9, from which it can be seen that the magnitude of individual gusts can be measured and the variations of wind in periods of a few minutes, e.g. in squalls and thunderstorms, can be studied in detail.

The main feature of the wind structure is that although both speed and direction often maintain the same average values during a period of, say, several successive hours, nevertheless from minute to minute both speed and direction show a considerable range, the effect of which is to broaden the traces of both speed and direction in the characteristic way shown by almost any anemograph record. A typical example is shown in Figure 3.9. In terms of the wind itself there is a continuous succession of gusts and lulls associated with equally rapid changes of direction over a range which may exceed 30°.

This effect is most marked over land, where the gustiness, as shown by the magnitude of the variations of wind speed and direction between successive gusts and lulls in the wind during a period of a few minutes, is much greater than over the sea. In terms of the anemograph record the typical trace from a land station shows a broad ribbon for both speed and direction, whereas over the sea at a typical station such as the Bell Rock Lighthouse, 12 miles off the east coast of Scotland, the trace is quite narrow, showing a small difference between gusts and lulls. In the case of Bell Rock there is a clear difference between winds from different directions: winds from the west, having travelled

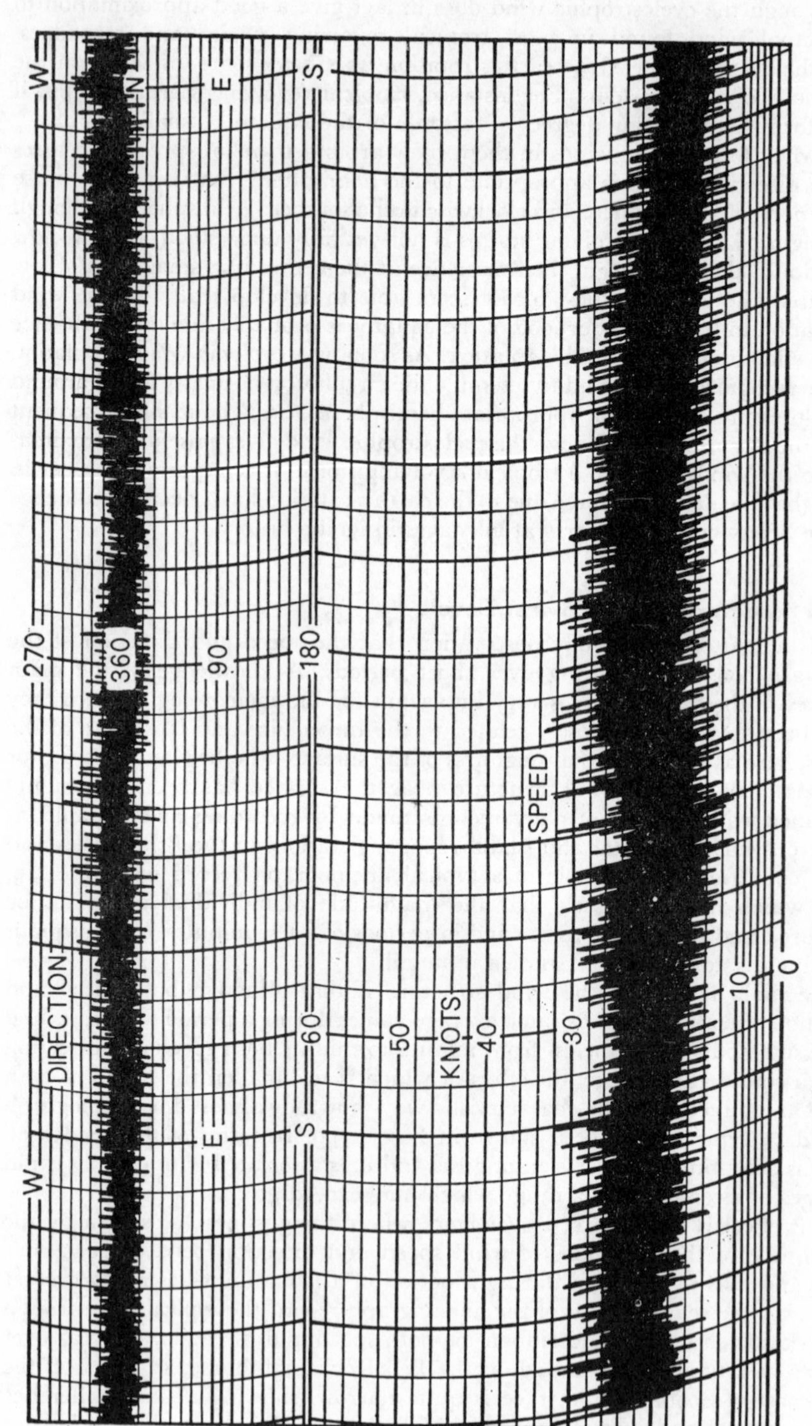

Figure 3.9. Wind direction and speed as recorded by an electrical anemograph

mainly over land, are noticeably more gusty than east winds, which have travelled a long distance over the North Sea.

Thus the wind is seldom steady but consists of a series of gusts and lulls at short intervals, and its speed and direction vary. Short-period wind variations, such as are found on any anemograph trace, are partly due to eddies caused by the roughness of the surface over which the wind travels. The mean wind speed over a period of time is therefore the mean of many gusts and lulls. Anemograph traces of wind at sea have a much narrower band of variations both of speed and direction, because, in general, the 'roughness' of the sea surface is much less than that of the land.

A gust is any sudden increase of wind of very short duration, commonly a few seconds. A measure of the intensity of a gust is given by the peak velocity, which is easily recognizable on the anemograph trace.

A squall comprises a rather sudden increase of the mean wind speed which lasts for several minutes at least, before the mean wind speed returns to near its previous value. A squall may include many gusts.

The cause of the gustiness of the wind also lies in the formation of eddies due to turbulence. In a very light breeze, wind speed and direction over a smooth surface are sometimes nearly constant, because the air flows quite smoothly. A small increase in wind speed is sufficient to break up the flow into eddies, or whirls of irregular shape and size which resemble in many ways the eddies visible on the surface of a fast-flowing river or in any river just below the piers of a substantial bridge. Through the agency of turbulence inequalities in the air are smoothed out faster than would otherwise be the case.

The increase of wind with height which is most pronounced near the earth's surface has been described earlier. It can equally well be looked on as a reduction of the wind speed below the geostrophic value from about 600 m to the surface, and which becomes most marked in the lowest layers. Irregularities in the land or sea surface bring about this reduction of wind speed, by causing turbulent eddies to form within the air flow near the surface. These eddies consume energy and thereby reduce the wind speed, making it seem as though the land or sea surface is exerting a frictional drag on the air flow near the ground. Over hilly country, forests or built-up areas many more eddies are produced and the gustiness and reduction of the wind speed below the geostrophic value is therefore much greater than over open level country and the open sea.

Within the layer affected by friction the wind direction veers with height in the northern hemisphere from the surface upwards until near 600 m it blows parallel to the isobars.

It was stated at the beginning of this chapter that wind structure in the lowest layers was related to the lapse rate as well as to turbulence. The reason for this can now be appreciated. Consider first, occasions when the lapse rate is large, i.e. steeper than the dry adiabatic, as often occurs when strong winds of polar origin are carrying air rapidly towards warmer latitudes and the air is much colder than the sea. Such conditions favour vigorous convection, which ensures that turbulent eddies formed in the strong wind prevailing are carried upwards rapidly. Turbulence is also responsible for bringing other eddies from a higher level downwards to a lower one. In the absence of turbulence and convection, the frictional effect of the sea surface would ensure that the horizontal velocity of the air at the surface would be less than that of the air at 600 m, i.e. the momentum of eddies near the surface would be less than that of eddies at 600 m.

However, in the case considered, surface layers are being rapidly mixed with other layers at higher levels; the sharing of momentum thus effected ensures that the mean surface wind over a short period will differ little from the geostrophic wind at about 600 m in speed and direction, although its gustiness is rather accentuated.

In the opposite case with a small lapse rate or an inversion in the lowest layers, say below 600 m, which is associated with warm air moving over a relatively cold sea surface, convection currents cannot develop. Thus turbulent eddies mainly develop, exist and decay at the level where they formed in each case. There is thus almost no exchange of momentum between layers, so that the frictional effect of the sea or land is almost wholly exerted upon the air closest to the surface, with the result that the surface wind is considerably less than the geostrophic wind at, say, 600 m.

Above 600 m the variation of wind speed and direction is largely dependent on the horizontal distribution of temperature. Within the region of 'westerlies' of both hemispheres winds normally increase in strength up to the height of the tropopause. (*See* page 2.)

Pressure Exerted by Wind

Figures for pressure exerted by the wind in millibars at standard air density are included in the Beaufort Force tables. This wind pressure is roughly proportional to the square of the wind speed. The mechanical effect of wind on any structure, however, depends not only on the mean wind speed but also on the short-term variations of speed and direction due to gusts and squalls. A structure, such as a bridge, may be perfectly safe while supporting a steady wind pressure but unsafe when subjected to a series of blows, as from a gusty wind, particularly if the frequency of blows approximates to the natural period of vibration of the structure. Similar reasoning will apply in considering the risk of a ship dragging her anchor in heavy weather in an exposed anchorage, or the more uncommon case of a ship listing due to shift of cargo or instability and experiencing a gusty wind on the beam, or in the case of a sailing-vessel which is exposed to sudden gusts.

Geographical Distribution of Winds

Other aspects of wind, particularly the general circulation of the wind over the oceans, the character of trade winds and monsoons and details of local winds, land and sea breezes, etc. are discussed in Chapter 7. Hurricane winds are dealt with in Chapter 11.

Waves

The mariner lives in intimate contact with the waves of the sea and is able to realize better than most people the extent to which their size and energy, as shown by their destructive power, are related to the speed of the wind. He is also accustomed by his training to make frequent estimates of wind force and to use various terms by which to describe the state of the sea surface.

The way in which the wind produces waves on the sea surface is still not completely understood. It is known that moving air exerts a drag upon a water surface over which it flows. Also we know that almost all air movement is to some extent turbulent, i.e. the motion is irregular with eddies and vortices complicating the flow in a manner similar to that observed in the flow of a stream of water. Due to the uneven pressures exerted by this irregular air flow, we can imagine that the water surface will itself become uneven, i.e. ripples will appear. As soon as this happens a new factor is introduced tending to increase the irregularity of the surface. For now, even assuming steady horizontal air flow, the air will exert a greater force on the windward slopes of the ripples than on their leeward side. This causes the ripples to be driven along in the direction of the wind and to increase in size as long as the wind is moving more rapidly than the water.

Elementary theory shows that, as a close approximation, small waves occurring at the boundary surface between two fluids (in this case air and water) will be such that the vertical cross-section along the line of propagation conforms to a sine wave (the curve obtained by plotting θ against sine θ).

In this case useful relationships can be established between the various measurements used in observing waves.

A simple wave is described in terms of the following measurements:

(*a*) SPEED, **C**, usually measured in knots,

(*b*) LENGTH, **L**, usually measured in metres,

(*c*) PERIOD, **T**, measured in seconds

and (*d*) HEIGHT, **H**, usually measured in metres.

In the case of simple sine waves it can be established that

SPEED (in knots) = 3·1 times PERIOD

LENGTH (in metres) = 1·56 times (PERIOD)^2

where the period is measured in seconds.

Using these equations, measurement of one of the first three of these variables can be used to calculate the other two. The following table gives the values of wave length and speed corresponding to specified wave periods.

Table 3.3. Wave length and speed in terms of period—for simple sine wave

Period (seconds)	Length (metres)	Speed (knots)
2	6·2	6·2
4	25·0	12·4
6	56·2	18·6
8	99·8	24·8
10	156·0	31·0
12	224·6	37·2
14	305·8	43·4
16	399·4	49·6
18	505·4	55·8
20	624·0	62·0

There is no theoretical relation between the fourth variable (wave height) and the other parameters just listed. Waves of a specified period (and corresponding length and speed) can be associated with a wide range of different heights

When the height of a wave is small compared with its length the wave profile can be adequately represented by a simple sine curve. As the height becomes relatively greater, however, it is seen that the crests become sharper and the troughs much more rounded, the precise profile being a curve known as a 'trochoid'. This is the curve that would be traced on a bulkhead by a marking point fixed to the spoke of a wheel, if we imagined the wheel to be rolled along under the deck-head.

In Figure 3.10 the large circle represents the wheel and **P** the marking point on a spoke **OP**, the distance from the axle being called the tracing arm. The arrow shows the direction in which the circle rolls and in which the wave is supposed to be travelling. **AB** is the base, i.e. the straight line under which the circle is to roll, the length **AB** being equal to the half-circumference of the wheel **AR**.

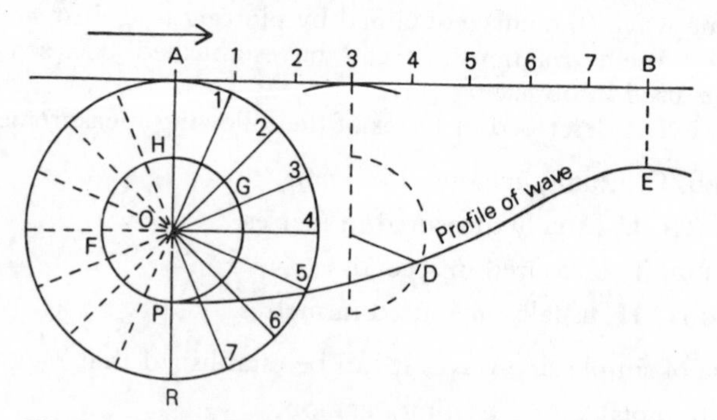

Figure 3.10. Representation of a trochoidal wave form

Now as the circle rolls, when position 3 of the circle reaches position 3 of the base, the semi-circle **FPG** will be in the position shown by the dotted semicircle; and the marking point **P** will coincide with the point **D**, having described part of a trochoid **PD**. When the circle has completed half a revolution, the marking point **P** will coincide with **E**, having described the trochoid curve **PDE** which is half a wave's length; the diameter **POH** represents the height of the wave. The nearer the marking point is to the axle of the wheel, the flatter will be the trochoid.

In an ideal wave each water particle revolves with uniform speed in a circular orbit, perpendicular to the wave ridge (the diameter of the orbital circles being the height of the wave) and completes a revolution in the same time as the wave takes to advance its own length. At a wave crest the motion of the particles is wholly horizontal, advancing in the same direction as the wave; at mid height on the front slope it is wholly upwards; in the trough it is again horizontal but in the opposite direction to the travel of the wave, and at mid height on the back slope it is wholly downwards. This motion may be seen by watching a floating object at the passage of a wave. The object describes a circle but is not carried bodily forward by the wave.

The disturbance set up by wave motion must necessarily extend for some distance below the surface; but its magnitude decreases very rapidly in accordance with a definite law, the trochoids becoming flatter and flatter as the depth increases, and the water particles revolving in ever-decreasing circles. At a depth of one wave's length the disturbance is less than a 500th part of what it is at the surface, so that the water at that depth may be considered undisturbed. The motion associated with the largest ocean waves is inappreciable at even moderate depths, as is demonstrated by experience in submarines.

Waves actually observed at sea seldom present the simple picture described above. Instead a complicated wave form such as is illustrated in Figure 3.11 more resembles what is commonly observed.

The profile shown in Figure 3.11 is the wave form recorded by a sophisticated instrument called a wave recorder. At present, wave recorders can only be effectively used from the shore or from stationary ships, buoys or drilling platforms, but within the relatively short period since their introduction they have already provided much useful information. The complex trace shown in the Figure can be regarded as due to the superposition of a number of simple regular wave motions having different wave lengths and periods.

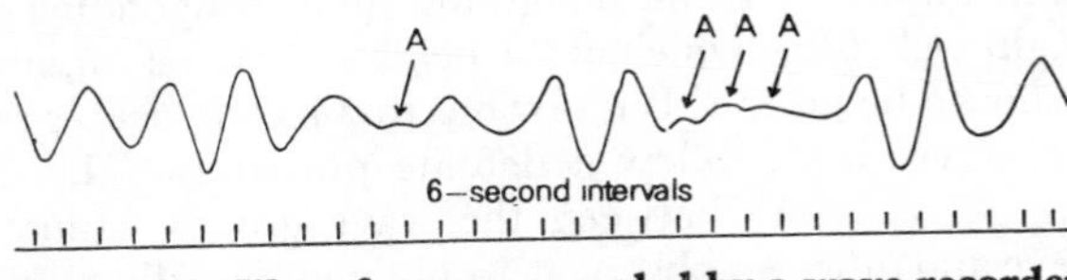

Figure 3.11. Wave form as recorded by a wave recorder

Wave Groups

Experience shows that waves generally travel in groups with patches of dead water in between, the wave height being a maximum at the centre of each group. We have said earlier that any observed wave motion can be regarded as built up from a number of simple wave forms. Let us consider, for example, the superposition of two simple wave motions having the same height but slightly different periods. If the crests of the two wave motions are made to coincide at the initial point of observation the height of the resultant wave will be twice that of each component wave. To each side of this point, however, owing to the difference of period, the additive effect becomes less until a point is reached where the heights of the component waves, being of different sign, completely annul each other's effect. Beyond this point the heights again become additive until the troughs of the component waves coincide. In other words, there is a variation of height superposed on the ordinary wave motion. It can also be shown that two simple wave trains moving in slightly different directions give a resultant pattern composed of 'short-crested' waves as distinct from the 'long-crested' waves of simple wave motions.

The speed of a wave group is not the same as that of the individual waves comprising it. Each individual wave in its turn emerges from the dead water in the rear of the group, travels through the group and subsides in the dead water ahead of it. The speed of the wave group must therefore be less than the speed of the individual wave. Both theoretical considerations and experience show that the wave group travels at one-half the speed of the individual wave.

Limiting Wave Height

We have seen how, given a wind blowing over the sea surface, waves develop and increase in height with time. There is a limit to the average height of the waves so produced because of the instability which arises when the steepness (the ratio of height to length) of the wave exceeds a certain value (about 1 in 13). When the limiting value is reached, surplus energy received from the wind is dissipated by the breaking of the waves at the crests (formation of 'white horses'). For this reason the mean maximum height of sea waves is roughly proportional to their length, for example wind-driven waves of length 130 metres would not be expected to have a mean maximum height greater than 10 metres. The foregoing limit in terms of steepness applies to persistent waves. It does not apply to the occasional transitory wave which arises from the chance coincidence of two or more crests.

Determining Wave Height, Significant Height

Determining wave height by visual observation is not a simple matter. Looking again at Figure 3.11 and imagining the corresponding appearance of the sea, one might well ask—'which of the heights are to be considered and what average value should be taken?' It is obvious that if comparable results are to be obtained the observer must follow a definite procedure. The flat and badly formed waves ('**A**' in 3.11) between the wave groups cannot be observed accurately by eye and different observers would undoubtedly get different results if an attempt were made to include them in the record. The method to be adopted, therefore, is to observe only the well-formed waves in the centre of the wave groups. Reliable average values can only be obtained by observing at least 20 waves. Of course these cannot be consecutive; a few must be selected from each succeeding wave group until the required number has been obtained. Further details regarding the observing of wave heights are given in the *Marine Observer's Handbook*.

In theoretical and statistical work on waves much use is made of the concept of 'significant height'. This is defined as the average of the highest one-third of the waves. It can readily be determined from the record of a wave recorder and has been found to agree fairly well with the wave height reported by experienced observers aboard ship who are endeavouring to estimate the average height of the higher well-defined waves. Of obvious interest to all concerned is maximum wave height. This, however, is difficult to define and to evaluate. Largely because of the occasional occurrence of 'freak' high waves due to the local superposition of the crests of a number of wave trains, the absolute maximum is many times in excess of the average. Also, to be meaningful, one must specify the length of the observing interval. A commonly quoted value is $\mathbf{H}_{max}$ (10 min) which is the highest wave to be expected over a 10-minute interval. According to Darbyshire and Draper,* the 10-minute maximum is about 1·6 times the significant height. It must not be assumed that the highest wave in each 10-minute interval will be 1·6 $\mathbf{H}_s$—it might be as low as 1·4 $\mathbf{H}_s$ or as high as 1·8 $\mathbf{H}_s$, but on the average it is reasonably close to 1·6 $\mathbf{H}_s$. Sometimes it can be twice the value of $\mathbf{H}_s$ and such a value can be expected once in every 3 hours. Figure 3.12 shows the relation between the expected 10-minute maximum, and the corresponding maximum of any longer period from 1 to 48 hours. Thus if the 10-minute maximum wave height is 5 metres, the maximum height that might be expected over a 12-hour period (in coastal waters) would be 6·95

* M. Darbyshire and L. Draper, Forecasting wind-generated sea waves. *Engineering*, April 1963.

metres. Where depths are greater than 100 metres the corresponding maximum height would be 7·1 metres.

In terms of numbers of waves rather than of the sampling interval, it has been shown that: 1 wave in 23 is twice the average height,
1 wave in 1175 is 3 times the average height
and 1 wave in 300 000 is 4 times the average height.

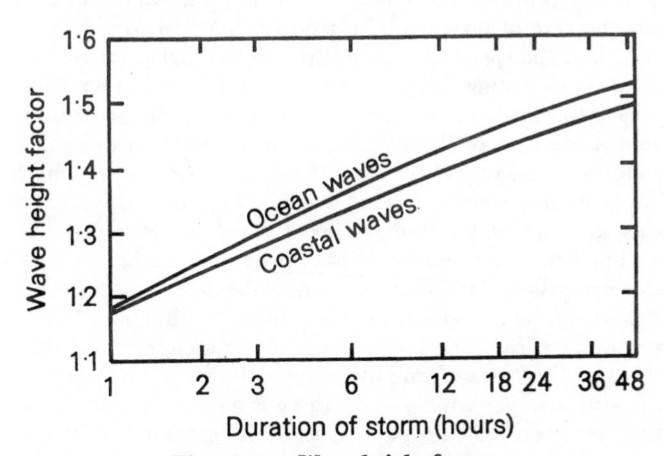

Figure 3.12. Wave-height factor
This gives the factor by which the 10-minute maximum wave height must be multiplied to give the probable height of the highest wave in a storm of a given duration.

It may be of interest to note that the highest wave known to have been recorded to date (1976) by a wave recorder is 25 metres. It is to be expected that when such instruments have been employed for longer periods and in more varied locations this value will be appreciably exceeded.

Swell

Much of the foregoing discussion has been limited to the consideration of sea waves and it is now necessary to turn our attention to swell. The distinction between the two is as follows: sea waves are those generated by the wind blowing at the time, and in the recent past, in the area of observation. Swell waves are waves which have travelled into the area of observation after having been generated by previous winds in other areas. For example, in a given sea area, the wind at the time of observation might be calm. This means, from the above definition, that there will be no 'sea' waves. We know from experience, however, that there may well be an appreciable wave motion on the sea surface, typically having a long wave length in proportion to its height. This is 'swell'. An oily unbroken surface is another characteristic of the pure swell wave. These waves are commonly due to the action of strong winds in some distant area and may travel thousands of miles from their origin before dying away. The energy associated with waves of short wave length is more rapidly dissipated than that associated with longer waves. Hence, in general, it follows that swell waves are long in comparison with the wind-driven waves at the place of observation.

As swell travels its height decreases. Investigations by the Institute of Oceanographic Sciences show that if R is the distance, in nautical miles, from the point of generation, then the amplitude at distance R is $\left(\frac{300}{R}\right)^{\frac{1}{2}}$ of that point of generation. Thus, a swell would lose one-half of its height in travelling a distance of 1200 nautical miles. The long swells are the greatest travellers.

The onset of swell, particularly in tropical and subtropical latitudes is often a first indication of a storm or hurricane in the vicinity. In any calculations concerning the onset of swell, it must be borne in mind (as discussed under wave groups) that the rate of progress of a group of swell waves, from the point of origin, is only half the speed of travel of the individual swell waves.

While it is easy to distinguish between 'sea' and 'swell' in terms of definitions it is rarely easy to make the distinction in practice. The great majority of sea states present a mixture of the two phenomena which is difficult to unravel. Clearly if one experiences a succession of waves of long wave length and with height of say 3 metres when the wind has not exceeded 10 knots, these waves must be classed as swell because the local wind is not strong enough to be responsible. Their direction of motion may provide a further clue. If it does not accord with the local wind then the waves must be described as swell. Frequently two wave trains can be recognized, having different direction. Even if the wave heights are commensurate, the waves whose direction agrees with the local wind can safely be described as 'sea' and the other series should be reported as 'swell'. Particularly with stronger winds, when there is a considerable sea, it is difficult to distinguish between sea and swell if there is not much difference between their directions of motion. In cases in which it becomes too difficult to differentiate between the two wave types, it is best to regard the combined motion as being due to sea waves.

Waves in Shallow Water

All the previous remarks refer to waves in deep water. When a deep-water wave enters shallow water it undergoes profound modification. Its speed is reduced, its direction of motion may be changed and, finally, its height increases until, on reaching a certain limiting depth, the wave breaks on the shore. Water may be regarded as shallow when the depth is less than half the length of the wave.

The decrease in speed when a wave approaches the shore accounts for the fact that the wave fronts become, in general, parallel to the shore prior to breaking. Figure 3.13 shows a wave, approaching the shore at an angle, being refracted until it becomes parallel to the shore.

The same reasoning may be applied to explain how waves are enabled to bend round headlands and to progress into sheltered bays.

Forecasting Sea Waves

A forecast of the wave heights likely to be encountered on a voyage is of obvious value to the mariner. The meteorological services of several countries provide such forecasts either in terms of wave-height distribution forecasts for a given area and period, or in connection with a 'weather routeing' service wherein the vessel may be advised to proceed on a particular route in order to avoid the higher seas expected on the normal route. The way in which such forecasts are prepared is briefly summarized as follows.

The height of sea waves is determined by the following variables:

1. wind speed,
2. duration of the wind and
3. wind fetch.

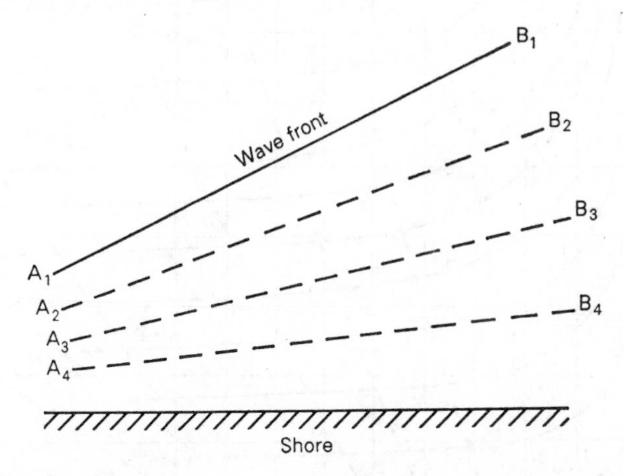

Figure 3.13. Refraction of a wave approaching the shore at an angle

Assuming a constant wind blowing for a long period—say more than 48 hours, then the wind speed determines the value of the significant wave height produced. If the wind is varying over the period an average value can be assumed. It is found in practice that, given an initial and final value of the forecast wind, it is better to take the final value less a quarter of the difference between the initial and final values, rather than to use the simple average.

Assuming that the initial conditions are a flat sea and no wind, then the introduction of the wind has a small effect initially which increases with time. In other words the height of the waves is also a function of the duration of the wind. A further variable is 'fetch'. This is defined as the distance up-wind from the point of observation over which the wind blows constantly and uninterruptedly over the sea. For example, with a steady westerly wind over the whole region, the fetch at a point 100 n. mile to the east of an extensive north-to-south coastline will be 100 n. mile. If there was no coastline in the picture, and the steady wind could be traced back in the direction from which it was blowing for 500 n. mile from the point of observation before reaching a position beyond which there was a major change in direction, then the fetch would be 500 n. mile.

Knowing how the waves depend on these three variables, one can produce a forecast of the wave height based on forecasts of the variables. Various authorities have produced nomograms which enable this to be done. Figure 3.14 is a reproduction of a nomogram due to Dorrestein.*

---

* Quoted in Korevaar, C. G., WMO Met Services to Marine and Coastal Activities, *Proc. Regional Seminar*, Rome 1974.

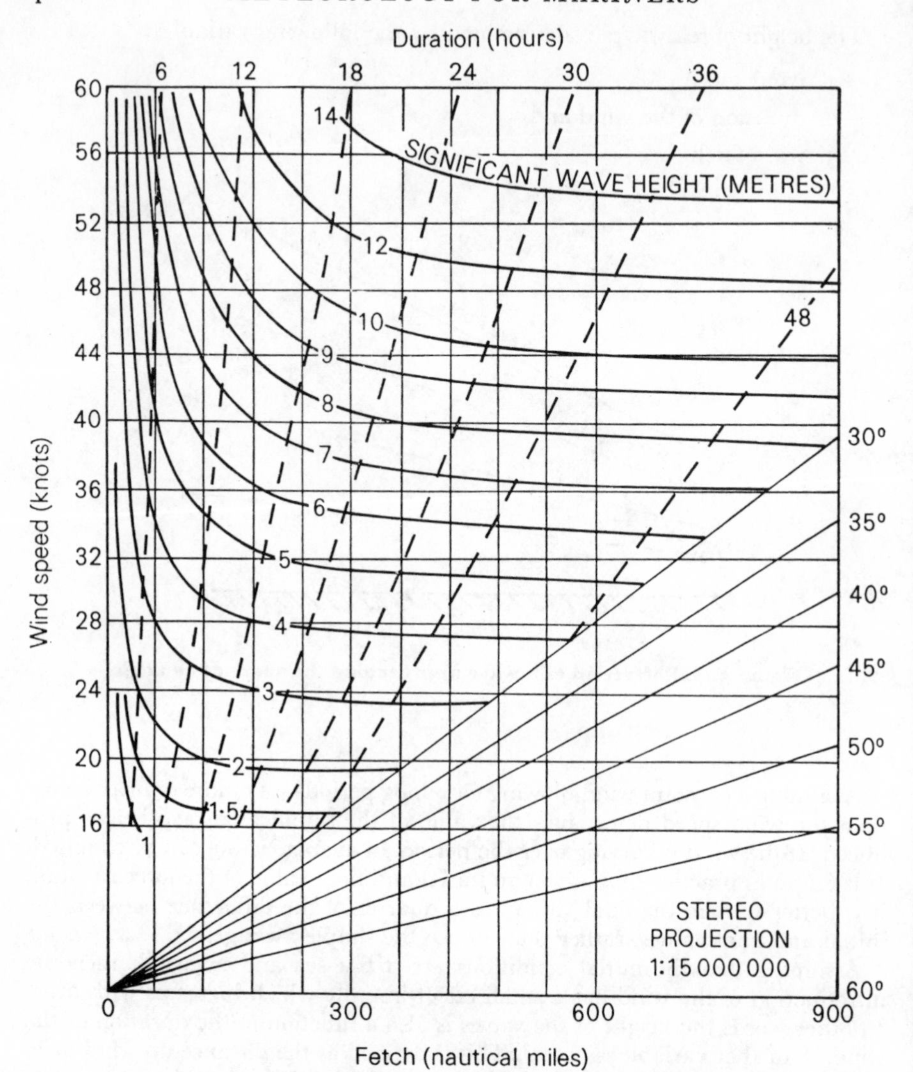

Figure 3.14. Dorrestein's nomogram for significant wave heights

Assuming a knowledge of the wind speed, duration and fetch, one can read off from the nomogram the corresponding significant wave height. Assuming a forecast wind speed of, say, 30 knots, the diagram shows that, at the right-hand margin of the diagram (corresponding with maximum wind duration and maximum fetch) the value of the significant wave height is about 4·9 metres. If the fetch were to be limited to 240 n. mile, this height would be reduced to 4·5 metres. Moreover, if the duration of the wind were only 12 hours the height would be reduced to 4·0 metres. In any specific case the height to be chosen is the lower of the two values obtained from reading off the height against duration and fetch. It can be seen from the diagram that the duration is an important factor up to a value of about 24 hours but for longer periods there is little additional effect. In the case of fetch, the effect increases significantly up to about 500 n. mile, but for greater distances there is little further effect.

Other authorities have produced similar diagrams which agree fairly well with the results just described.

Figure 3.15 is a diagram representing the variation of significant height with wind speed assuming fetch 1000 nautical miles and duration 50 hours (i.e. fetch and duration unlimited). This represents an average curve derived from those of a number of authorities which were in reasonably close agreement.

The Value of Wave Data

The development and introduction of wave recording instruments in recent years has considerably advanced our knowledge of waves. At the present time, however, their use is largely confined to inshore positions or to stationary vessels such as lightships, weather ships or research vessels, or on buoys. There is an obvious need for an inexpensive wave recorder which could be used on a typical commercial vessel, but at present this requirement has not been satisfactorily met. So in general the demand for wave information on the high seas still depends upon the supply of wave reports made 'the hard way' by visual observation. The need for these reports is twofold. Firstly, a forecast of the wave conditions in a particular position tomorrow depends, among other things, upon the waves occurring in the area today. This might be called the 'synoptic' requirement. Secondly, a very great number of individual reports of waves occurring in a given region under different conditions, at different times, is required before an effective 'wave climatology' can be established. Such a wave climatology is required to answer such questions as 'what is the average, maximum and minimum wave height in a particular region at a particular time of year?' From data already accumulated we can provide some sort of answer to such questions in the more frequented parts of the ocean. But a much more precise answer could be given if it were based on a greater number of observations. Even on the more frequented routes we are still hampered by too few observations. In some of the less frequented areas the data are so sparse as to be inadequate to supply the basic requirements of average and extreme values.

The synoptic requirement is particularly important in connection with the weather routeing of ships (*see* page 199). The climatological need for wave data concerns a large variety of purposes including the design and behaviour of ships at sea. The behaviour of an individual ship is governed to a considerable extent by the period of her roll and pitch in various conditions of loading in relation to the period of the waves she encounters, and her longitudinal and transverse strength calculations must inevitably take account of similar factors. Other purposes include the design and orientation of harbours and the construction of breakwaters; problems of coast erosion and silting; discharging of ships in open anchorages; landing operations on exposed beaches. Mention has already been made of the value of noting the onset of a swell which might serve as a clue to the approach of a tropical revolving storm. Wave reports can also provide valuable information in connection with inquiries into storm damage, whether in the tropics or elsewhere.

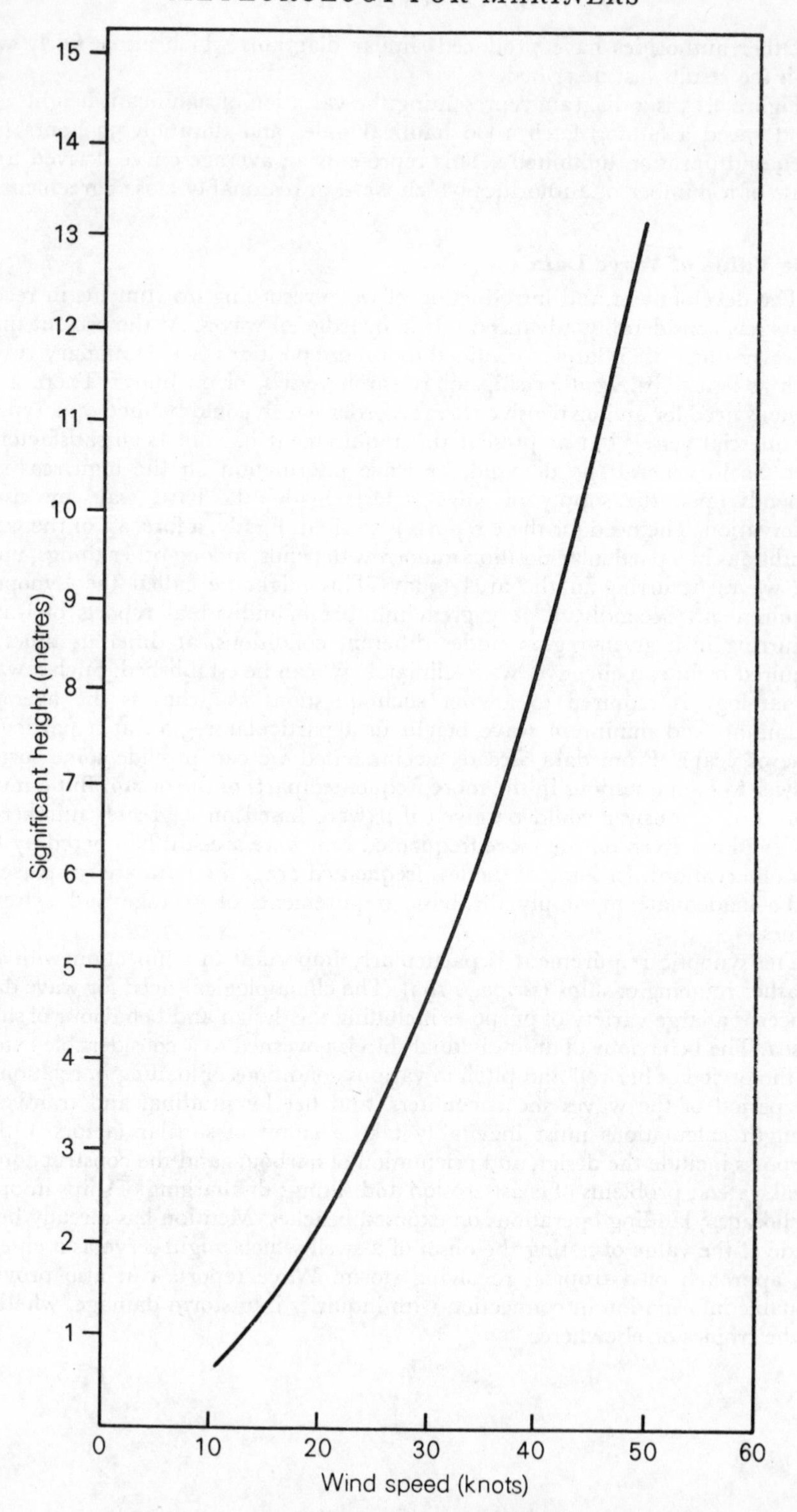

Figure 3.15. Significant wave height in terms of wind speed, with fetch 1000 nautical miles and duration 50 hours

CHAPTER 4

CLOUD AND PRECIPITATION

Formation of Cloud

The formation of cloud results from the condensation of water vapour in the atmosphere. When such condensation happens near the surface, mist or fog is produced and this commonly builds upwards and may extend to something like 1000 feet (300 metres) above the surface. The only distinction between fog and cloud composed of water droplets is that cloud has a base above the earth's surface. The distinction is arbitrary in that, if one imagines a layer of stratus extending across a valley to surrounding hills, then an observer at the bottom of the valley would report cloud, while another observer on the hillside who was within the cloud would report fog. The formation of fog will be discussed in greater detail in Chapter 5.

Clouds occur almost wholly within the lowest layer of the atmosphere, the troposphere (*see* Figure 1.1), and can form only in air which has been cooled sufficiently to bring about condensation. Ascent in convection currents, up-sliding at a frontal surface, forced ascent over high ground and, in certain circumstances, direct cooling of air by radiation, are all processes which lead to a fall of temperature. Condensation can only occur if the air contains sufficient water vapour for it to become saturated before the fall in temperature ceases.

The presence of minute particles or nuclei is also necessary to initiate the formation of water droplets or ice crystals which must precede the visible products of condensation such as cloud, rain, drizzle and snow. Condensation nuclei in the atmosphere result from a number of processes but chiefly from the action of strong winds on salt spray from breaking waves, from the products of industrial processes and, less frequently, after volcanic eruptions or when strong winds blow over desert regions.

Laboratory experiments have shown that the smallest droplets of pure water can be cooled to near $-40°$C before they freeze, and the presence of water-droplet clouds at these temperatures has been inferred from observations of coronae and 'glories'. It is concluded that direct condensation from vapour to ice crystals does not occur in the atmosphere until the temperature approaches $-40°$C, and that at higher temperatures condensation takes place on suitable nuclei which are always present in sufficient numbers.

In the stratosphere the amount of water vapour is very small. Thin high cloud occasionally forms in the lower levels of the stratosphere but, apart from the comparatively rare nacreous or 'mother-of-pearl' clouds, average altitude 82 000 feet (25 km) and noctilucent clouds, average altitude about 280 000 feet (85 km), condensation is seldom observed at higher levels.

In temperate latitudes cloud can form at altitudes ranging from near the ground to about 43 000 feet (13 000 metres) but, as the troposphere extends to greater heights in the tropics, the highest cloud there may reach about 49 000 feet (15 000 metres). The tropopause is higher in summer than in winter at all latitudes, allowing convection to lift water vapour to greater heights in the summer half of the year; the highest clouds are therefore somewhat higher in summer than they are in winter.

The production of water droplets or ice crystals takes place when temperatures fall below the dew-point or frost-point. The air may cool in a number of ways, but the formation of practically all cloud, as distinct from fog or mist, results from adiabatic cooling when a parcel of air ascends to regions of lower pressure. There are four ways in which such an ascent may be brought about:

1. By turbulence. The formation of turbulence cloud is illustrated in Figure 4.1. Water vapour is carried up and down within the turbulent layer, cooling as it rises and warming as it falls. If, during the ascent, it cools to below the dew-point, condensation will occur, this level marking the base of a layer of turbulent cloud.

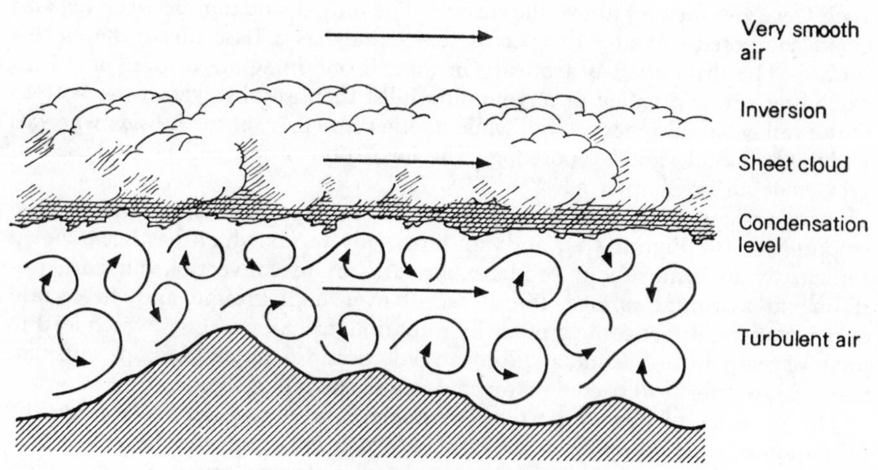

Figure 4.1. Formation of a layer of turbulence cloud

This type of cloud is common when tropical-maritime air reaches temperate latitudes. Such air has a high moisture content and on reaching cooler regions its relative humidity becomes even higher. Under these conditions a layer of cloud can be widespread and persistent over the oceans at any season, and over land areas for much of the year, although inland in summer the cloud will often become broken or even disappear altogether during the day.

Over land the most rapid changes in turbulence and relative humidity take place in the early evening and again after dawn; it is at these times that changes in turbulence cloud are most likely with fresh clouds being formed and/or existing cloud dissipating. Over the open sea (in the absence of air-mass changes) there is little diurnal variation of temperature, humidity or turbulence, and therefore there are fewer changes in cloud types and amounts.

2. By orographic ascent. When moist air is blown against rising ground such as a range of hills or a mountainous island the forced ascent and adiabatic cooling can lead to condensation and the formation of orographic cloud (Greek *oros* = mountain) as illustrated in Figure 4.2.

Orographic clouds formed on a small scale, as over an isolated ridge or a small hilly island, often exhibit a characteristic 'laminated' shape resembling a flat plate. The island or hill then has a cap of cloud which may envelop the summit or lie above it, but not move with the wind. In effect, the cloud is continually renewed as air enters it on the windward side and continually evaporating as air leaves it on the leeward side. The 'table-cloth' of Table Mountain and the 'levanter' cloud over Gibraltar are examples of orographic clouds.

When warm, moist air is blown against an extensive mountain range the orographic cloud may cover a wide area and be accompanied by heavy and prolonged precipitation.

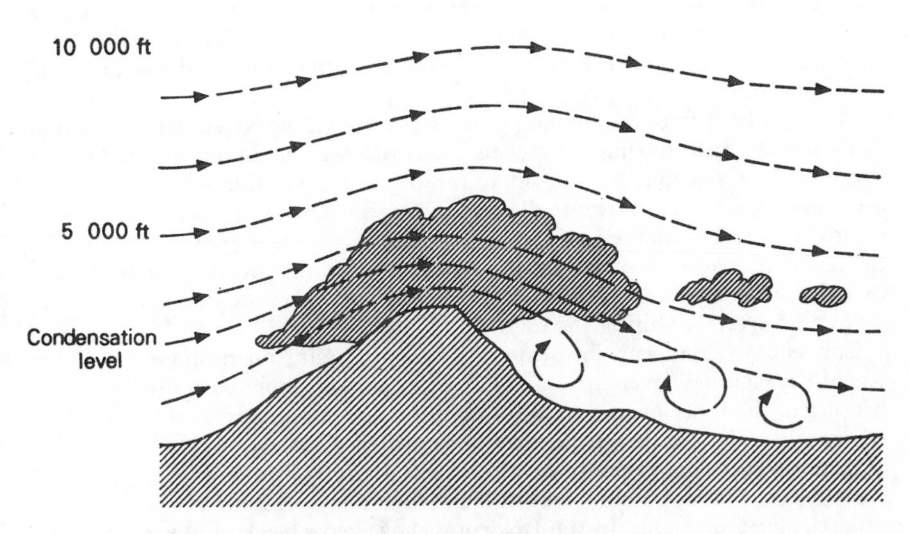

Figure 4.2. Formation of orographic cloud

3. By general ascent or upsliding over a wide area. Ahead of a frontal depression, when warm air is overtaking cold air at the warm front, the warm air slides up over the cold air all along the surface of separation. In the rear of a depression, another cold air mass follows and undercuts the warm air at the cold front, forcing the warm air upwards. (It will be shown in Chapter 9 how these air movements occur.)

General ascent of air due to large-scale convergence also occurs in polar and tropical depressions which do not possess fronts. Many extensive belts of more or less continuous cloud and precipitation are produced in this way.

4. By convection. When a portion or 'parcel' of air is heated to a temperature higher than that of the surrounding atmosphere, the parcel will rise freely if the environmental lapse rate exceeds the dry adiabatic up to the condensation level and will continue to rise, accompanied by cloud development, if the environmental lapse rate exceeds the saturated adiabatic above this level.

Convection may be caused by local heating of the air in contact with the ground as on sunny days over land, or by cool air moving over progressively warmer sea, as occurs in the trade-wind belt or when polar air flows to more temperate latitudes. This warmed air over land or sea cannot rise as a single mass over a large area because other air must fall to take its place; it therefore ascends in distinct columns, hence the well-known pattern of individual clouds with clear spaces between where the air is descending. Such columns or 'heaps' of cloud are known as 'cumulus' (*see* also following section on cloud forms).

The formation of these individual columns is facilitated over land by irregularities in the surface and its vegetation; over the sea a small island is often sufficient to start a convection cell. On sunny days such an island can become capped with cumulus cloud as the day advances and, if a breeze is blowing, a succession of cumuli may develop and drift for a considerable distance downwind before they disperse. Whether the day is calm or windy the typical cumuliform convection clouds are readily distinguished from the flat or lens-shaped orographic clouds which remain in the same position under all conditions of wind.

Over the land fine-weather cumulus clouds show a marked diurnal variation. They usually form during the morning and reach a maximum in number and size by mid afternoon; they dissipate rapidly in the evening when the ground cools and convection currents die out. Over the sea the diurnal variation of clouds is much less marked and tends to be the reverse of that over land because the sea temperature remains nearly constant while the air aloft cools at night by radiation; convection and cloud formation may therefore increase a little by night and decrease during the day.

Convection cloud can form along a coastline with an onshore wind even when it is cloudless over the sea. This can occur on sunny days owing to solar heating of the land, or as a result of uplifting if there are cliffs or hills near the coast, in which case the clouds may be regarded as partly orographic.

In the trade-wind regions over the oceans and on summer days in temperate latitudes over land, cumulus clouds increase in depth only slowly unless the air is particularly unstable. In the Doldrums, however, once cumulus clouds have started to develop they often do so rapidly.

When a steep lapse rate exists through a deep layer, convection will be vigorous, resulting in large cumulus or cumulonimbus. When the rapidly rising air reaches a level where a more stable lapse rate makes further ascent impossible the water vapour spreads sideways to form the characteristic layer of ice crystals known as 'anvil cirrus'.

Cloud Form and Height

Clouds may assume an almost infinite variety of forms, but for purposes of description and identification they are classified as simply as possible. The present international cloud classification, which distinguishes 10 main forms, has been in use since 1895. Although the condensation of water vapour which produces clouds may occur in different ways the international classification is based solely on the clouds' appearance.

For simplicity, clouds are grouped broadly according to their heights into high (or upper), medium (or middle) and low clouds. A fourth category 'clouds of marked vertical extent' have their bases at 'low cloud height' and their tops at 'medium' or 'high' cloud height.

Because the levels at which each cloud type can form are subject to considerable variations the heights given above should not be taken as precise limits.

Clouds may be composed of water drops, ice crystals, or a mixture of the two, according to their temperature. High cloud consists mainly of ice crystals, in medium cloud water drops predominate and low cloud usually consists entirely of water drops. Clouds of marked vertical extent frequently contain water drops near their base and ice crystals in their tops.

The following table shows the main cloud forms and the average range of heights of the four categories.

Table 4.1. Cloud forms and heights (temperate latitudes)

High Clouds	Medium Clouds	Low Clouds	Low clouds of marked vertical extent
45 000 feet (14 000 m) to 18 000 feet (5500 m)	23 000 feet (7000 m) to 6500 feet (2000 m)	6500 feet (2000 m) to close to ground	45 000 feet (14 000 m) to close to ground
Cirrus (Ci) Cirrocumulus (Cc) Cirrostratus (Cs)	Altocumulus (Ac) Altostratus (As)	Nimbostratus (Ns) Stratus (St) Stratocumulus (Sc)	Cumulus (Cu) Cumulonimbus (Cb)

Cloud Classification

Cloud forms are divided into 10 main groups, called genera, which are set out in Table 4.1.

The definitions of genera, given below, are limited to a description of the main types and the essential characteristics necessary to distinguish a given genus from genera having somewhat similar appearance.*

1. CIRRUS (Ci). Detached clouds in the form of white, delicate filaments or white or mostly white patches or narrow bands. These clouds have a fibrous (hair-like) appearance, or a silky sheen, or both.

Cirrus appears in the most varied forms such as isolated tufts, lines drawn across a blue sky, branching feather-like plumes, or curved lines ending in tufts; it is often arranged in bands which cross the sky like meridian lines, and which, owing to the effects of perspective, converge to a point on the horizon or to two opposite points (cirrostratus and cirrocumulus often take part in the formation of these bands).

2. CIRROCUMULUS (Cc). Thin white patch, sheet or layer of cloud without shading, composed of very small elements in the form of grains, ripples etc., merged or separate, and more or less regularly arranged; most of the elements have an apparent width of less than one degree.

In general, cirrocumulus represents a degraded state of cirrus or cirrostratus both of which may change into it. In this case the changing patches often retain some fibrous structure in places. Real cirrocumulus is uncommon. It must not be confused with small altocumulus or the edge of altocumulus sheets.

3. CIRROSTRATUS (Cs). Transparent, whitish cloud veil of fibrous (hair-like) or smooth appearance, totally or partly covering the sky, and generally producing halo phenomena.

Sometimes it is quite diffuse and merely gives the sky a milky look; sometimes it shows a more or less distinctly fibrous structure with disordered filaments.

* Photographs of cloud forms can be found in the *Marine Observer's Handbook*.

4. Altocumulus (Ac). White or grey, or both white and grey, patch, sheet or layer of cloud, generally with shading, composed of laminae, rounded masses, rolls, etc., which are sometimes partly fibrous or diffuse and which may or may not be merged; most of the regularly arranged small elements usually have an apparent width of between one and five degrees.

Altocumulus may sometimes be confused with cirrocumulus but, if the clouds have shading, they are by definition altocumulus, even if their elements have an apparent width of less than one degree.

5. Altostratus (As). Greyish or bluish cloud sheet or layer of striated, fibrous or uniform appearance, totally or partly covering the sky and having parts thin enough to reveal the sun at least vaguely, as through ground glass.

This cloud is like thick cirrostratus but without halo phenomena. Sometimes the sheet is thin with forms intermediate with cirrostratus (altostratus translucidus); sometimes it is very thick and dark (altostratus opacus), sometimes even completely hiding the sun or moon. In this case differences in thickness may cause relatively light patches between very dark parts, but the surface never shows real relief, and the striated or fibrous structure is always seen in the body of the cloud.

6. Nimbostratus (Ns). Grey cloud layer, often dark, the appearance of which is rendered diffuse by more or less continuously falling rain or snow, which in most cases reaches the ground. It is thick enough throughout to blot out the sun.

It appears as though feebly illuminated, seemingly from inside. When it gives precipitation this is in the form of intermittent or continuous rain or snow. But the precipitation alone is not sufficient criterion to distinguish the cloud which should be called nimbostratus even when no rain or snow falls from it. There is often precipitation which does not reach the ground; in this case the base of the cloud is always diffuse and looks 'wet' on account of the general trailing precipitation (virga) so that it is difficult to determine the limit of its lower surface. Low ragged clouds frequently occur below the lower surface, with which they may or may not merge.

7. Stratocumulus (Sc). Grey or whitish, or both grey and whitish, patch, sheet or layer of cloud which almost always has dark parts, composed of tessellations, rounded masses, rolls, etc., which are non-fibrous (except for virga) and which may or may not be merged; most of the regularly arranged small elements have an apparent width of more than five degrees.

Stratocumulus may sometimes be confused with altocumulus. If most of the regularly arranged elements, when observed at an angle of more than 30 degrees above the horizon, have an apparent width of more than five degrees, the cloud is stratocumulus.

8. Stratus (St). Generally grey cloud layer with a fairly uniform base which may give drizzle, ice prisms or snow grains. When the sun is visible through the cloud its outline is clearly discernible.

Stratus does not produce halo phenomena except, possibly, at very low temperatures. Sometimes stratus appears in the form of ragged patches or irregular shreds which are designated as stratus fractus (St fra).

Stratus is distinguished from stratocumulus by the fact that it shows no evidence of the presence of elements, merged or separate. Stratus fractus is distinguished from cumulus fractus in that it is less white, less dense and shows a smaller vertical development.

9. Cumulus (Cu). Detached clouds, generally dense and with sharp outlines, developing vertically in the form of rising mounds, domes or towers of which the bulging upper part often resembles a cauliflower. The sunlit parts of these clouds are mostly brilliant white; their base is relatively dark and nearly horizontal. Sometimes cumulus is ragged.

Since cumulonimbus generally results from the development and transformation of cumulus, it is sometimes difficult to distinguish cumulus with a great vertical extent from cumulonimbus. The cloud should be named cumulus as long as the sprouting upper parts are everywhere sharply defined and no fibrous or striated texture is apparent. If it is not possible to decide on the basis of other criteria whether a cloud is to be named cumulus or cumulonimbus, it should by convention be called cumulus if it is not accompanied by lightning, thunder or hail.

One of the species of cumulus, cumulus congestus, can produce abundant precipitation in the tropics. Cumulus fractus (Cu fra), formerly called fractocumulus (Fc), is a small cumulus cloud with very ragged edges and with outlines continuously undergoing changes which are often rapid.

10. Cumulonimbus (Cb). Heavy and dense cloud with considerable vertical extent, in the form of a mountain or huge towers. At least part of its upper portion is usually smooth, or fibrous, or striated and nearly always flattened; this part often spreads out in the shape of an anvil or vast plume.

Under the base of this cloud, which is often very dark, there are frequently low ragged clouds (cumulus fractus or stratus fractus), either merged with it or not, and precipitation, sometimes in the form of virga.

Cumulonimbus clouds generally produce showers of rain or snow and sometimes of hail or snow pellets, and often thunderstorms as well.

Cloud Movements and Changes

Clouds move with the wind at their own level but, because winds are seldom constant throughout the depth of the troposphere, clouds at different heights may be moving in different directions, which may differ again from the direction and speed of the surface wind. The apparent or angular speed of low clouds may be greater than that of higher clouds even if their true speed is less, because the low clouds are so much nearer to the observer.

Clouds are continually changing in form, and often in height also. Changes in developing cumuliform clouds are rapid and obvious, but it is more difficult to see changes in structureless overcast skies. Development may be due to growth or decay without change of type, to change of cloud from one type to another with or without growth or decay, and to deformation by the wind.

A commonly observed cloud sequence is from cirrus to cirrostratus, altostratus and, finally, nimbostratus as a warm front approaches, but in this case the clouds form part of a large system which is moving past the observer rather than changing in type.

Precipitation

Precipitation is the general term given to water drops or ice particles formed at a higher level and falling to the ground. It includes rain, drizzle, sleet, snow, snow pellets, snow grains, ice pellets, hail and ice prisms. Surface condensation phenomena such as dew, hoar frost and rime are not classified under the heading of precipitation.

Formation of Precipitation

The water droplets or ice crystals comprising cloud or fog are very small in comparison with the drops which fall as drizzle or light rain; condensation must therefore be assisted by some other processes before appreciable precipitation can fall.

One such process requires that water droplets and ice crystals should coexist, at least in the upper parts of the cloud. Because the saturation vapour pressure is lower over ice than over water, the ice crystals will grow at the expense of the water drops until they are large and heavy enough to start falling through the cloud. This process of growth will continue during the descent until they finally emerge from the base of the cloud and reach the ground, either as snowflakes if the temperature is low enough, or as drops of rain or drizzle if the melting level is sufficiently high. Most precipitation in temperate and polar latitudes can be explained by this process since the tops of precipitating clouds usually reach levels where the temperature is low enough for ice crystals to form.

In tropical regions heavy showers are quite common from clouds whose tops are known to be too warm for the formation of ice crystals; numerous observations have been made of showers falling from clouds which were warmer than 0°C throughout. It is believed that this precipitation depends on the existence, at an early stage in the growth of the cloud, of a few droplets much larger than the majority which have formed on a small number of nuclei that are themselves very much larger than the average size of all nuclei present. These large drops are able to grow by collision with and absorption of smaller-cloud droplets, the process being self-sustaining as the larger drops fall faster than the smaller ones; updraughts also assist in the collisions by carrying all but the largest drops up through the cloud and by reducing the fall-speed of the larger drops so that the time during which collision can occur is increased.

These theories explain the main observational evidence concerning the formation of precipitation; thin layers of cloud seldom precipitate except in the form of drizzle, in which the drops are very small, and light rain or snow rarely falls from cloud that is less than 3000 feet (1000 metres) thick. Clouds can measure as much as 10 000 feet (3000 metres) vertically without giving precipitation and are often observed to be deeper than this before heavy rain starts to fall.

Snow

Once an ice crystal has formed on a condensation nucleus further growth will be in the solid state, provided the temperature remains below 0°C. The initial ice crystal is very small and falls correspondingly slowly, giving plenty of time for growth of the feathery crystalline structure which is a snowflake; this structure, when examined under the microscope, is seen to be of infinite variety and great beauty. Dry snowflakes do not readily combine; this explains why, at temperatures well below freezing, individual snowflakes are small and why freshly fallen snow is easily lifted by the wind to be blown along in a fine cloud as in a blizzard. Large snowflakes are either aggregates of crystals or result from the coalescence of crystals and water droplets at temperatures around the freezing point. When the ground and surface-air temperatures are below 0°C, snow will remain unmelted where it falls, but if the air near the ground is more than a few degrees above freezing the slowly falling snowflakes will have time to melt, and precipitation will reach the surface in the form of rain.

Partially melted snow or a mixture of snow and rain can occur when surface-air temperatures are within a narrow range, a few degrees above freezing. In the British Isles this mixture is known as sleet, but it should be noted that this word is often used in American terminology to signify ice pellets.

Hail

The formation of hail can be described only in outline. For a hailstone to remain in the cloud and grow to any size, rather than fall to the ground, very strong updraughts must be present. These updraughts can develop only when the atmosphere is very unstable, as in active cumulonimbus cloud, hence the association of hail with thunderstorms. If a hailstone is cut open it shows an open soft core surrounded by a shell of clear ice; larger stones may exhibit several concentric layers. The soft core of a hailstone is formed in the upper part of a cumulonimbus by the freezing of supercooled water droplets on to the original ice crystal. When it has grown heavy enough to overcome the force of the ascending air currents the enlarged hailstone will start to fall through the cloud, picking up a coating of clear ice from the partial freezing of raindrops which it encounters on the way. At this stage most hailstones will drop out from the cloud base and reach the ground but, if the updraughts are particularly violent, the hailstones caught in them can be carried up again and the process repeated, perhaps several times, to form the concentric shells of soft and clear ice sometimes found in larger hailstones.

Thunderstorms

The mechanism of thunderstorms is complex and, as with the formation of hail, only a brief explanation will be attempted.

When the atmosphere is very unstable throughout a considerable layer of depth and a plentiful supply of moisture is available for the production of cloud and precipitation, towering cumulonimbus clouds are likely to develop. In addition to the formation of heavy rain and perhaps hail, the violent convection can lead to the separation of positive and negative electric charges, so that large potential gradients are set up within the cloud. When these potential differences have become large enough the positive and negative charges will be neutralized by a charge of electricity between them. This may take place between two clouds, between different parts of the same cloud or between a cloud and the earth.

The discharge will result in two observable effects: by ionizing the atmospheric gases it will make its path visible as a flash of lightning, and by setting up a shock wave it will cause a peal of thunder to be heard. Light travels at 3×10^8 metres per second ($5{\cdot}8 \times 10^8$ kn) so that an observer will see a flash of lightning virtually at the instant when the discharge took place, whereas the sound of thunder travelling at between 330 and 340 metres per second (640–660 kn) (depending on temperature) will take about six seconds to cover each nautical mile between flash and observer.

Whilst the duration of a flash of lightning is extremely short, that of a peal of thunder may extend over many seconds; this is due partly to sound having to travel over varying distances from different parts of a lightning flash which may be one n. mile or more in length, and partly to echoes as the sound is reflected from surrounding hills and other obstructions.

Information concerning the effect of cloud and precipitation on radar will be found in Chapter 5.

CHAPTER 5

VISIBILITY

General Remarks

Visibility, in the meteorological sense, is a measure of the transparency of the atmosphere and may be defined as the greatest horizontal distance at which an object of specified characteristics can be seen by a person of normal vision under conditions of average daylight illumination. Whether an object can be seen at a distance depends, among other things, upon the nature and size of the object. Accordingly, the sort of object to be used in estimating visibility needs to vary with its distance from the observer. On land there is usually no shortage of suitable objects. For distances of a few hundred metres a relatively small object is appropriate, such as a telephone kiosk or small tree. For distances of around a kilometre (about ½ n. mile), a house or large tree might be suitable. For distances of 30 kilometres (about 16 n. mile) or more, a larger feature becomes necessary, for example a large clump of trees or a prominent hill. Assuming the distance of each such object is known, whether, and how well, it can be seen enables an estimate to be made of the visibility on a particular occasion.

In a large vessel the lowest ranges present no difficulty because objects at known distances on deck can be used. Visibility at the higher end of the scale can be determined as follows. When coasting and when fixes can be obtained, the distances of points when first sighted, or last seen, may be measured from the chart. In the open sea, when other ships are sighted, visibility may be estimated by noting the radar range when the vessel is first sighted visually, and again when it disappears from view. The horizon, whether, and how well, one can see it, is also used to estimate visibility. This, however, must be used with caution. The distance of the horizon is obviously a function of the height of the observer above sea level, which could be considerable from the bridge of a large ship. Also there are occasions of abnormal refraction which give a false impression of the visibility. With a vessel close to a precipitous shore there might be occasions when an elevated coastal tower could be seen at 2 n. mile but owing to a shallow fog layer the horizontal visibility might be only ¼ n. mile.

At night, visibility is difficult to estimate. The figure to be entered in a meteorological report is the distance which could be seen assuming normal daylight. Whether or not there is bright moonlight will considerably affect how far one can see, but it should make no difference to the meteorological 'visibility' which is concerned with the atmospheric clarity or obscurity. If there is no change in the meteorological conditions, the visibility just after dark will be the same as that just before dark. Caution is needed, however, in that, near to land, changes in thermal conditions which commonly occur at sunset tend to favour a deterioration of visibility at this time. The appearance of lights provides the most important clue to the visibility at night. Lights on the observer's vessel will usually serve as a basis for estimating at the lowest end of the visibility scale. Lights from other vessels, in conjunction with a radar range measurement, can sometimes be used for estimating at medium and higher ranges. Lights from shore establishments can also be useful but in the case of very bright lights, such as that of a powerful lighthouse, lights may be seen in conditions of obscurity which would not permit the observer to see the lighthouse under normal daylight illumination. The presence of a 'loom' around the vessel's navigation lights is frequently a guide to deteriorating visibility.

Visibility depends chiefly on the number of solid or liquid particles held in suspension in the air. It may vary in different directions because the concentration of particles varies. This may occur in the neighbourhood of a large port or other industrial centre due to the variation in the concentration of solid particles, or due to variations in the concentration of water droplets, as in the case of patchy fog.

The main causes of atmospheric obscurity are:

(*a*) Visible moisture in the atmosphere. Under this heading are cloud, mist or fog consisting of water droplets, precipitation (i.e. drizzle, rain, sleet or snow, etc.) at the observer's level and spray blown up from the sea. Water vapour is a transparent gas and so does not affect visibility.

(*b*) Solid particles such as those produced by factories, domestic fires and forest fires, by sea spray and by sand and dust due to strong winds in desert regions, or as the result of volcanic eruptions.

Where particles from these sources are absent the atmosphere is nearly transparent, although the smaller nuclei when present in large quantities in a layer not necessarily near the surface cause a scattering of the light. This accounts for the pale blue or white skies sometimes seen in waters adjacent to desert areas.

Although there are several agencies whereby visibility in the atmosphere is reduced, over most of the world the commonest factor producing low visibility is the presence of water droplets. When atmospheric visibility is reduced below 1 km (about ½ n. mile) due to water droplets, fog is said to occur. It frequently happens during duststorms, sandstorms and in industrial towns in winter time, that the suspended matter in the air reduces the visibility below 1 km (about ½ n. mile). Although, strictly speaking, these conditions are not aptly described as 'fog', for making statistical summaries of visibility at a particular station all such occasions when visibility was below 1 km (about ½ n. mile) are returned as fog or thick dust haze. In a synoptic weather report the visibility is indicated in metres irrespective of the cause. Thus a reduction of the visibility to 200 metres may be caused by fog, snow or heavy rain; the cause will be indicated in the weather report under the heading of present weather.

Formation of Fog

A fog composed of water droplets may also be described as a cloud on the surface. Over high ground, fog may be merely one of the cloud types formed because of cooling by adiabatic ascent. In other cases, the condensation in a fog is almost entirely produced by the direct effect of a relatively cold surface. Two main types may be distinguished:

(*a*) RADIATION FOG, due to cooling of the ground and the air in contact with it, by radiation. It forms almost entirely at night and only over land, since the sea surface retains a fairly constant temperature. Radiation fog is, however, liable to drift to seaward and is frequently experienced in rivers and harbours.

(*b*) ADVECTION FOG, forming rapidly when warm moist air moves over a colder surface of land or sea.

There are also two less common types of fog:

(*c*) MIXING FOG, which forms at the boundary layer of two completely different air masses. This is also known as 'frontal fog' because it often occurs during the passage of a front.

(*d*) SEA SMOKE, which occurs when very cold air flows over relatively warm water.

Radiation Fog

The development of radiation fog depends upon the cooling of the ground at night. It is therefore a land fog, but it may drift over coastal waters with a slight wind. Very dense fogs are often formed in this manner, and those occurring in the eastern English Channel and in the Thames Estuary are mostly fogs of this type.

The cooling of the ground at night is communicated to the air in contact with it and the cooling effect is spread upwards by turbulent mixing. Since the cooling takes place at the ground, an inversion tends to develop, with the lowest temperatures on the ground; the dew-point is therefore first reached on the surface itself, and considerable moisture may be extracted from the air and deposited as dew.

On a clear night when the air is absolutely calm, the cooling is most intense close to the ground and extends upwards very slowly, and the temperature of only a shallow layer near the ground may fall below the dew-point so that fog forms within it. If there is just a little turbulence, which will be the case if there is a very light breeze, mixing occurs and the cooling spreads to higher layers, say up to 150 metres (500 feet) or more. Fog may then form up to that level. With stronger winds at low levels turbulence will extend through several thousand feet, and the cooling effect will be distributed through these layers so that the fall of temperature will be very small and the dew-point will not be reached at any level. Since turbulence tends to establish a dry adiabatic lapse rate through a layer which is free of cloud initially, thereby steepening the lapse rate, the temperature will fall at the top of this layer and an inversion will frequently be found just above the layer. After turbulence has been distributing moisture upwards through the layer for some time, the condensation level will be found near the top of the layer, resulting in the formation of a layer of stratus or stratocumulus cloud just below the inversion, while no fog occurs at the surface.

From this it can be understood that the conditions favouring formation of radiation fog are:

(*a*) large moisture content in the lowest layers,

(*b*) little or no cloud,

(*c*) light breeze at the surface,

(*d*) surface of the ground initially cold and wet.

Diurnal and Seasonal Variation of Radiation Fog

The minimum night temperature occurs about dawn and the highest frequency of radiation fog occurs about an hour after sunrise. The slight increase in turbulence due to the first heat from the sun reaching the lowest layers often causes a sudden formation of fog or results in a thickening of any fog already formed in the night. Further heating by the sun is needed to clear this fog. In the British Isles, in summer, radiation fogs rarely last more than a few hours, and it is the long nights of autumn and winter which provide conditions most favourable for radiation fog. In these islands the late autumn is more subject to fogs of this type than the winter. This is because the moisture content of the air masses arriving from the Atlantic is higher in autumn than in winter, which is associated with the sea being warmer in the former season. Radiation fogs may occasionally persist all day during the midwinter months. Then, the short period of heating by day is insufficient to offset the long period of cooling by night. In the absence of wind such fog may sometimes last for several days.

Local Effects on Radiation Fogs

The topography and ground conditions are responsible for the local nature of radiation fog which is, however, occasionally widespread. It has a tendency to collect first in valleys, due to the greater cooling that takes place therein and to the katabatic (*see* page 37) draining of cool air into low-lying places. Although water-logged ground cools more slowly than dry ground, the effect of the increased humidity is the more important, and fogs are more likely to occur over wet ground if other conditions are suitable. Similarly, fogs are particularly likely when the sky clears at night after rain.

Under suitable conditions the fog may be only about a metre in thickness, constituting ground fog. Most radiation fogs have a depth of about 150 metres (500 feet) and rarely more than 300 metres (1000 feet). The upper surface is usually sharp. Owing to the fact that dust and moisture cannot be lifted through an inversion by turbulence, the clearness of the air above is usually in marked contrast to that of the air below.

Advection Fog

This type of fog occurs when warm, damp air moves over a surface which is cooler than the air dew-point. It may occur over either land or sea. Ashore, it is particularly likely when in winter, after a cold spell, a supply of milder air arrives from the sea. The air is cooled by the colder land surface over which it passes and fog is formed. If the wind is more than moderate, turbulence lifts the condensation level above the surface and low cloud forms instead of fog. Over the sea, this type of fog occurs when warm, damp air moves from the land over a colder sea, or from a region of fairly warm sea water to one of colder sea water. The frequent fogs on the Grand Banks of Newfoundland are formed when the warm damp air overlying the Gulf Stream is blown northwards to the region over the cold Labrador Current. The fogs along a narrow strip of the Californian coast, which are most frequent near San Francisco, are due to the cooling of the maritime air during its passage across the cold inshore waters of the California Current.

Mist, Haze, Dust* and Smoke

Mist is similar in cause and character to fog, but the visibility is not so seriously affected. By international agreement, visibility which is impaired, but is not less than 1 km (about ½ n. mile) is described as mist when the obscurity is caused by water particles, and as haze when the obscurity results from smoke, dust particles or other impurities in suspension in the atmosphere. The usual criterion for deciding between mist and haze is whether or not the relative humidity is above 95%.

Off large industrial centres near the shore, industrial haze from the numerous factories may be encountered some distance to seaward in certain wind conditions. The smoke from forest fires such as are experienced in the United States and Canada can similarly cause considerable haze when it drifts to seaward.

In desert or other arid regions, the visibility may be greatly reduced by dust or sand in the atmosphere. Dust or sand is raised from the ground by the wind and carried upwards to a greater or lesser degree according to meteorological conditions and the nature of the dust particles. This phenomenon is known as a sandstorm or a duststorm. The effects of such storms may be observed well out to sea when arid regions border the coast. Off the West African coast, dust haze from the Sahara is experienced far to seaward during certain seasons of the year, such as when the Harmattan wind is blowing.

In and near large towns and industrial areas smoke can seriously reduce visibility. The larger particles of soot and smoke settle easily under gravity and do not drift far, but much of the pollution is in the form of minute particles, comparable in size with the water droplets in a cloud, which may remain suspended in the air for an indefinite period.

The thickness of smoke haze depends largely on the rate at which it is dispersed through the air, and the dispersion may occur in two ways:

(*a*) horizontally, by being transported in the wind (advection) and

(*b*) vertically, in rising currents of air (convection).

Convection currents are most effective in moving the smoke to higher levels, to be dispersed in the stronger winds aloft. Even when humidity is insufficient to produce a 'water-drop fog', smoke alone may reduce visibility below the value at which fog is defined as occurring, if the wind is light and other conditions are suitable. Since, however, the favourable conditions for producing a high concentration of smoke are, in general, similar to those favouring the formation of water-drop fog, the two commonly occur together, with the smoke adding to the obscurity due to the water drops. This may lead to the very unpleasant phenomenon of very thick fog, usually of a dirty yellow or brown colour and an unpleasant smell and taste which has been termed 'pea-soup fog' or 'smog'. At least in the United Kingdom, the frequency of the latter has appreciably decreased in recent years due, it is thought, to Government and other measures to reduce smoke pollution and the general trend towards 'smokeless' fuel. Even so, all things being equal, one may still expect fog to be more frequent in industrial areas than in non-industrial areas.

---

* *See* also *Marine Observer's Handbook*, Chapter 11.

Mixing Fog

Mixing fog is liable to occur when air streams of widely different origin meet. For example, if a cold air current meets a warm moist air current, the latter will be cooled at the boundary and fog may form there. Fog near a warm front or occlusion is quite common over the sea in temperate and high latitudes, and may be persistent in an area when a front becomes nearly stationary. Mixing fog is often known as frontal fog.

Sea Smoke

The description 'sea smoke' is given to a peculiar kind of surface mist or fog observed close to the open sea surface when the air temperature is very low. As a rule this layer of mist or fog is shallow and visibility within it is rather variable. There is some evidence that sea smoke does not form unless the air temperature is at least 9°C below the temperature of the sea surface, which explains why it is more commonly observed in the polar regions and off the east coasts of the continents than elsewhere, since cold offshore winds are frequent there in autumn and winter. Sea smoke results from the rapidity with which cold air becomes saturated by evaporation from a relatively warm sea surface. The sea also supplies a large amount of heat, in addition to moisture, to the lowest layers of air, with the result that such a large air–sea temperature difference can only persist while strong or gale force winds continually renew the supply of cold air.

Sea smoke is most common in Arctic and Antarctic waters and in areas such as the Baltic, but it can also occur elsewhere. For example, in the Newfoundland region and the Gulf of St Lawrence dense sea smoke (in which visibility was below 1 km (about $\frac{1}{2}$ n. mile) and which sometimes extended to a height of 1500 metres (5000 feet) has been observed by ships and aircraft during the winter, with surface air temperatures below −9°C and gusty winds between W and N, often exceeding gale force. This kind of extensive deep-sea smoke is given the local name of 'ice-crystal fog', and constitutes a serious hazard to navigation. As a rule sea smoke is shallow and patchy and does not impede navigation to any extent. It has been observed occasionally in this form locally in the Mediterranean, at Hong Kong and once from an ocean weather ship in the north-east Atlantic, also in the Gulf of Mexico during the occurrence of a Norther (*see* page 88). Sea smoke is sometimes described as frost smoke, or Arctic sea smoke.

Visibility Meters

In view of the difficulty of estimating visibility at night, instrumental measuring devices have been introduced and are in common use ashore. Visibility meters such as the 'Meteorological Office Visibility Meter, Mark 2' depend for their working on using a simple device known as an 'optical wedge', which is used in conjunction with one or more lights of known brightness and at known distances from the observer. An optical wedge is a strip of glass made completely opaque at one end, the opacity decreasing uniformly to the other end where it is quite transparent. A variable degree of obscurity can be introduced between the observer's eye and a distant fixed light by sliding the wedge across the line of sight. Moreover this degree can be read off on a scale on the instrument. In its simplest mode of use the chosen light is viewed at night under conditions of very good visibility when the obscurity introduced by the atmosphere over the distance in question can be regarded as negligible. By sliding the wedge until the light is only just visible a reading is obtained and noted. Then, on a later occasion, assuming that the visibility is not so good, repeating the procedure with the same light will give a different reading. The difference between the two readings gives a measure of the obscurity due to the intervening atmosphere on the second occasion. The instrument needs to be calibrated for each observer and for each light, the calibration depending upon the distance of the light and its intensity. In using the instrument it is necessary to keep the wedge and other glass surfaces absolutely clean and the voltage of the lamps constant.

Audibility of Sound Signals in Fog

It is rightly said in the books of seamanship that 'sound is conveyed in a capricious manner through the atmosphere'. It is important to realize this when considering the distance at which one may expect to hear a sound signal in fog. In calm, or relatively calm, weather the strength at which sound signals are heard from a stated distance in fog may vary from day to day, or on a given day the signal audibility may vary on different azimuths from the observer. Sound waves may at times be reflected or refracted, and hence attenuated, due to the presence of the excessive number of water particles present in a foggy atmosphere. The echo which is often heard from a ship's whistle in fog is evidence of this. In a breeze it is fairly obvious that sound signals originating from a source situated to leeward of the observer, will not normally be heard at as great a distance as from a source to windward. When the wind is blowing at all hard it is often difficult to hear fog signals owing to the noise of wind and sea. The mariner needs to be constantly on his guard when estimating distance from an object in foggy weather by the strength of its fog signal or its apparent azimuth.

Use of Radar in Fog

When using radar for navigation in conditions of low visibility, the navigator should bear in mind the meteorological factors which affect the performance of the instrument. A reduction in the expected range may be because of conditions favourable to sub-refraction or because the radar impulses are attenuated due to absorption by water particles in the atmosphere. On the other hand, ranges may be considerably greater than expected if conditions are favourable for super-refraction. Standard atmosphere conditions may be defined broadly as follows:

(*a*) barometric pressure at sea level 1013 mb, decreasing with height at a rate of 12 mb per 100 m,

(*b*) temperature at sea level 15°C, decreasing with height at a rate of 0·7°C per 100 m, and

(*c*) relative humidity 60% (remaining constant with height).

Under such conditions the radar range should be normal, i.e. it should somewhat exceed that of the optical horizon. If the temperature decrease with height is more than standard, or if humidity increases with height, then sub-refraction may be expected and the radar range may be considerably reduced. If the temperature decrease with height is less than the standard, or if humidity decreases with height, then super-refraction may be expected and the radar range may be considerably increased. Another form of super-refraction known as 'ducting' is also experienced due to irregular change in the temperature–humidity gradient in height between the observer and the target, and this may give excessive radar ranges. It is perhaps unfortunate that super-refraction is most frequent when light winds carry warm dry air across a relatively cool sea, that is, when the weather is clear. Although the observer aboard a ship cannot be sure when either condition is present without an accurate knowledge of the vertical gradients of temperature and humidity, he can at least obtain some indication as to when conditions are very favourable for sub-refraction or super-refraction by taking readings of dry-bulb and sea temperature. When the air temperature is considerably below the sea temperature (by 5°c or more) it is likely that sub-refraction is present. Alternatively, when the air is 5°c or more warmer than the sea it is almost certain that super-refraction conditions exist. In both cases this is only true with light and moderate winds when turbulence is not marked. It has also to be remembered that the radar range regarded as normal for one region would be abnormal for another region, and that the range in many localities varies between the seasons.

Reduction of radar range by attenuation may be caused by the presence of fog, rain, ice particles, hail or snow due to the concentration of water particles therein. Generally the larger the water particles and the closer they are concentrated the greater the attenuation, so that the effect in heavy rainfall is probably greatest and that due to fog is least. The effect of snow and hail can normally be expected to be less than that of rain at an equivalent rate of precipitation. Thus, in the conditions which are the greatest menace to navigation (fog and snow) the probability is that the attenuation will not be as great as in heavy rain. Some attenuation may be experienced due to the dust, sand or smoke particles present in haze. The meteorological aspects of the use of radar at sea are more fully dealt with in the book *The Use of Radar at Sea*, published by the Royal Institute of Navigation.

PART II. CLIMATOLOGY

CHAPTER 6

METEOROLOGICAL CHARTS OF THE OCEANS

Use made of Ships' Meteorological Observations

Meteorological observations, whether made on land or at sea, serve several purposes. When transmitted by radio through a coastal radio station to a central meteorological office and there plotted on a synoptic chart, they provide an up-to-date picture of the weather over a large area. From a study of such a chart, together with other synoptic charts plotted several hours earlier, forecasts can be made of the weather for the following 12 to 24 hours. As much of western Europe's weather comes from across the Atlantic, the importance of ships' observations for synoptic purposes in that area is obvious. Similar considerations apply in other parts of the world. As nearly three-fourths of the earth's surface is ocean, synoptic charts would be very incomplete without the help afforded by those ships' officers who voluntarily provide information about the weather at sea.

These observations are also used for climatological purposes. If enough observations can be collected from one place, or from a limited area, they will show, among other things, the most frequent type of weather, and within what limits any element such as temperature, wind or wave height may be expected to vary, at any time of the year. Such information is of value in circumstances where daily forecasts are unobtainable or inadequate as, for example, when it is required to know the most favourable time of year at which to perform a difficult operation such as towing an offshore drilling platform. Statistics of sea and air temperature and humidity can be valuable in dealing with practical problems such as offshore mining, the construction of harbours and the design of specialized ships.

Some observations, such as barometric tendency and characteristic, and cloud types, are used chiefly for forecasting; others, such as wind, temperature and humidity observations, are very necessary for forecasting and are used also for climatological and research purposes.

There is one essential difference between land and marine climatology. On land there are regular series of observations, made several times a day over long periods of years at fixed locations, from which a moderately complete picture of the climate can be built up. At sea the situation is different; ships are still practically the only source of observations and even on the main shipping routes the chance of getting many observations from one spot, or even from a comparatively small area of, say, ten square miles, is only slight. The exceptions are in the North Atlantic and North Pacific oceans, where the existence of ocean weather stations since 1947 has provided regular series of observations from within relatively small fixed areas in the ocean. It is necessary, therefore, to divide the ocean into relatively large areas, at the very least into 1° 'squares', that is, areas bounded by two meridians and by two parallels 1° apart, while 2°, 5° and even 10° squares are also used. All observations taken in such a square are grouped together as if they had been taken at the same location.

Systems of defining Ocean Areas

For marine climatological purposes the surface of the earth is divided into 10° squares bounded by meridians and parallels at intervals of 10°. The Marsden system, named after a former Secretary to the Admiralty who introduced it in 1831, of numbering these squares and dividing them into smaller squares was used by the Meteorological Office for many years and is still being used for some purposes. In this system, in both the northern and southern hemispheres the numbers start at the intersection of the Greenwich meridian and the Equator and increase westwards around the globe. In the northern hemisphere the first square to the west of the Greenwich meridian is square 1, in the southern hemisphere it is square 300. To find the position of a given square it is desirable to have a world map showing the numbers (although this can be deduced by dividing the number by 36, i.e. the number of squares between any two parallels). This system has the advantage that the numbering of adjacent squares is progressive which facilitates the processing of data for groups of squares. Figure 6.1 illustrates the Marsden system of numbering.

An alternative system recently introduced by the World Meteorological Organization (WMO) and agreed internationally has the advantage that the ten-degree square number can be deduced directly from the reported position. If a ship's position is 41·2°N, 165·3°W, the ten-degree square number can be evaluated as 146, where 1 is the octant of the globe, as shown in Table 6.1, 4 is the 'tens' figure of the latitude, and 6 is the 'tens' figure of the longitude.

Table 6.1. Octant of the Globe

Octant	Greenwich longitude	Hemisphere	Octant	Greenwich longitude	Hemisphere
0	0°– 90°W	North	5	0°– 90°W	South
1	90°–180°W	North	6	90°–180°W	South
2	180°– 90°E	North	7	180°– 90°E	South
3	90°E– 0°	North	8	90°E– 0°	South

The 10° squares are subdivided into 1° squares as shown in Figure 6.2 using exactly the same system as is used for subdividing the Marsden squares. It may be noted that in the International reporting code the position of an observation is given to the nearest one-tenth of a degree. The two-digit number of the 1° square is given by the units figure of latitude, followed by the units figure of longitude.

The fact that observations from a fairly large area are considered representative of one spot is not such a disadvantage over the open sea as it would be on land, for climate does not generally change as rapidly with position over the sea as it does over land. Near coasts, however, and at the boundaries separating ocean currents with distinctive temperatures, the climate may change rapidly within short distances, and observations in these regions may thus require special treatment.

Observations at sea are not made with anything like the same regularity as those made on land, except in the special cases of observations from ocean weather ships when on station, or those from light-vessels. As a ship passes through a particular square, a series of observations is obtained at successive intervals of 6 or perhaps 12 hours, but that square may contain no further

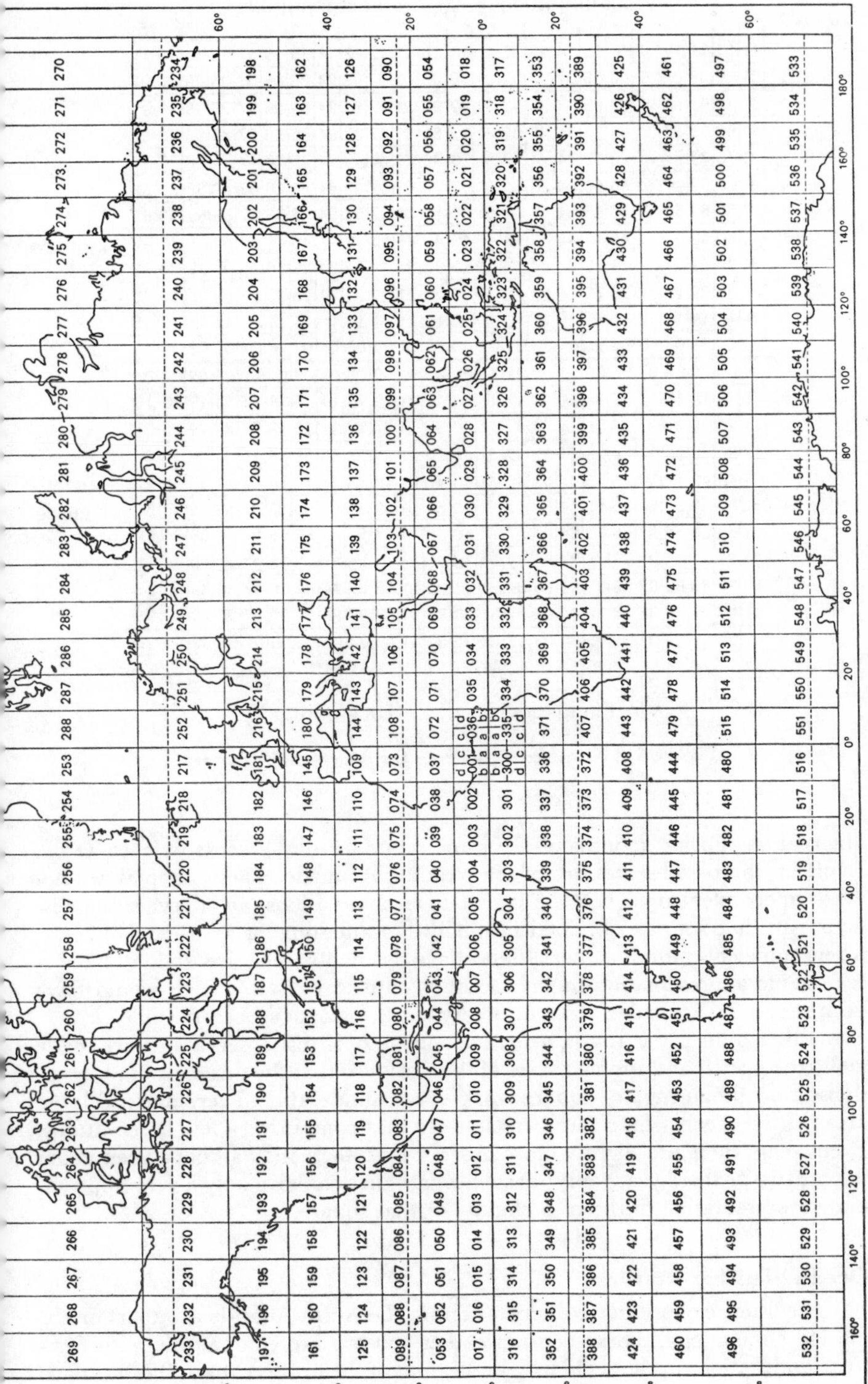

269	268	267	266	265	264	263	262	261	260	259	258	257	256	255	254	253	288	287	286	285	284	283	282	281	280	279	278	277	276	275	274	273	272	271	270
233	232	231	230	229	228	227	226	225	224	223	222	221	220	219	218	217	252	251	250	249	248	247	246	245	244	243	242	241	240	239	238	237	236	235	234
197	196	195	194	193	192	191	190	189	188	187	186	185	184	183	182	181	216	215	214	213	212	211	210	209	208	207	206	205	204	203	202	201	200	199	198
161	160	159	158	157	156	155	154	153	152	151	150	149	148	147	146	145	180	179	178	177	176	175	174	173	172	171	170	169	168	167	166	165	164	163	162
125	124	123	122	121	120	119	118	117	116	115	114	113	112	111	110	109	144	143	142	141	140	139	138	137	136	135	134	133	132	131	130	129	128	127	126
089	088	087	086	085	084	083	082	081	080	079	078	077	076	075	074	073	108	107	106	105	104	103	102	101	100	099	098	097	096	095	094	093	092	091	090
053	052	051	050	049	048	047	046	045	044	043	042	041	040	039	038	037	072	071	070	069	068	067	066	065	064	063	062	061	060	059	058	057	056	055	054
017	016	015	014	013	012	011	010	009	008	007	006	005	004	003	002	001	036	035	034	033	032	031	030	029	028	027	026	025	024	023	022	021	020	019	018
316	315	314	313	312	311	310	309	308	307	306	305	304	303	302	301	300	335	334	333	332	331	330	329	328	327	326	325	324	323	322	321	320	319	318	317
352	351	350	349	348	347	346	345	344	343	342	341	340	339	338	337	336	371	370	369	368	367	366	365	364	363	362	361	360	359	358	357	356	355	354	353
388	387	386	385	384	383	382	381	380	379	378	377	376	375	374	373	372	407	406	405	404	403	402	401	400	399	398	397	396	395	394	393	392	391	390	389
424	423	422	421	420	419	418	417	416	415	414	413	412	411	410	409	408	443	442	441	440	439	438	437	436	435	434	433	432	431	430	429	428	427	426	425
460	459	458	457	456	455	454	453	452	451	450	449	448	447	446	445	444	479	478	477	476	475	474	473	472	471	470	469	468	467	466	465	464	463	462	461
496	495	494	493	492	491	490	489	488	487	486	485	484	483	482	481	480	515	514	513	512	511	510	509	508	507	506	505	504	503	502	501	500	499	498	497
532	531	530	529	528	527	526	525	524	523	522	521	520	519	518	517	516	551	550	549	548	547	546	545	544	543	542	541	540	539	538	537	536	535	534	533

Subdivision labels in squares 001/036 and 300/335: d c c d / b a a b; b a a b / d c c d

Latitude: 60° 40° 20° 0° 20° 40° 60°

Longitude: 160° 140° 120° 100° 80° 60° 40° 20° 0° 20° 40° 60° 80° 100° 120° 140° 160° 180°

Figure 6.1. Marsden chart of the world

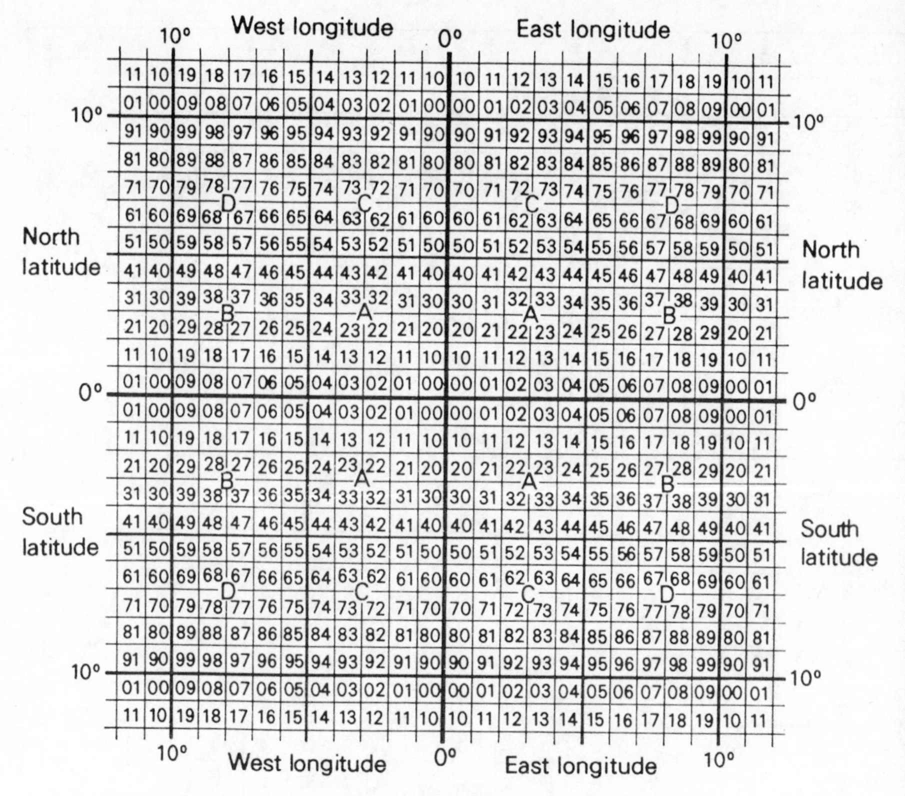

Figure 6.2. Numbering scheme for degree sub-squares

observations during months of the same name for days, weeks or even years. In effect, sample observations of the weather are made. These samples are not even independent ones because consecutive observations, at intervals of a few hours, are liable to be obtained in much the same type of weather. However, when sufficient sample observations have been obtained we should have a reasonable approximation to the monthly statistical results which would have been shown by a regular series of observations, had such been possible.

Another difficulty is that a great proportion of ships' observations are made along the principal shipping routes of the world, so that the numbers of observations in neighbouring 10° squares may vary considerably. There are still large areas of the ocean where there are squares that contain very few observations. Even where there are slightly more observations, these may all have been taken in one, two or three years only; in such a case they can give only a very approximate representation of the true climate in that square.

Ocean Weather Stations

The existence of ocean weather stations in the North Atlantic is important for marine climatology. These ocean weather stations are comparable with first-class land stations in the regularity and extent of their observations, both surface and upper-air, taken at a 'fixed' location. Many years of surface and upper-air data are now available from these fixed ship stations; these data supplement those received from other ships and provide valuable information which has already been of use in many investigations and published papers dealing with both forecasting and climatological problems.

The presence of ocean weather stations by no means lessens the value of regular weather observations from other ships. The expense of providing and maintaining ocean weather ships is considerable and they are consequently few in number. From a meteorological standpoint it is almost as though some new islands had appeared in the North Atlantic and had been manned, but information from the large areas between them is just as necessary as before and must be obtained from merchant ships in order to obtain a comprehensive picture.

Manipulation of Data

When observations recorded in meteorological logbooks are received in the Meteorological Office they are scrutinized and classified and then recorded in a form suitable for computer processing. At the time of writing, magnetic tape is the normal medium for recording the data. Until recently punched cards were used. They are still used to some extent for the international exchange of data but the tendency is for magnetic tape to replace them to an increasing extent. After quality-control and correction procedures have been carried out the data are sorted into ten-degree squares and into months. They can then be readily processed by the computer to produce the statistical summaries required. The results are most usually expressed in the form of monthly means or as monthly percentage frequencies for squares of various sizes, depending on the element(s) concerned and the number of observations available. Squares of 1° or 2° are normally used for temperature observations, but for winds, currents and waves, 5° squares are often used. The results may be presented either in tabular form, e.g. if they relate to ocean weather stations or to small selected areas, or in chart form if a fairly large region is being studied. On charts the values are often portrayed by means of isopleths. An isopleth is a line drawn so as to pass through all the points at which the element concerned has the same value. Examples are isobars, which join points at which the atmospheric pressure is the same, and isotherms which join points with the same temperature.

Portrayal of Distribution of Vector Quantities

Wind and current are vector quantities because they both possess two properties, those of magnitude and direction. Isopleths are not suitable for the portrayal of vectors except when one constituent only is to be shown and the other ignored; for instance, isopleths may show the frequency of gales, irrespective of their direction.

The most important vector quantity in marine climatology is the wind. Wind observations are usually summarized in the form of wind roses. These are diagrams showing, for each area, the percentage frequency of winds from each direction by means of arrows, the length of each arrow being proportional to the percentage frequency. Every arrow is also divided into segments, each segment representing winds between certain limits of strength.

Figure 6.3 illustrates the type of wind rose used in marine charts prepared by the Meteorological Office. The directions used are the cardinal and intercardinal compass points; all directions between NNE and ENE are classed as NE and so on. Within 30° of the equator, however, the intermediate points are used because the trade winds and monsoons are more constant in direction than the winds at higher latitudes, and the use of only eight points might blur this constancy. For each direction the percentage frequencies of Beaufort force groups 1–3, 4, 5–6, 7 and 8–12 are shown by the lengths of sections of the arrows. The thickness and shading of the sections indicate which group of wind forces they represent.

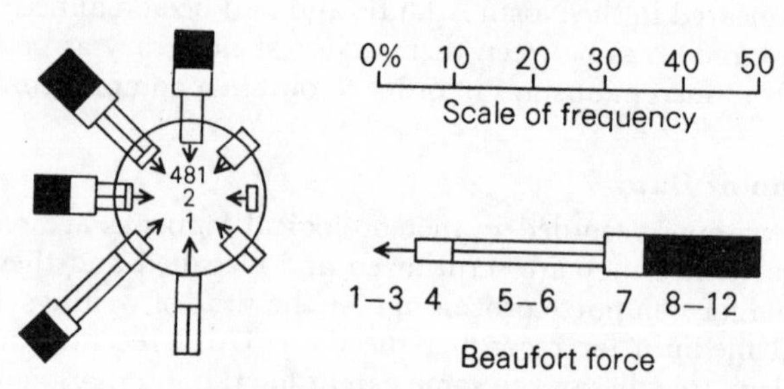

Figure 6.3. A typical wind rose

From the wind rose a rough estimate can be made of the direction from which the wind blows most frequently. It is, however, better to show this explicitly on a separate chart, in the form of arrows which indicate the direction of the prevailing or predominant winds. Such charts, when combined with an indication of the average speed in that direction, are useful as representing the most likely wind in a given area. For areas of high variability they are more meaningful for most purposes than charts showing vector mean values.

Another method of summarizing wind data is to show the vector mean of all the winds observed in a given small area.* This is of value in showing the general flow lines of the main air currents, and for investigations into the relation between these and ocean currents. Vector mean winds, however, give no idea of the mean wind strength. For example, if during half of a month there are easterly winds of force 5, and for the other half westerly winds of the same force, the two half-months will cancel out in the vector mean wind for the whole month, which will therefore be force 0, yet the average wind during the month, irrespective of direction, is force 5. This is an extreme case, although it is always found that the strength of the vector mean wind is less than the wind strength averaged irrespective of direction.† In the northern hemisphere the

* To find the vector mean of a number of separate wind observations first resolve each observation into two components at right angles. It is usual to regard flows to the north and east as positive and those to the south and west as negative. The algebraic sums of the separate components and their average values are then computed. The vector mean wind is formed by combining the two mean components by simple trigonometry or the use of tables. The vector mean wind thus takes account of the direction, as well as the velocities, of all the observed winds.

† It is of interest to note that a measure of the constancy of winds is formed by expressing the ratio of the vector mean wind to the average wind strength irrespective of direction, as a percentage. Care is needed, however, to distinguish this constancy of winds from the constancy of ocean surface currents which is defined differently (*see* page 214).

difference is greater to the north of 30°N, where wind directions are conspicuously variable, than it is to the south where the comparatively constant trade winds and monsoons blow.

Marine Meteorological Atlases

Some of the earliest examples of climatic charts which showed the wind and currents of the oceans were those prepared by Lieutenant Maury of the United States Navy in the 1850s. With the aid of these charts, the sailing-ship passage to Australia, which had previously occupied an average of 124 days, was reduced to an average of 97 days.

Since those early days marine atlases covering the various seas and oceans of the world have been prepared by the meteorological services of several countries. These atlases have, for the most part, been based on observations reported from ships of the countries preparing them. In recent years it has been recognized that much better results would be obtained at no greater overall cost by an international arrangement whereby the work would be shared and each set of charts would be based on all available observations, whatever the nationality of the ships providing them. To this end certain member countries of the World Meteorological Organization have accepted responsibility for specific ocean areas. They have agreed to act as collecting centres for surface marine observations of all nations made within their areas from 1961 onwards and to prepare routine climatological summaries for a number of selected representative areas within their areas of responsibility. In this way the groundwork has been laid for the eventual preparation, when sufficient data have been accumulated, of a marine section of the World Climatic Atlas, planned by WMO. The area for which the United Kingdom has accepted responsibility is the North Atlantic north of 20°N and east of 50°W, excluding the Mediterranean and Baltic Seas.

CHAPTER 7

THE WIND AND PRESSURE SYSTEMS OVER THE OCEANS

General

In Chapter 2 the main features of the pressure systems shown on synoptic charts were described. Such charts represent the distribution of pressure and wind on a particular occasion. On another occasion the synoptic chart for the same region may be quite different. The degree of variation in the synoptic chart for a given region from one occasion to the next varies considerably in different parts of the world. In some areas, notably in the tropics, one day is usually very similar to the next so that the actual conditions on an individual occasion do not normally differ much from the average conditions. In other areas, such as the temperate regions in general, and the seas around the British Isles in particular, the variation from one day to another is so great that the average conditions constitute a much smaller proportion of the whole range and in consequence become something of an abstraction. In spite of this, much can be learnt about the general distribution of pressure and wind around the globe, and the seasonal variation of this distribution, by compiling and studying charts showing long-term averages of these elements, based on the accumulated reports of vessels over a long period of years, as described in Chapter 6. Most useful for this study are charts evaluated on a monthly basis which show monthly mean values averaged over a long period of years (preferably 30 or more). World charts prepared on this basis show a decided pattern of distribution of both pressure and wind which changes progressively from month to month. The greatest contrast is usually apparent in the charts for summer and winter. Figures 7.1 and 7.2 illustrate the distribution of mean monthly pressure at mean sea level in January and in July respectively. Corresponding charts showing the distribution in these months of vector mean wind* are shown in Figures 7.3 and 7.4.

A comparison of the vector mean wind chart with the mean-pressure chart for the same month shows the wind to be related to the pressure in accordance with Buys Ballot's Law (*see* page 30 Chapter 3). These charts are especially useful in indicating the general circulation of air over the surface of the globe.

The General Circulation

It will be seen that the pattern of wind circulation is rather complicated. One of the most obvious features is that the winds are comparatively uniform in latitudinal belts extending round the world. This is broadly true in spite of local exceptions. Thus, on or near the equator, there is a belt of rather variable and mainly light winds (the doldrums). To the north of this belt, roughly between lat. 10°N and 30°N lies the belt of the north-east trades. These fall off northwards in a relatively narrow belt of light and variable winds which correspond in position with the subtropical high-pressure belt. This in turn gives place to a wider belt of mainly westerly or south-westerly winds extending over most of the

---

* *See* page 72 near end of Chapter 6 for definition and details.

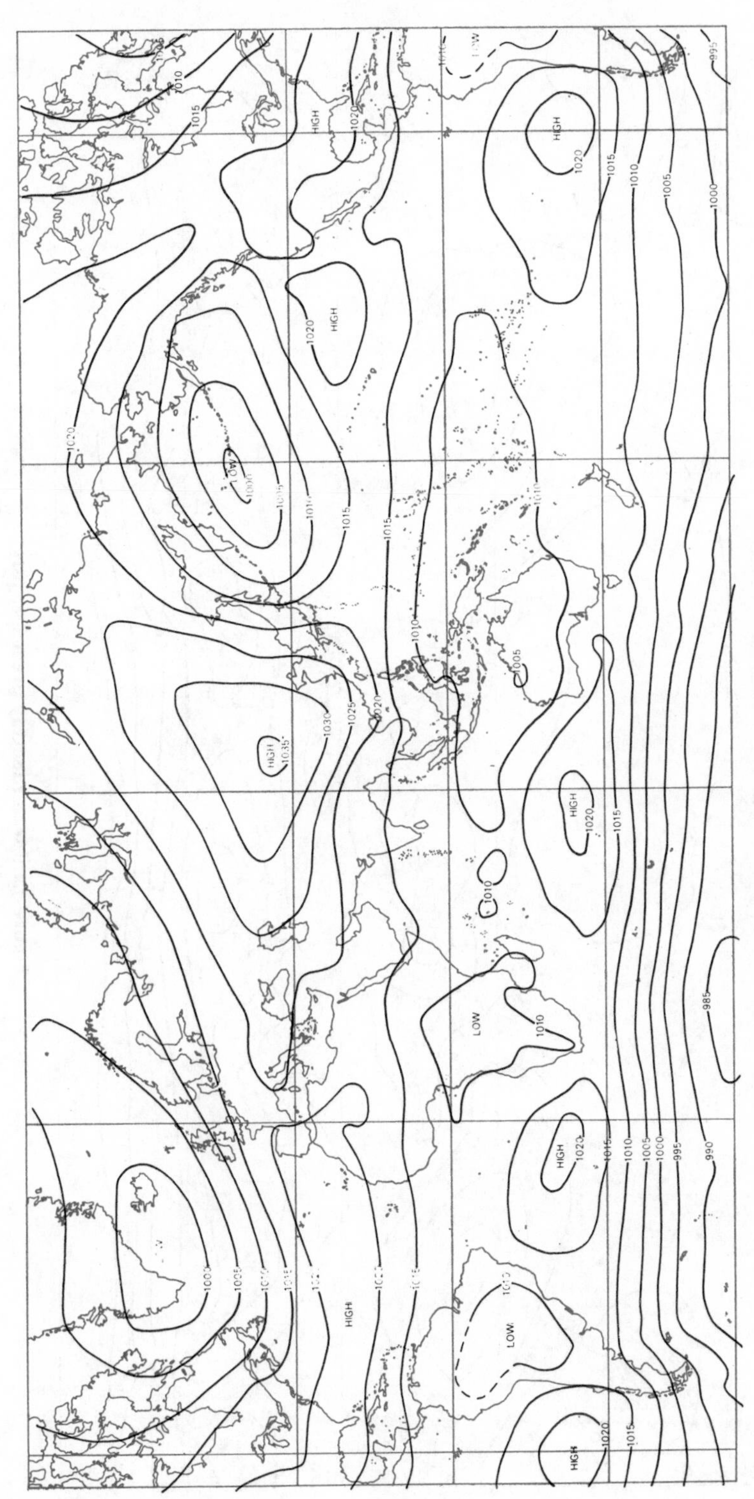

Figure 7.1. Mean monthly pressure, January

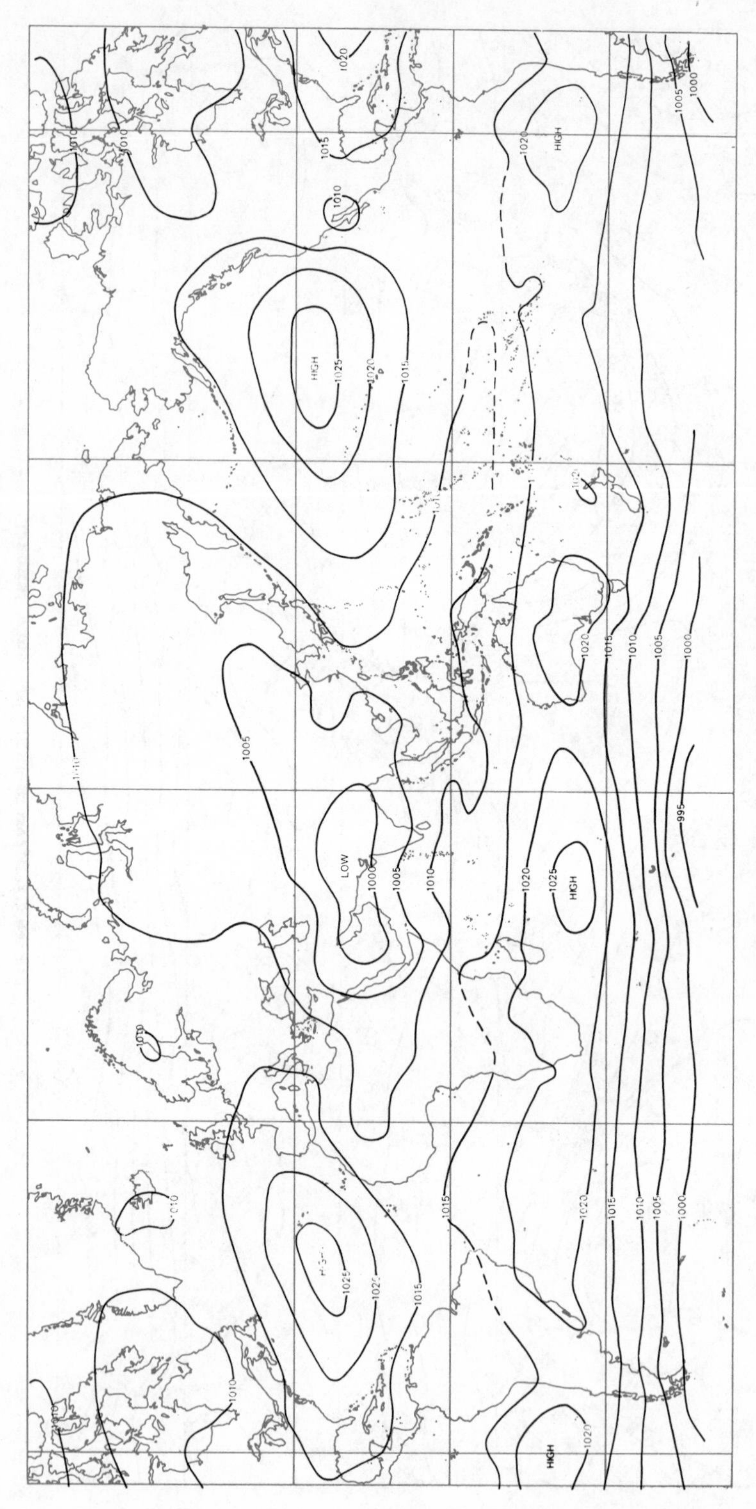

Figure 7.2. Mean monthly pressure, July

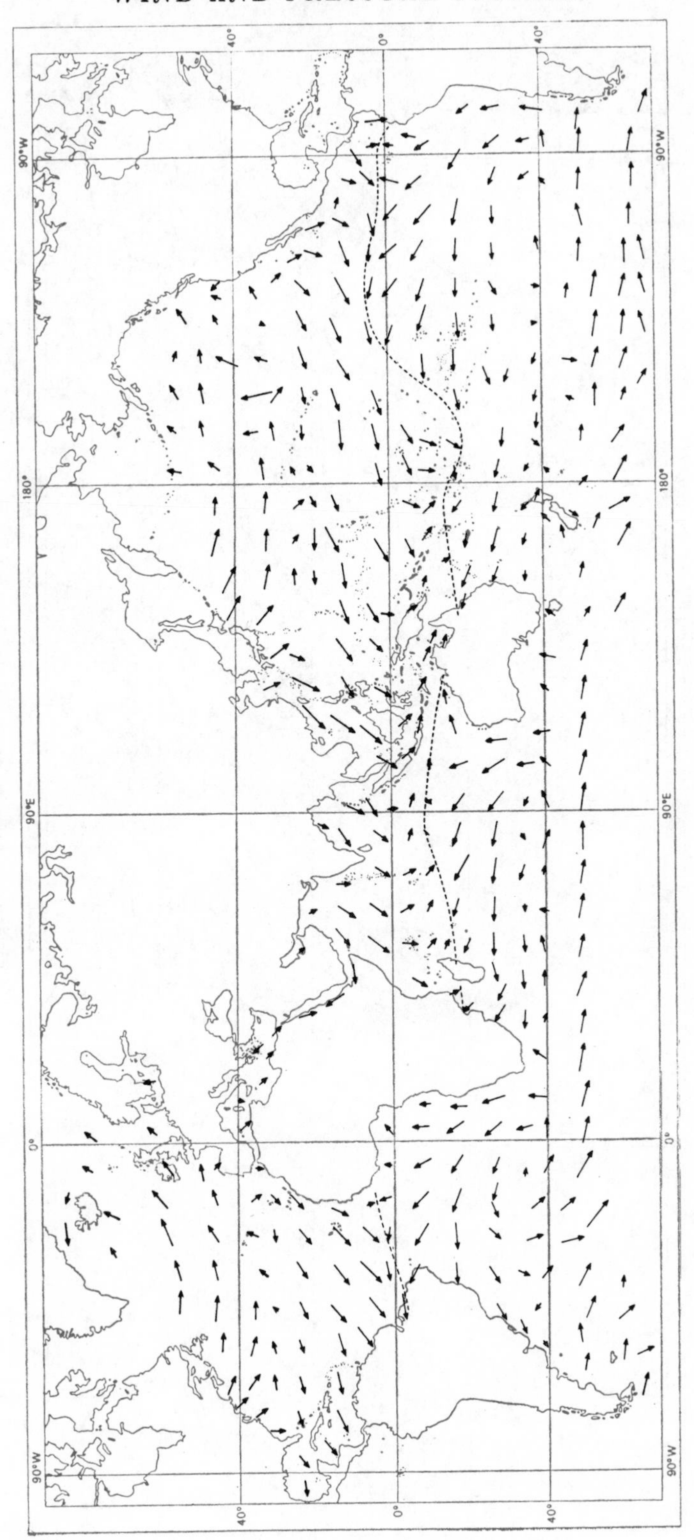

Figure 7.3. Vector mean winds, January

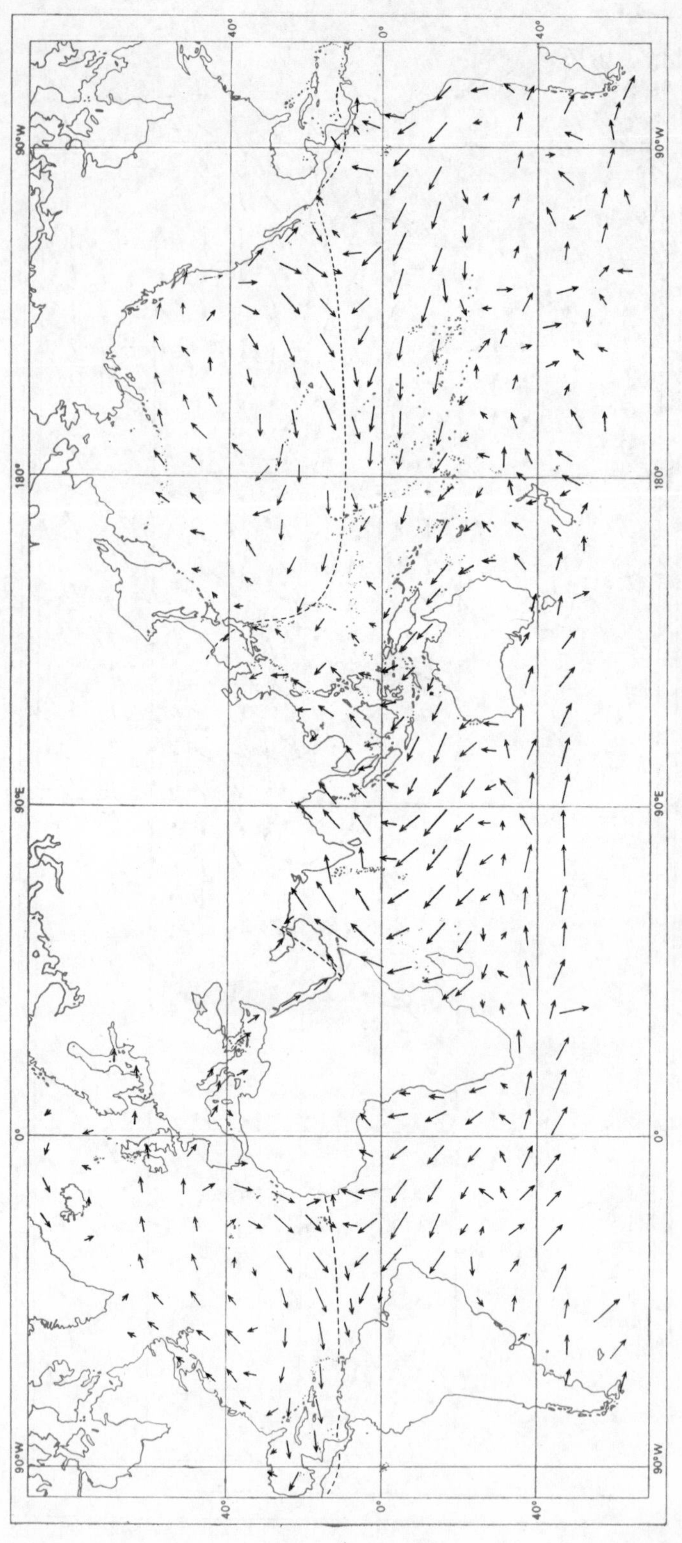

Figure 7.4. Vector mean winds, July

region between about lat. 40°N and 63°N. Between about lat. 63°N and 65°N there are local areas of variable winds conforming with the average positions of the Icelandic and Aleutian Lows. Further north the wind observations become scanty but in the region north of Iceland the winds are mainly north-easterly. In some sectors, however, other directions are reported, for example southerly off the central section of the north coast of central Siberia.

In the southern hemisphere there is a similar sequence of latitudinal belts. To the south of the doldrums lie the south-east trades between about lat. 10°S and 30°S. After a belt of light winds corresponding with the subtropical high-pressure belt near lat. 30°S, westerlies become predominant in a wide belt from about lat. 35°S to about lat. 63°S (the 'roaring forties'). Then, variable winds mark the position of the low-pressure belt off the coasts of Antarctica and give place polewards to south-easterlies blowing outward from the Antarctic continent.

There have been many attempts to produce a simple explanation of the general circulation of the atmosphere. So far the attempts have only been partially successful. George Hadley in the 18th century produced a theory to explain the trade winds. This postulated a circulation in the vertical wherein surface heating in the equatorial region produced low surface pressure and rising air in that region. The trade wind flow (NE'ly in the northern hemisphere) fed in surface air from higher latitudes. Above the equatorial surface low was a high-pressure region at high level from which air flowed outwards in a south-westerly direction. The circulation was completed by downflow in a more northerly latitude where high pressure occurred at the surface and low pressure at high level above. Figure 7.5 illustrates the circulation of the simple Hadley cell.

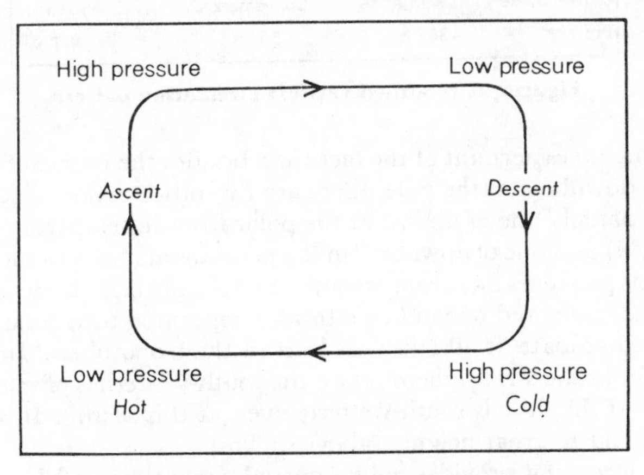

Figure 7.5. Circulation in a Hadley cell

An attraction of this circulation pattern is that it fits the theoretical expectation of the circulation on a uniform rotating earth heated at the equator and cooled at the poles. In these conditions one would expect that air would rise at the equator and sink at the poles, giving rise to a surface flow from the pole towards the equator and a return current flowing polewards at height. The effect of the geostrophic force would be to convert the surface northerly flow (in the northern hemisphere) into a north-easterly flow, and the return flow at height would be south-westerly.

This simple pattern, while fitting some of the facts, does not fit all of them. There are understandable reasons, concerned with the uneven distribution of land and sea in the northern hemisphere, why the observed flow does not fit the simple theory in the North Polar regions. In the southern hemisphere, where the presence of land at the pole keeps the region of lowest temperature near to the pole, the circulation in the polar region is as expected (south-easterlies at the surface overlain at height by north-westerlies). In the tropical region, too, the observations are in accord with the simple theory. The south-east trades are overlain at height by north-westerlies. What makes the simple theory untenable is the observed fact that between the polar south-easterlies and the south-east trades lies the broad belt of north-westerly winds (the 'roaring forties').

To accommodate this and other known facts it has been suggested that the original single-cell concept should be modified as in Figure 7.6.

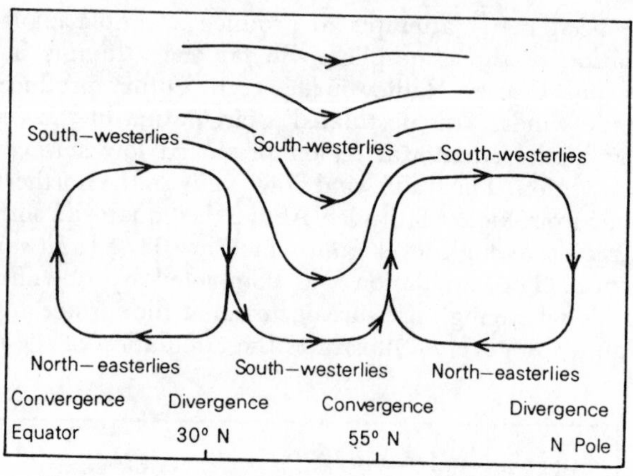

Figure 7.6. Modified vertical circulation pattern

This scheme takes account of the fact that, besides the region of upflow at the equator and downflow at the pole, there are two other major regions of vertical air currents, namely one of upflow at the polar front in temperate latitudes (say about lat. 55°N) and one of downflow in the permanent anticyclonic belt in about lat. 30°N. This produces an arrangement of two simple cells, as described earlier, one in high latitudes and one in low latitudes, separated by a zone where south-westerlies predominate at all levels, at least in the troposphere. In summer one finds easterlies in the stratosphere above the south-westerlies of middle latitudes but the bulk of the flow is south-westerly even at this season. In winter south-westerlies extend to great heights (above 65 km).

While the foregoing provides only a partial explanation of the circulation at higher levels, the low-level circulation can be usefully summarized in the manner shown in the idealized global circulation of surface winds shown in Figure 7.7 which fits most of the facts.

This shows the surface winds arranged in latitudinal belts which are symmetrical with respect to the two hemispheres. To the north of about lat. 65°N the winds are shown as north-easterly. Between about lats. 60°N and 40°N south-westerlies predominate, and between about lat. 30°N and 10°N north-easterlies (the trades) predominate. A similar pattern characterizes the flow in the southern hemisphere.

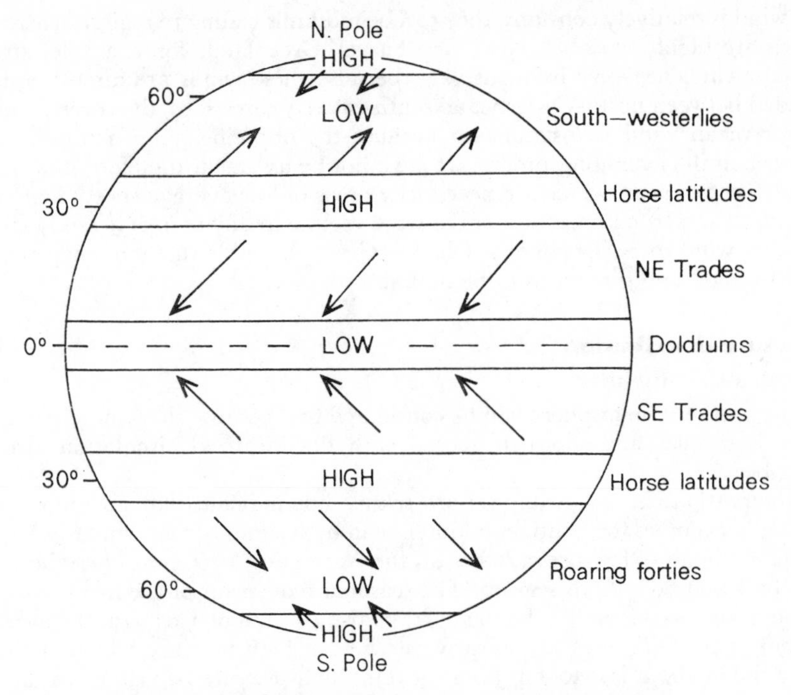

Figure 7.7. Idealized global circulation of surface winds

As far as the southern hemisphere is concerned the observed winds fit the idealized pattern quite well. In the northern hemisphere the fit is not so good on account of the irregular distribution of land and sea. Because the North Polar regions are occupied by sea, and the vast land-mass of Asia is remote from the pole, the centre of lowest winter temperature is displaced from the polar region to Siberia. Accordingly the main centre of subsidence and outflow is displaced from the polar region to the winter high-pressure centre in the Asian interior (Mongolia). In consequence, the high-latitude north-easterlies of the idealized pattern are observed only in certain sectors. Similarly, in summer, the Asian land mass gives rise to the extremes of heat which produce the monsoon low-pressure area (centred near the Himalayas). The converging circulation towards this centre suppresses the north-east trades in the Asian sector and replaces them with south-westerly winds during the summer season. Consequently in the summer months, one can recognize the northern hemisphere trade winds in the Atlantic and east Pacific but not in the Indian Ocean or China Seas.

A complication which must be borne in mind is that all the latitudinal belts of wind undergo a seasonal shift northwards and later southwards following the declination of the sun, with a slight delay. The amount of the displacement varies from place to place but generally amounts to about 5–8 degrees of latitude. The charts in Figures 7.3 and 7.4 represent an approximation to the furthest north and furthest south positions reached by the various wind belts.

When considering the winds shown on the vector mean wind chart, it is important to bear in mind that, while these winds give a reasonable approximation to the prevailing wind in the trade-wind belts and in other regions where the wind is relatively constant, they can be most misleading in regions where the winds are highly variable. Near Cap Farvel, Greenland, for example, strong to gale winds are very frequent. Yet, because these winds are almost equally divided between north-westerly and south-easterly directions, the corresponding vector mean wind is insignificant because the opposing gales neutralize one another in the averaging process. It must not be assumed, therefore, that areas of low vector mean winds are necessarily areas of low average speed. The only way to gain a true impression of the wind regime in any area is to study charts showing wind roses (*see* page 72, Chapter 6) which enable the frequency of both speed ranges and directions to be determined.

Pressure distribution

Southern Hemisphere

The southern hemisphere will be considered first because the wind circulation there is simpler and more in accord with the idealized circulation already described.

Observations in the Antarctic are scanty but indicate that an anticyclone normally occupies the continental interior in both summer and winter, while an annular trough of low pressure lies off the Antarctic coasts somewhere between lats. 60°s and 70°s at all seasons. The seasonal movement of the mean position of the trough axis is small. To the north of this circumpolar trough, the average pressure rises fairly steadily towards the belt of high pressure which encircles the globe in about lat. 30°s in January. This high-pressure belt shifts seasonally, following the sun, and is some 3–5 degrees farther north in July, than in January. The considerable pressure gradient indicated between about lat. 40°s and 60°s corresponds with the strong westerly winds of the 'roaring forties'. North of about lat. 40°s, the gradient falls off and the winds become light in the central region of the high-pressure belt. It will be noted that the cells of high pressure are best developed over the oceans whereas over land in the same latitude pressure tends to be lower. North of the high-pressure belt lies a moderate gradient for south-easterly winds (the trades) which falls off in the proximity of the equator where, except in the Indian ocean, there is a pressure minimum. This equatorial low-pressure belt, in the Atlantic and Pacific Oceans, moves north and south, following the sun, with a maximum amplitude of about 5–8 degrees of latitude. The situation in the Indian Ocean is described in the following section.

Northern Hemisphere

In the northern hemisphere, the comparatively simple distribution of pressure in latitudinal belts, as found in the southern hemisphere, is prevented by the more irregular distribution of land and sea. In particular the absence of a land mass in the North Polar region means that the winter temperature in this region does not fall as low as it does in the interior of Siberia. In consequence the main centre of high pressure in winter lies over central Asia instead of near the pole. From this high centre a ridge extends to north-east Siberia and as a comparatively weak ridge across the Canadian side of the pole to link up with the high over north Canada.

In summer, however, the pressure in the polar regions is higher than in the north temperate latitudes. The north temperate low-pressure belt is identifiable in all seasons but is not so well-marked or as constant a feature as in the corresponding zone in the southern hemisphere. The Icelandic low, with its more or less latitudinal trough, remains a feature throughout the year although it is less intense in summer than in winter. The other main cell, the Aleutian low, so called from its position near the Aleutian Islands of the Pacific, is well marked in the winter but practically disappears in summer, when there is a continuous latitudinal belt of relatively low pressure lying near the Arctic Circle.

In winter the low-pressure regions of the Iceland and Aleutian lows are separated by high-pressure areas in central Asia and again over the North American continent. In summer there is a more or less continuous belt of relatively low pressure in north temperate latitudes (60–65°N) in all sectors except that of central Asia. There, the axis of lowest pressure is displaced much further south and lies over the Himalayas (the monsoon low).

The subtropical high-pressure belt is well in evidence in the Atlantic and Pacific areas throughout the year. In the Asian region, however, the massive seasonal pressure fluctuations associated with the Asian monsoon obliterate all signs of this belt. In Asia there is a gradual change during the year from very high pressure centred over Mongolia in midwinter to very low pressure centred near the Himalayas in midsummer. Both in the Atlantic and in the east Pacific the pressure decreases southwards from the subtropical high-pressure belt to a minimum in a roughly latitudinal belt lying between about the equator and lat. 10°N, shifting with the sun. In the neighbourhood of India and south-east Asia in summer, however, this tropical low-pressure area is suppressed by the presence of the monsoon low near the Himalayas. In winter the monsoon high-pressure area over Mongolia has the effect that the tropical trough, which in the other oceans lies north of the equator at this season, is displaced southwards to about lat. 5°S.

Individual Wind Systems

Some of these systems such as the trades and monsoons have already been mentioned in connection with the general circulation but will now be reviewed systematically in greater detail.

Doldrums, Intertropical Convergence Zone

The doldrums is the name given to the zone of light and variable winds, often associated with heavy rain or thunderstorms, which form a narrow, roughly latitudinal, belt occupying a position between the equator and about lat. 12°N, which varies with longitude and with season. In the days of sailing-vessels the zone was dreaded on account of the risk of being becalmed. This is the zone where the trade-wind systems of the northern and southern hemispheres converge together and where, after prolonged heating over the equatorial ocean, some of the air finally ascends vertically.

The term INTERTROPICAL CONVERGENCE ZONE (ITCZ) concentrates on the latter concept, i.e. on the meeting place of the converging trade-wind belts of the northern and southern hemispheres whereas the term doldrums refers particularly to the tropical region of light winds. Apart from this difference of emphasis, they refer to the same phenomenon.

The ITCZ was formerly known as the INTERTROPICAL FRONT by analogy with the fronts of higher latitudes but the name was changed because of the rather vague and variable character of the accompanying weather characteristics.

On a world map of mean winds such as Figure 7.3 it is comparatively easy, with the definition of the ITCZ in mind, to trace its position as a line extending across the oceans around the world. We do not, however, find a corresponding continuous belt of heavy cloud and rain marking the position of the line. In the relatively few cases where there are sufficient observations to support a detailed analysis, the situation is revealed as being more complex. The line is then seen as a belt or zone whose width varies considerably from place to place and to some extent from one occasion to another. Sometimes the northern boundary of the zone will be marked by rain storms, squalls or thunderstorms, and sometimes it is the southern boundary which is so marked. On one occasion one boundary of the zone will be marked by bad weather with frontal characteristics, but a vessel two or three days later will report nothing untoward in the same region. It is as though the zone is one where rain, thunderstorms and squalls are liable to occur locally, and at times, rather than as a firm expectation.

The mean position of the ITCZ in January and July is indicated as a dotted line in Figures 7.3 and 7.4. Only the small scale of these figures justifies the use of a line to represent this belt of varying width. In the January chart the zone can be traced across each ocean. In July the zone can be recognized in the Atlantic and in the east Pacific. In the China Seas and Indian Ocean, however, the zone loses its identity. This is a result of the action of the monsoon low-pressure centre which dominates the air circulation of this region and season. Its effect is to suppress the northern-hemisphere trade winds in this sector. Because of the monsoon low, south-westerlies occupy all the region of the Arabian Sea and Bay of Bengal and the South China Sea in the summer months. Since there are no north-easterly trades there can be no ITCZ by definition. If the definition of the ITCZ is extended to cover convergence between air masses of southern- and northern-hemisphere origin (rather than limiting it to the trade winds) then its position can be traced further. From the tropical central Pacific it can, with difficulty, be traced towards Japan. In this sense a corresponding convergence line can sometimes be traced over South Arabia marking the boundary between north-westerlies in the Red Sea and Persian Gulf, and the south-westerlies in the Arabian Sea.

Between the extreme positions of summer and winter the ITCZ moves, more or less progressively, in a manner which can be studied from the mean charts for individual months which are available in climatic atlases.

In places, notably in the eastern parts of the Pacific and Atlantic, the trade winds of the opposing hemispheres converge at a comparatively high angle but in others, particularly in the western parts of these oceans, the winds of different origin flow alongside each other, both blowing from roughly an easterly direction. The ITCZ can be expected to be comparatively innocuous in this latter case.

Trade Winds

Except in the Arabian Sea, Bay of Bengal and China Seas, the trade winds are clearly prominent over all the major sea areas in both summer and winter as north-easterlies or south-easterlies (in the northern and southern hemispheres respectively) with a comparatively high vector mean wind speed. They blow more or less constantly through the year from about lat. 30° towards the equator. In each case they extend furthest from the equator on the east side of the relevant high-pressure area and blow round that area towards the doldrums or the equator, changing direction from north to north-east (south to south-east), to east-north-east (east-south-east) or even to east.

The trade-wind areas tend to shift northwards and southwards following the sun. In the southern hemisphere the movement is small, but in the northern hemisphere the zone of the highest trade-wind speeds moves through 8 to 10 degrees of latitude in the Atlantic and Pacific Oceans. In the Atlantic the mean wind speed of the trades is about 13 to 15 knots, the higher value occurring in the north-east trade. The highest mean value anywhere (18 knots) is found in the south-east trades of the Indian Ocean. In general the trades in each hemisphere blow most strongly at the end of the winter.

In the Arabian Sea, Bay of Bengal and the China Seas, the trade winds are suppressed and the wind regime is dominated by the Asian monsoons as described below.

Monsoons

The word monsoon is derived from an Arabic word meaning 'season'. It is used equally to describe a wind which is characteristic of a season, or to describe a season so characterized.

The simple sequence of latitudinal wind belts discussed in connection with the idealized general circulation of the globe is suppressed in the Asian theatre on account of the thermal effects of the presence of the great Asian land mass. The intense cooling which occurs in the interior of this area in winter and the extreme heating in summer has the effect of producing a great anticyclone, normally centred over Mongolia, in winter, and an intense depression, normally centred near the Himalayas, in summer. These two great pressure systems dominate the whole Asian region and extend roughly to the equator in the Indian Ocean and over all the China Seas and the seas around Japan. Because of them, the winds blow with great regularity in a clockwise circulation around the Mongolian high in winter and with similar regularity in an almost opposite direction around the continental low-pressure area in summer. Although there is a considerable shift in the position of the centre of lowest and highest pressure from summer to winter, the more important consideration for the mariner is that winter implies high pressure inland on the continent and low pressure at sea. Summer reverses this position so that the low pressure is now inland and the high pressure at sea. In accordance with this distribution the winds roughly follow the main lines of the Asiatic land mass, blowing slightly off the land in winter and towards the land in summer. Thus, for example, between south Japan and the Philippines the winds are predominantly north-north-easterly in winter and south to south-south-westerly in summer. In the Bay of Bengal and the Arabian Sea the winds are mainly north-easterly in winter and south-westerly in summer. In these seas the south-west monsoon is stronger and more stormy than the north-east monsoon. In the China Seas, however, the winds are stronger and steadier in the north-east or winter monsoon. In general the monsoon winds when fully established tend to be somewhat stronger than the trade winds. The south-west monsoon of the Indian Ocean brings copious rainfall, particularly to the windward coasts of India and south-east Asia, and is sometimes associated with poor visibility, which is not as a rule persistent.

For the most part the weather associated with the north-east monsoon is fine and clear. An exception occurs near the coasts of South China and Vietnam where spells of overcast, drizzly weather with poor visibility are frequent. Such spells may persist for up to ten days along the south China coast in February to April, where they are known locally as crachin and are an important climatic feature in that season.

The Horse Latitudes (Subtropical High-pressure Belt)

These are zones of light and variable winds and fine, clear weather marking the central regions of the subtropical high-pressure belt. They are mainly located in about latitude 30–35°N and S. The name refers to the occasional necessity, in sailing-ship days, of throwing overboard horses being carried to America and the West Indies, because of shortage of water and fodder, when the ship's passage was unduly prolonged due to the light winds.

Winds of the Temperate Zones

On the poleward side of the subtropical high-pressure belts are those disturbed regions where the mobile depressions and anticyclones of the temperate zone are found. As these pressure systems move, generally from a westerly direction, they cause considerable variations in wind direction and force at any given place. On the whole there is a predominance of westerly winds and a relatively high mean rate of travel of the surface air masses in some easterly direction. In the southern hemisphere these winds so often reach gale force that they are known as the Roaring Forties, after the latitudes wherein they are generally located. Figure 7.3 shows that in the southern summer they form a continuous belt around the globe; in winter the continuity of this belt is broken in the South Pacific where, according to Figure 7.4, the Roaring Forties appear to be less in evidence. It is likely that this effect is real, in so far as the wind directions in this region are extremely variable and thereby the magnitude of the vector means is reduced, but it is important to remember that the information in this region is based upon scanty data.

Polar Winds

The wind systems of the polar regions have not yet been studied in detail. In north polar regions the winter winds are in most areas very variable both in force and direction; quiet periods associated with developing independent anticyclones over the polar icefields alternating with disturbed periods due to incursions of depressions formed on the Atlantic and Pacific Arctic fronts. In summer winds are generally lighter, as depressions on the Arctic fronts are less frequent and less active.

In Antarctica, winds from the south-eastern quarter predominate both in the coastal region and, so far as is known, inland. They become more variable with increasing distance from the coast as a vessel approaches the low-pressure trough which surrounds the Antarctic continent.

Katabatic winds (*see* page 87) are an important feature of local weather in many places in Arctic and Antarctic regions. They are particularly frequent and severe where high ice- or snow-covered land lies close to a straight coast as in Adélie Land in the Antarctic and in east Greenland. The former area is one of the windiest regions of the world, and the Mawson Expedition experienced a mean annual wind speed near gale force.

Land and Sea Breezes

The main wind systems of the world may be modified by local causes, of which differential heating of land and sea is one of the most important. This gives rise to land and sea breezes which, by their reaction on the general wind distribution, may modify it appreciably.

During daylight hours the land warms up much more rapidly than the sea, partly because the specific heat of soil and rock is less than that of water and partly because the sun's rays penetrate to a greater depth in water than on land, and therefore have to heat a greater mass of water than of land. Thus the air near the surface warms up more rapidly and rises more easily over land than over water. Air flowing in from seaward to replace this rising air forms the sea breeze which under favourable conditions may set in well before midday, but often does not appear till the afternoon.

At night the reverse action takes place. The air over the land cools more rapidly by radiation and, becoming heavier, flows down the slope to the surface of the sea. It displaces and forces upwards the air already over the sea, which flows landwards at a higher level to complete the circulation. The land breeze, which is impeded by inequalities in the ground, trees, houses, etc., is much weaker than the sea breeze.

Sea and land breezes, especially the former, appreciably modify the climate near the coast. The diurnal range of air temperature on the coast is usually appreciably smaller than further inland.

Sea breezes are important near coasts where their effect is usually confined to within some 10–20 n. mile. They are most noticeable in fair or fine conditions when the general winds over a larger area are light. It is then quite common to observe force 4 winds on the coast when the winds further inland are only force 2. The direction of the coastal winds may be quite different from that of the inland winds.

In the tropics the effect of the land breeze is often felt several nautical miles to seaward, and the sea breeze can affect wind direction over the sea for 20 n. mile off the coast.

Anabatic Winds (Greek *ana* = up, *baino* = to move)

This is the name given to winds which blow up the slopes of a mountain or the sides of a plateau in calm, sunny conditions. They are found most commonly on steep-sided islands and bare mountain slopes in the cloudless arid regions of the tropics and subtropics. Anabatic winds can be considered as part of the convection currents which form in the day-time over a mountain heated by sunshine. At a later stage clouds may be formed over the mountain and showers may occur.

Katabatic Winds (Greek *kata* = down, *baino* = to move)

During the night, particularly with a clear sky, heat is radiated from the surface of the earth, which cools and consequently cools the air immediately above it. Where the ground is sloping, gravitation causes this cooler, denser air to flow down the slope, forming a katabatic wind, which may have no relation to the distribution of atmospheric pressure. In mountainous countries these winds can be violent; for example, in Greenland katabatic winds up to storm force locally, sometimes blow down the slopes to the sea.

Local Winds

The following are some brief notes on important local winds.

BORA. A cold, dry, katabatic wind which blows, from directions varying between north and east, down from the mountains of the north and east shores of the Adriatic. It is often dangerous, as it may set in suddenly and without much warning, the wind coming down in violent gusts from the mountains. It may occur behind the cold front of a depression, or when there is an intensification of the winter anticyclone over the continent to the north. It occurs chiefly in winter, when it may attain gale force. It is liable to be especially severe at Trieste. Similarly dangerous winds are reported at Novorossiysk on the Black Sea coast.

ETESIAN. The summer winds of the Aegean Sea and eastern Mediterranean blow with considerable constancy from a northerly direction. These seasonal winds, called *etesian* (Greek) or *meltemi* (Turkish), blow in response to the pressure gradient between relatively high pressure over the western Mediterranean and low pressure inland over Iraq.

GREGALE. A strong NE wind, found in the central and western Mediterranean. It has a significance for the mariner at Malta and on the east coast of Sicily, where the chief harbours are open to the NE. It occurs mainly during the winter. At Malta it generally occurs with pressure high to the north and low to the south, but may also occur after the passage of a depression.

HARMATTAN. A dry E wind, which blows on the west coast of Africa, between Cabo Verde and the Gulf of Guinea, in the dry season (November to March). Coming from the Sahara it brings clouds of fine dust and sand, which may be carried hundreds of nautical miles out to sea. Being dry and comparatively cool, it is a pleasant change from the normal damp tropical heat of this district in spite of the dust it brings, and is known locally as the 'Doctor'.

KHAMSIN. This is a southerly wind which blows in Egypt and in the Red Sea ahead of eastward-moving depressions. It is liable to be strong and, as it blows from the African interior it is hot, dry and often dusty. It occurs from February to June and is most frequent in March and April.

LEVANTER. An E wind in the Strait of Gibraltar. It brings excessive moisture, cloud, haze or fog, and sometimes rain. It is usually of only light or moderate strength, and under such conditions the 'Levanter cloud' usually stretches from the summit of the Rock for 1 n. mile or so to leeward; when it is fresh or strong, violent eddies, which are troublesome to sailing craft, are formed in the lee of the Rock.

MISTRAL. The Mistral is a strong N or NW wind which blows over the Golfe du Lion and adjoining coastal districts, particularly the Rhône Valley. It usually occurs with high pressure to the north-west, over France, and low to the south-east, over the Tyrrhenian Sea, but the wind due to this pressure distribution is strengthened by the katabatic flow of air down the mountains, and by the canalizing of the air down the narrow, deep Rhône Valley. In spite of warming caused by compression during this descent, the Mistral is usually a cold wind, but it is dry, so that the weather is usually sunny and clear. The Mistral often reaches the sea as a gale and rapidly produces a rough sea.

NORTHER. The Norther of Chile is a northerly gale, with rain, which occurs usually in the winter, but occasionally at other times. It generally gives good warning of its approach, i.e. a falling barometer, a cloudy or overcast sky, a swell from the northward and water high in the harbours, while distant land is unusually visible.

The term Norther is also applied to strong, cool, dry N winds which blow over the Gulf of Mexico and western Caribbean, chiefly in the winter. These Northers occur when there is an intense anticyclone over western North America and one or more depressions off the eastern seaboard of the United States. The Norther frequently sets in suddenly, with no warning. These winds sometimes reach gale force in the Gulf of Mexico, but as they travel southwards into the Caribbean their strength diminishes.

PAMPERO. This is the name given, in the Rio de la Plata area, to a line-squall occurring at the passage of a sharp cold front. It is usually accompanied by rain, thunder and lightning, while the wind backs suddenly from some northerly direction to a south or south-west direction.

Pamperos are most frequent between June and September. They can generally be foretold. Pressure falls slowly, while the wind blows at first from some northerly direction, with high temperature and humidity. As the cold front approaches the winds become NE, strong and gusty. Other signs of the approach of the Pampero are said to be a rising of the water level in the river, and increase in the number of insects in the air and the extreme clearness of the atmosphere.

The wind dies away for a short time prior to the arrival of the SW squall. This is usually accompanied by large Cu or Cb cloud, torrential rain and a fall in temperature. The SW wind may be very severe, up to 70 knots or more, during the first squall, but moderate afterwards.

SIROCCO. This name is given to any wind from a southerly direction in the Mediterranean. Since it originates from the deserts of North Africa it is hot and dry at most times of the year when it leaves the African coast. On its way north this air picks up a large amount of moisture from the sea, and is regarded as a disagreeable wind by the inhabitants of many countries of the northern Mediterranean on account of its enervating qualities. It frequently causes fog in these areas.

SHAMAL. In the Persian Gulf, the Gulf of Oman and along the Makran coast, the term Shamal denotes any NW wind, whether it is the normal prevailing wind or a gale associated with a depression. The average direction of the Shamal is NW, but this varies from place to place, according to the trend of the coast, and may be W or even SW. During a Shamal the air is generally very dry and the sky cloudless, but visibility is bad because of the masses of fine sand and dust blown from the desert. During summer months the Shamal seldom exceeds force 7, but in winter it often reaches force 8, sometimes force 9, and at this season may be accompanied by rain squalls, thunder and lightning. The barometer does not, as a rule, give any indication of the approach of a Shamal.

SOUTHERLY BUSTER. On the south-east coast of Australia the S wind behind the cold front in a trough of low pressure often starts with a violent squall. It is then known as a Southerly Buster. It occurs mainly in summer, but also in spring and autumn. As a rule, warning of a Buster is given only an hour or so beforehand by the appearance of Cu and Cb clouds in the south or south-west, sometimes accompanied by lightning. A long Cu roll then appears on the horizon. The wind drops to a calm and then becomes southerly, frequently blowing with gale force, accompanied by a rapid fall in temperature.

SUMATRA. Squalls from the SW known as Sumatras occur several times a month between May and October in the Malacca Straits and the west coast of Malaya, but less frequently in the vicinity of Singapore. The squalls develop over the

Straits and their approach can usually be seen in advance as they are almost invariably accompanied by a thunderstorm. At the onset the wind abruptly increases in a squall with gusts sometimes well over gale force, quickly followed by torrential rain. Sumatras generally occur between late evening and soon after sunrise, and are a distinct menace to small vessels.

TORNADO. This term is applied to two wind phenomena of different natures.

In the West African area it refers to the squall which accompanies a thunderstorm. This generally occurs with heavy rain and often lasts only a short time, but may do much damage. Tornadoes of this sort usually move from east to west and occur in the transition periods between the wet and dry seasons.

The other type of tornado is a violent whirl of air which may be about a hundred metres in diameter in which cyclonic winds of 150 knots or more blow near the centre. The whirl is broadest just under the cloud base and usually tapers down to where it meets the ground. In appearance it is mostly described as a dark funnel-shaped cloud like a long rope or snake in the sky, its darkness being due to the combination of thick cloud, rain, dust and other debris resulting from the destruction caused along its path. These tornadoes occur most frequently in the Middle West and central plains of the United States. The chance of any one locality being struck is small as the path of a tornado is usually less than 15 n. mile long, and the average number over the whole United States is 140 a year. Conditions most favourable for their occurrence are when maritime polar air from a north-westerly direction overruns maritime tropical air from the Gulf of Mexico, leading to a steep lapse rate and great instability, and then formation of tornadoes is most likely some distance ahead of the surface cold front. They can also occur in other localities, chiefly in temperate zones, about 50 having occurred in England in the past 80 years but these were all less intense than the United States variety. When a tornado moves from a land to a water surface it quickly takes on the characteristics of a waterspout, though it seldom maintains the intensity which it had over land.

Besides the tornadoes which have travelled from land to a water surface, there are numerous WATERSPOUTS* which originate over the open sea. The conditions favouring their formation over the sea are similar to those over land, namely, proximity of a frontal surface, marked changes of air and sea temperature over a short distance and above all great instability in the lower layers. Waterspouts are more frequent in the tropics than in temperate latitudes. Although they are less violent than tornadoes ashore, waterspouts are nevertheless a general hazard to all shipping and constitute a real danger to small craft.

A waterspout forms under the lower surface of a heavy Cb cloud. A funnel-shaped cloud appears, stretching downwards towards the sea. Beneath this cloud the water appears agitated and a cloud of spray forms. The funnel-shaped cloud descends till it dips into this spray cloud; it then assumes the shape of a column, stretching from sea to cloud.

The diameter of a waterspout may vary from about 1 metre to over 100 metres. The height, from the sea to the base of the cloud, may be as little as 50 metres but is usually about 300–600 metres (1000–2000 feet).

A waterspout may last from 10 minutes to half an hour; it travels quite slowly and its upper part often travels at a different speed from its base, so that it becomes oblique or bent. It finally becomes attenuated and the column breaks at about one-third of its height from the base, after which it quickly disappears.

---

* *See* also *Marine Observer's Handbook*, Chapter 11.

Frequency of Strong Winds

The frequency distribution of winds of Beaufort force 7 and higher in January and July over the oceans is shown in Figures 7.8 and 7.9.

The greatest frequencies of strong winds occur mainly in the temperate latitudes. In the northern hemisphere the incidence of strong winds is noticeably greater in the winter months but in the southern hemisphere the difference between summer and winter is less well marked. These strong winds are associated with travelling depressions and can blow from any direction, although winds with westerly components predominate.

Strong winds are also quite frequent in the Arabian Sea from June to August. They are associated with the south-west monsoon and, in contrast to the variable westerlies of the temperate latitudes, these strong winds are fairly constant in direction, blowing mainly from the south-west.

In general Figures 7.8 and 7.9 show relatively low frequencies of strong winds in tropical regions but there are a few small regions showing frequencies in excess of ten per cent. The strong winds of these regions are associated with tropical storms, (typhoons, hurricanes, etc.) which are described in Chapter 11.

Upper-air Circulation

Observations of wind and barometric pressure in the upper air are naturally more scarce than corresponding observations at the earth's surface. The main source of upper-level observations consists of radiosonde and radar soundings which are carried out at widely spaced stations which provide a world-wide network of observations. The coverage, while mostly adequate over many land areas is sparse or non-existent over large areas of the ocean. Soundings from weather ships play an important role, and aircraft provide a further source of observations.

Of recent years radiometer measurements from satellites have been increasingly used to provide temperature soundings and, indirectly, upper-wind data over non-tropical ocean areas. It is hoped to improve these techniques so that they can provide data similar to that presently supplied by radiosondes. This would have the advantage of giving a better coverage over the oceans than is practicable with radiosondes.

A knowledge of the wind and pressure fields at various levels throughout the atmosphere is important not only for the navigational needs of aircraft but also in connection with forecasting which depends not only upon the physical processes occurring in the lower levels, but also on all the changes occurring throughout the vertical extent of the atmosphere.

The average pressure pattern becomes progressively more simple at higher levels in the atmosphere. The closed circulations of individual lows and highs of the average surface chart become less marked and ultimately disappear. At high levels in the troposphere (e.g. about 16 km) the average northern hemisphere circulation in winter (January) takes the form of a simple vortex wherein westerly winds circulate around a centre located somewhere near the pole, and only two or three widely separated meridional troughs cause local deflections (to NW or SW) in the westerly flow which extends to all latitudes north of about 10°N.

In summer (July) the upper winds are much lighter. The centre of low pressure (at 16 km) is now centred near Baffin Land. The winds at this level are mainly between north-west and south-west to the north of about lat.

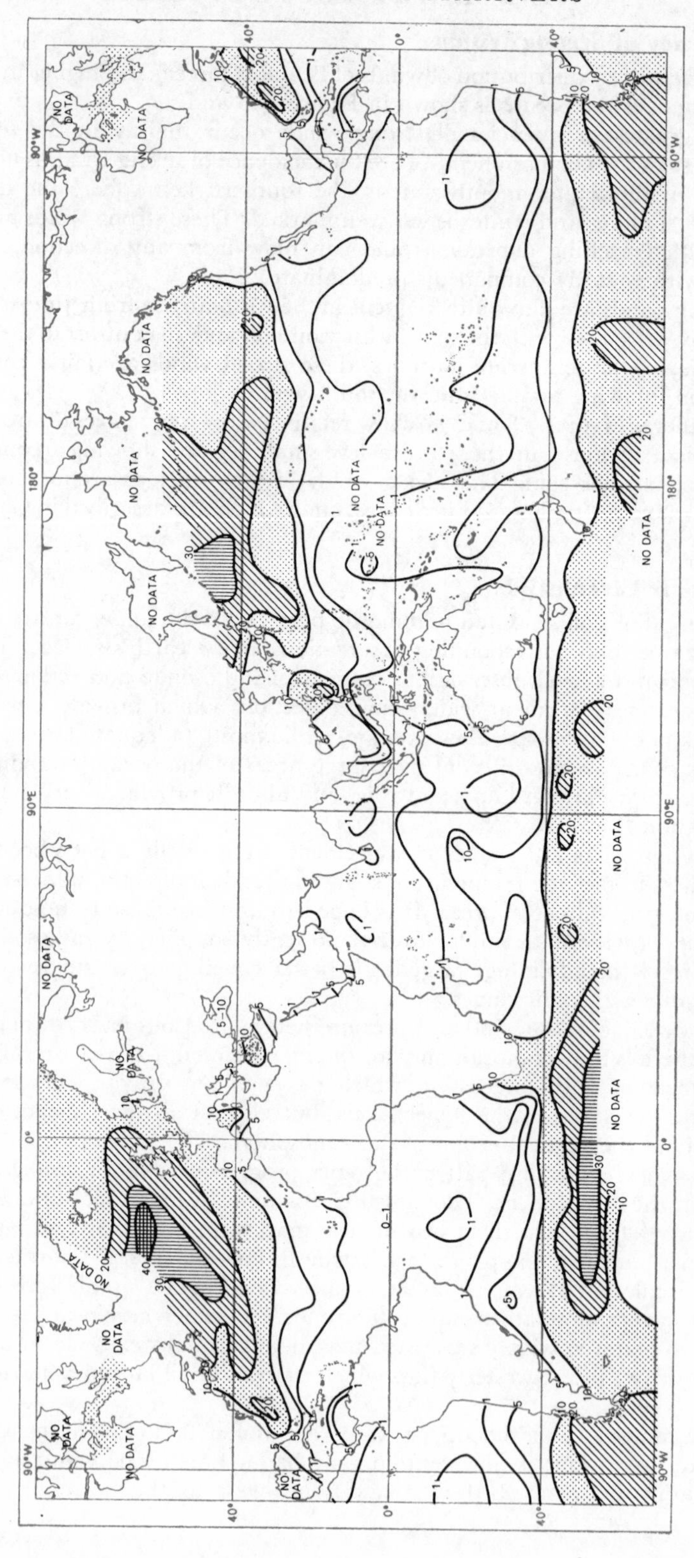

Figure 7.8. Percentage frequency of winds of Beaufort force 7 and higher, January

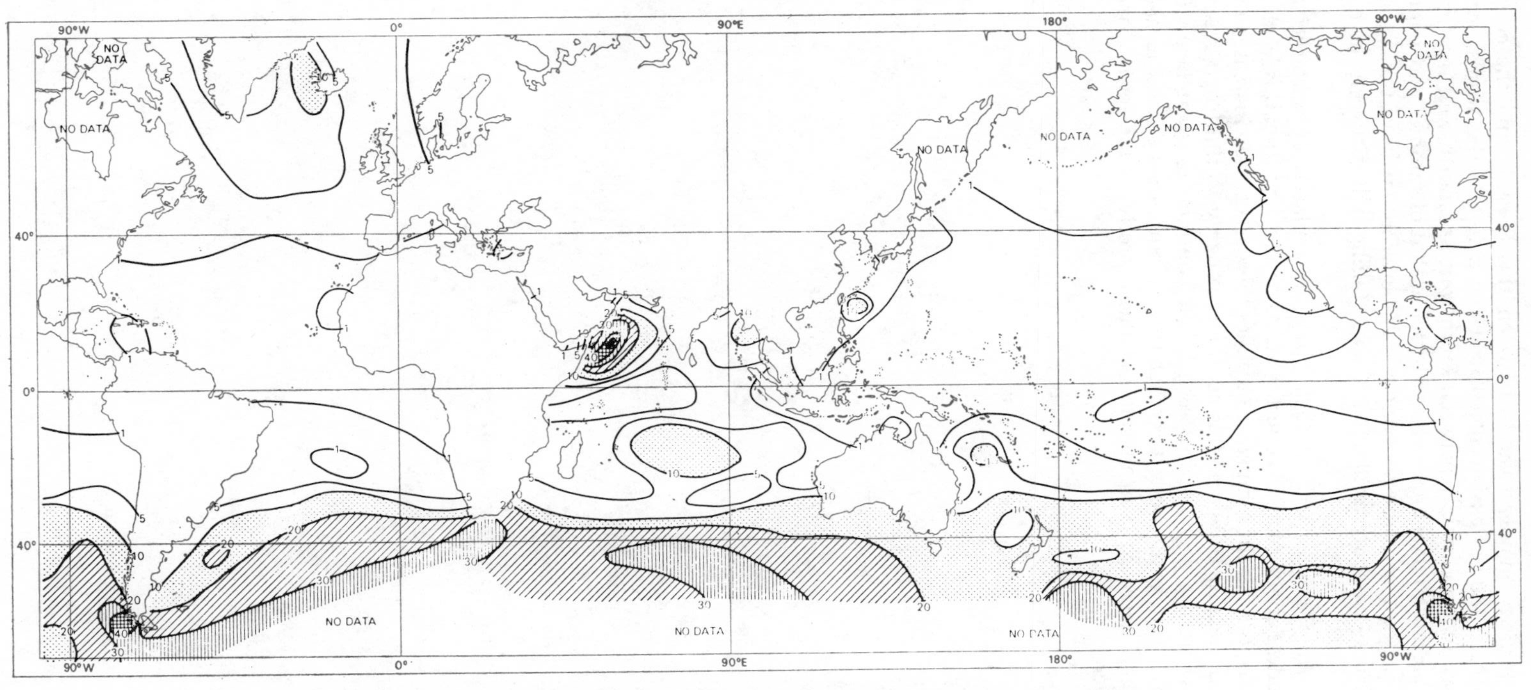

Figure 7·9. Percentage frequency of winds of Beaufort force 7 and higher, July

35–40°N but are mainly easterly between this latitude and the equator. In spring (April) the vortex is centred near the pole and the winds (less strong than in winter) are mainly between south-west and north-west in all parts. In autumn the situation is intermediate between that of summer and winter. The vortex is centred near the pole and winds between north-west and south-west are general as far south as about lat. 20°N, falling light between there and the equator.

In the Antarctic, easterly surface winds are replaced by westerlies at fairly low altitudes (about 3 km). Westerlies are also prevalent from this level upwards in arctic regions. In the temperate latitudes there is a steady increase in westerly winds with height. In winter this continues roughly to the tropopause. Above this the westerlies decrease in strength but continue to great heights (above 65 km). In summer the westerlies reach their maximum at a lower level (around 12 km) and decrease thereafter to become easterly in the lower part of the stratosphere.

CHAPTER 8

GENERAL CLIMATOLOGY OF THE OCEANS

This chapter reviews the remaining climatological elements following the discussion of wind and pressure distributions in the previous chapter.

Air and Sea Temperature

Except near the coasts, the air temperature over the oceans is largely related to the temperature of the sea surface. In consequence, the long-term average isotherms of air and sea surface temperature are somewhat similar in shape.

Figures 8.1 and 8.2 show the distribution of average sea surface temperature, and Figures 8.3 and 8.4 the corresponding average air temperature for February and August. These, in most regions, are the months when the long-term average sea surface temperatures are lowest, and highest, according to the hemisphere.

As a first approximation, the distribution of sea surface temperatures may be described as latitudinal. This is more nearly true in the southern hemisphere where, apart from a few corrugations, the isotherms roughly follow the parallels of latitude.

In the northern hemisphere a roughly latitudinal distribution obtains over most of the Pacific south of about lat. 45°N, and in the central parts of the Atlantic south of about lat. 40°N. Further north there are large departures from the latitudinal distribution.

In keeping with this distribution, the highest average sea temperatures are found in the tropics where there are extensive areas with values in excess of 28°C. There are relatively few areas wherein the average sea temperature exceeds 29°C. The highest value appearing on the charts of average sea temperatures is 33°C which occurs over a small area in the central part of the Persian Gulf in August.

The lowest sea temperature is limited by the freezing point of sea water which is normally about −2°C, but under conditions of low salinity the freezing point approaches more nearly to 0°C. Such temperatures are largely limited to Arctic and Antarctic regions but are locally experienced in lower latitudes, for example, near Kap Farvel, and in the Davis Strait and in the inner parts of the Baltic in February. Sea temperatures not far removed from 0°C are to be expected in the proximity of extensive sea ice, but one cannot assume a precise value for the water temperature. In some circumstances pack ice may have been driven by the winds from its place of origin into appreciably warmer water. With active melting taking place, the sea temperature near the ice edge may be well in excess of 0°C.

Departures from the latitudinal orientation of sea isotherms occur near some coasts in the southern hemisphere and to a much greater extent in the northern hemisphere. These more meridional trends in the isotherms are better understood if examined in conjunction with the chart of surface currents of the oceans (facing page 218).

One of the most striking examples of temperature irregularities is in the North Atlantic. The Gulf Stream, moving up the American coast and then across the Atlantic as the North Atlantic Drift, causes the isotherms to run not east and west, but almost SW to NE. The cold Labrador Current, moving southwards between the Gulf Stream and the American coast, complicates the isotherm pattern still more and creates, south of Newfoundland, the steepest sea-temperature gradients in the world.

In the North Pacific there is a similar phenomenon. The warm Kuro Shio Current, after flowing north-eastwards along the coast of Japan, moves eastwards across the ocean as the North Pacific Current, while the cold Kamchatka Current comes southwards to meet it off Japan. However, the resulting distortion of the isotherms east of Japan is not so marked as in the North Atlantic.

The effect of ocean currents on isotherms is also shown along the east coast of South America, where the warm Brazil Current moves south to meet the cold Falkland Current; and on the west coast where the relatively cool Peru Current flowing to the northward mixes with colder water which has upwelled from below (*see* Chapter 16).

Other examples of upwelling occur off the coast of South West Africa where, for some distance off shore, the isotherms run almost parallel to the coast; off the coast of Mauritania and Senegal and, in summer, off the coast of California and Oregon.

For the most part, the long-term average values of air temperature are within about 2°C of the underlying sea surface temperature for the same place and month. There is a tendency for the water to be slightly warmer than the overlying air, rather than vice versa. From the mariner's point of view regions where the sea is colder than the air are more important than those where the opposite is true. This is because any moist air advected over such regions may be cooled below its dew-point with the resultant formation of fog. The most notable area where this occurs is to the south of Newfoundland in late spring and early summer. There, in May, the average temperature of the water may be as much as 7°C below that of the overlying air. The frequent fogs in this region are largely due to the condensation resulting from the chilling of moist air by this cold water.

Because of the south-west to north-east orientation of the isotherms over the North Atlantic, winter sea and air temperatures in the triangle Norway–Iceland–United Kingdom are very much higher than the corresponding latitudinal mean. For example, mean February air temperatures around the Shetlands (+4°C), in about lat. 60°N, are comparable with the values in about lat. 38°N off the east coast of the United States. Alternatively one may compare this February temperature of +4°C in the Shetlands area with the corresponding value of about −20°C for the coastal air temperature in the same latitude on the other side of the Atlantic. In a similar way winter air temperatures along much of the coasts from north California to south Alaska are much higher than those in corresponding latitudes on the Asiatic side of the Pacific.

The foregoing description of air and sea temperatures has been limited to monthly mean values. On individual occasions more extreme values are liable to be encountered and, in particular, air/sea temperature differences are liable to be much greater. This is especially true near the coast, where a change from an onshore to an offshore wind can cause a large change of air temperature in a short time. Since sea temperature is more conservative, a large air/sea temperature difference can also develop. The proximity of large gradients of sea temperature such as are experienced at the inner edge of the Gulf Stream also favour large changes of air temperature within a short time. Here, a change of wind may displace air which has been in contact with one sea temperature to a position where the underlying water is much colder. The consequent chilling is then liable to lead to fog formation.

Extremes of air temperature such as those reported on land are unlikely to be experienced at sea. However, when sailing close to desert coasts, with offshore winds, very high values may be experienced. Still higher values are likely when in port. For example, 51°C has been reported at Abadan. At the time of writing the highest recorded air temperature on land is thought to be that of 57·8°C at Azizia, Libya.

It must be remembered that these temperatures, and the values that may be quoted in meteorological reports, are shade temperatures. The temperature attained by a steel deck exposed to the sun may rise well above these levels. In the Red Sea, for example, temperatures of 70–75°C have been reported on exposed steel decks.

The lowest air temperature known to have been reported on land is −88·3°C at Vostok, Antarctica. Such an extreme could not be experienced at sea, but if beset by ice in a cold port, a vessel could be exposed to very low temperatures. For example, an air temperature of −51°C has been recorded at the Antarctic coastal station McMurdo.

Humidity

Data in respect of humidity at sea are not readily available and many marine atlases exclude this element from their coverage because of insufficient data. However, some data are already published and no doubt will be extended in future publications. Monthly maps showing the distribution of average dew-point (°F) are given in *Climatological and Oceanographic Atlas for Mariners*, Vol. 1—North Atlantic, and Vol. 2—North Pacific, published by the United States Weather Bureau and United States Navy, Hydrographic Office. Another useful source which provides coverage for most of the world's oceans is the British Standards Institution's *Guide to hazards in the transport and storage of packages*, Part 2 (Maps and Diagrams), BS4672, 1971. This gives charts of maximum and minimum dew-points (°C) for the months of February and August.

Precipitation over the Oceans

The recording of precipitation amounts at sea presents considerable difficulty and the problem has not yet been solved satisfactorily. The occurrence of precipitation (rain, snow, hail) has, however, been regularly recorded aboard observing ships, and charts have been drawn showing the number of times precipitation of all types has been recorded in a given area, expressed as a percentage of the total number of observations. Charts for January and July are shown as Figures 8.5 and 8.6.

In the Atlantic there is an area of high rainfall frequency in the doldrums, caused by the convergence of the NE and SE trades. This is centred about long. 25°W and has a movement north, between January and July, of about 5° in latitude. North and south of this area the precipitation frequency decreases, especially off the coast of Africa; the frequency is also quite low in the Gulf of Guinea, well within the tropics. As we pass from the regions of the subtropical highs to those of the mobile depressions of the temperate zone, the frequency increases again, particularly in winter.

In the eastern Pacific there is an area of high rain frequency in the doldrums,

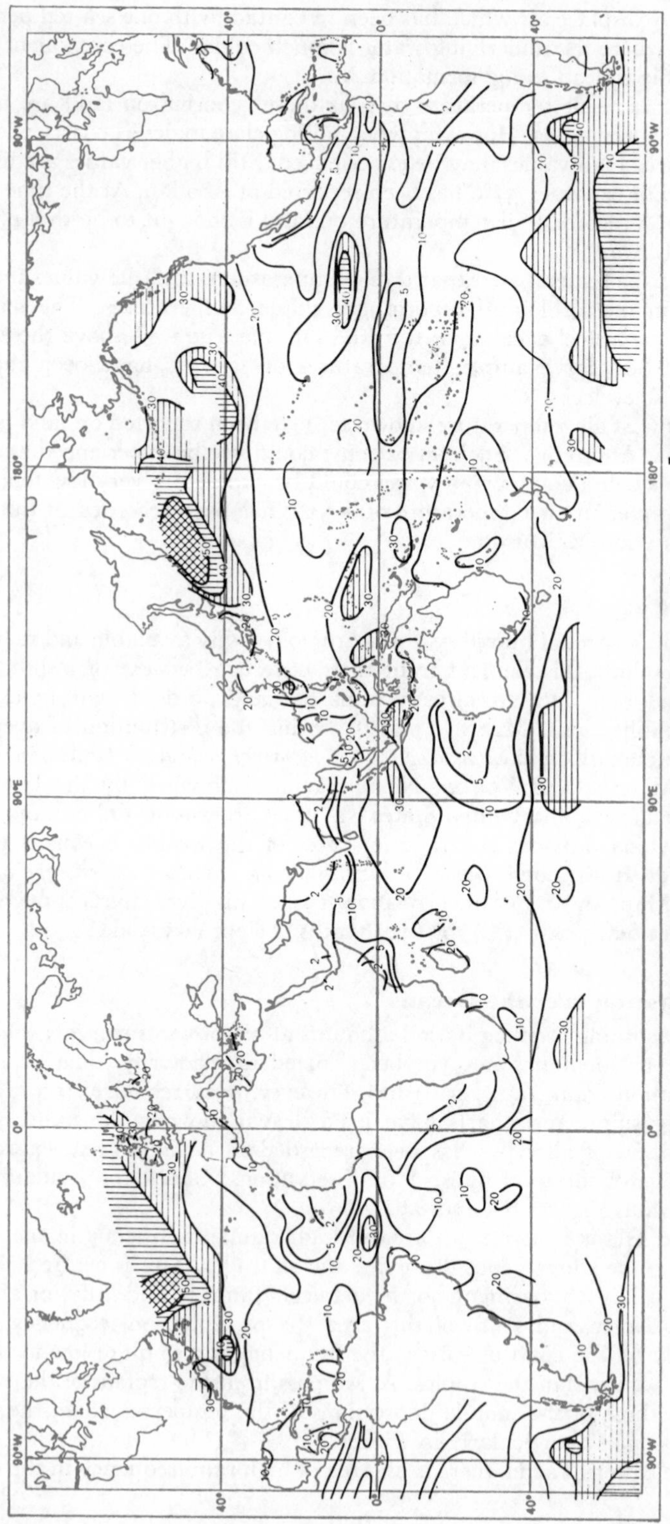

Figure 8.5. Percentage frequency of precipitation over the oceans, January

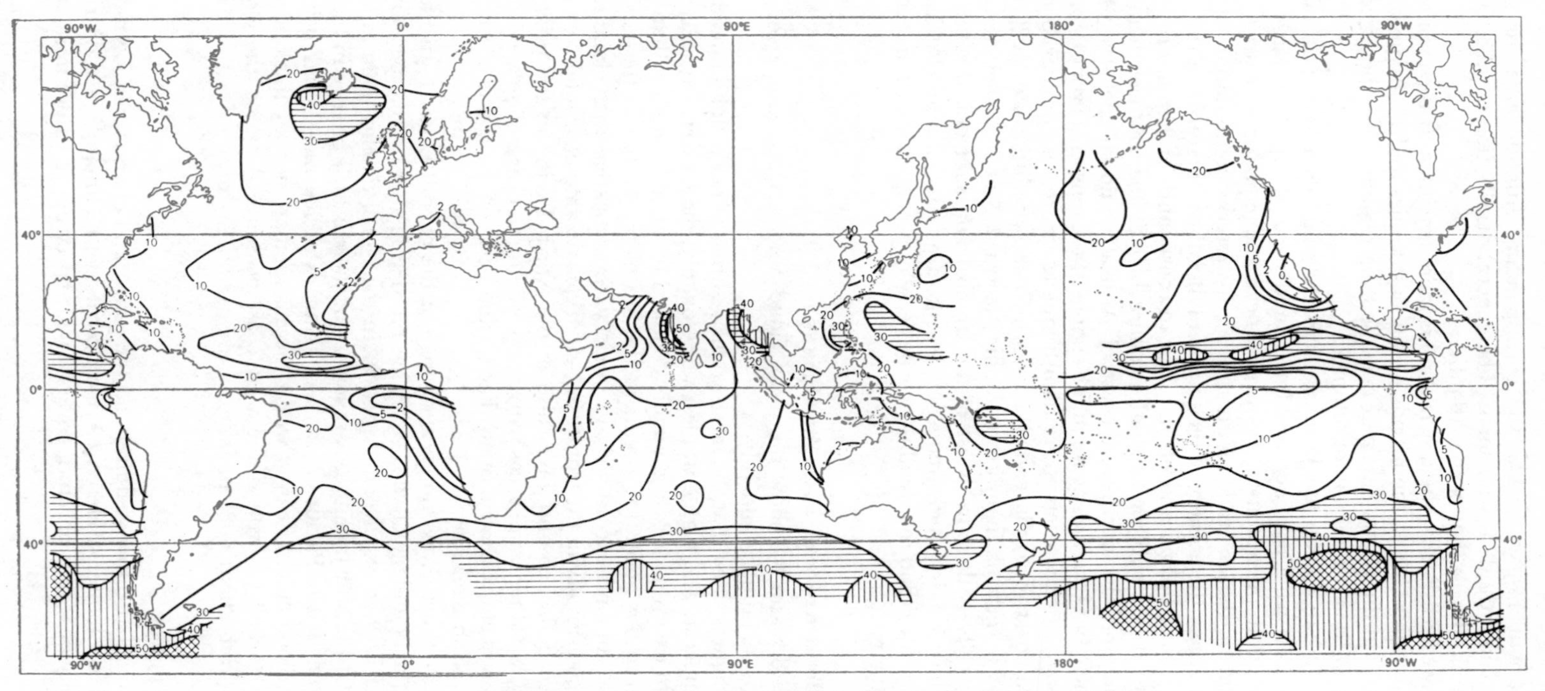

Figure 8.6. Percentage frequency of precipitation over the oceans, July

centred about 125°W. Unlike the corresponding Atlantic area, its seasonal movement in latitude is very small. The general distribution of precipitation is otherwise similar to that of the Atlantic.

In the western Pacific the distribution is more irregular. It is complicated by orographic rains due to the islands of the East Indies and Philippines and by the change, in the China Seas and neighbouring regions, from the NE monsoon of winter to the SW monsoon of summer.

In the Indian Ocean the January chart may be taken as typical of the NE monsoon months, December to March, and even of April. A large area of relatively high precipitation frequency near the equator is caused by the interaction of the SE trades with the north-westerly continuation of the NE monsoon across the equator. Precipitation falls off considerably in the Arabian Sea and Bay of Bengal and off the African coast south of 8°N. To the south it diminishes in the subtropical high-pressure belt of the southern hemisphere, particularly off the north-west coast of Australia and to the west of South West Africa, but increases again in the regions of the temperate zone depressions.

As the SW monsoon sets in, rainfall frequency increases considerably on the windward coasts of India and Burma, which show values up to 50% in July. The rainfall there is mainly orographic; leeward coasts or sheltered regions have generally little rain. Most of the tropical and subtropical south Indian Ocean shows a fairly high frequency, except near Australia and Africa, while further south we find the usual winter increase of precipitation in the temperate and subantarctic zones.

Mean Cloud Amount over the Oceans

Figures 8.7 and 8.8 show the mean cloud amount (percentage cover) over the oceans in January and July.

These figures show that the most extensive areas of large cloud amount occur in the temperate and high latitudes. Very roughly, these extensive areas of cloud amount in excess of 70% are mainly confined to latitudes higher than 40°. An exception to this generality occurs in the Sea of Japan and in the East China Sea where amounts exceed 70% in January. Smaller areas with cloud in excess of 70% occur locally in the tropics in January, for example, around New Guinea and between Borneo and Java, also in the North Pacific in the proximity of long. 135°W. Other small areas with cloud amount in excess of 70% occur in the tropical parts of the central South Pacific and between the coasts of Angola and the Greenwich meridian.

The July chart shows no great change in the overall pattern. In the North Atlantic the boundary of 70% cover has receded further north especially in the vicinity of Europe and North America. In the North Pacific cloud amounts in the East China and Japan seas are lower but there is little change in the central part of the North Pacific, and there is an extensive meridional belt in the neighbourhood of long. 135°W where there is even an increase in cloud amount (from 70 to 80%). Around the west coast of South America the cloudiness increases from about 60% in January to 70–80% in July.

Visibility, Fog

The factors affecting visibility and the causes and character of fog have been discussed in Chapter 5. Here we are concerned with climatological distribution.

Visibility at sea, as on land, covers a wide spectrum of values from excellent

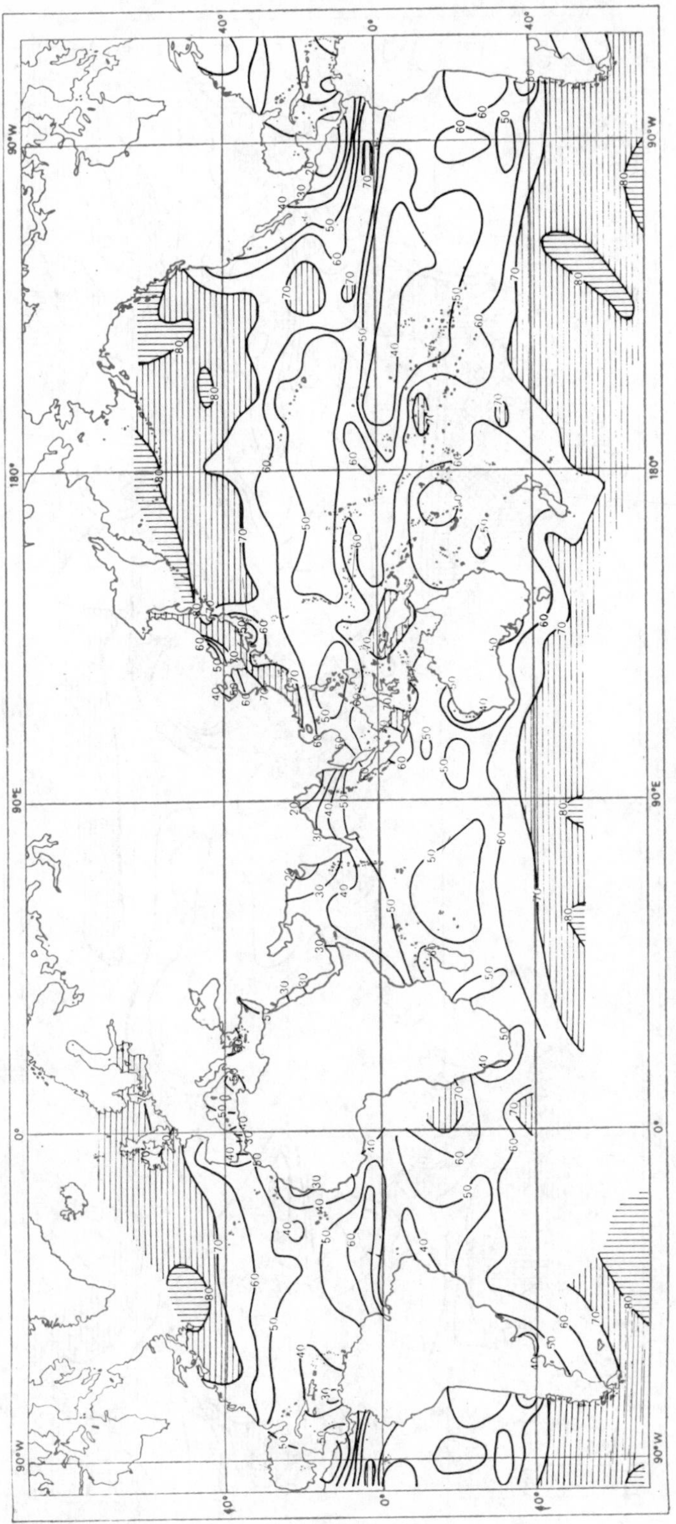

Figure 8.7. Mean cloud amount over the oceans (percentage cover), January

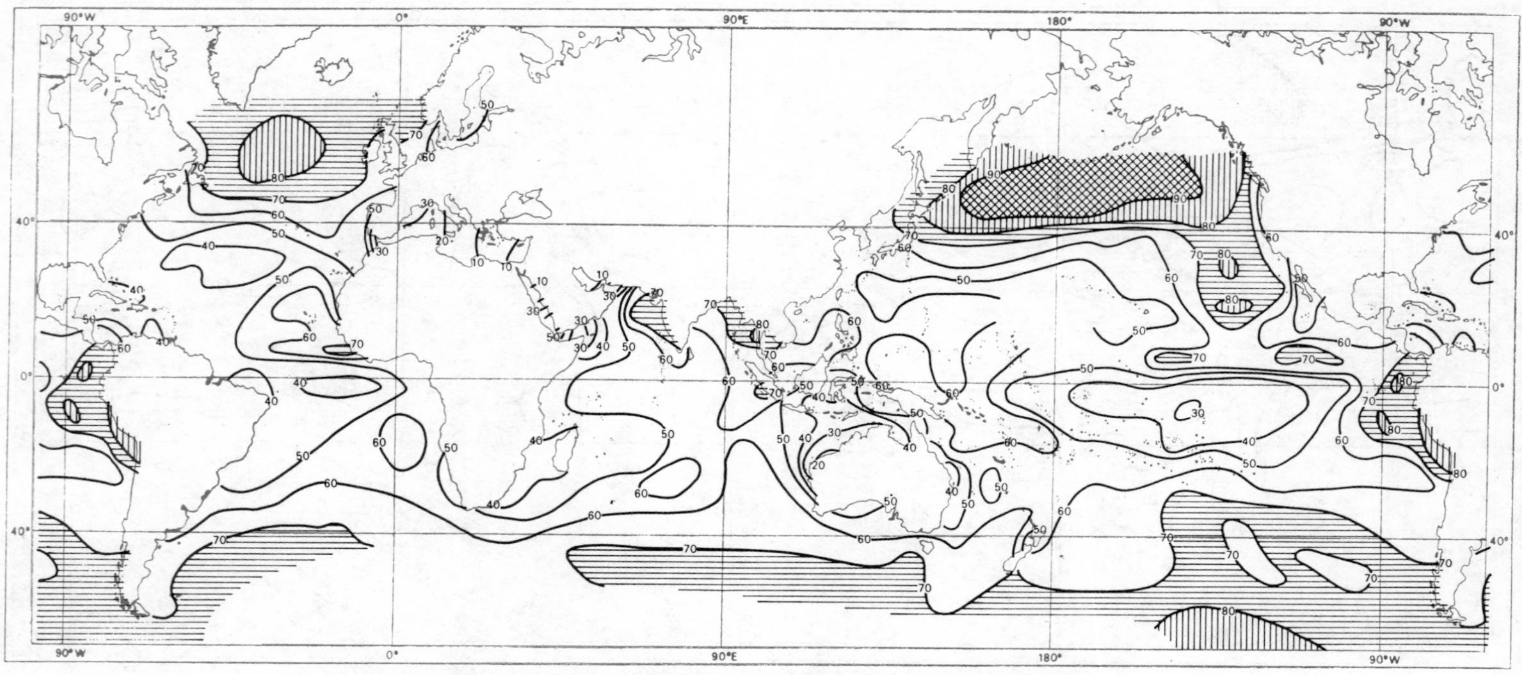

Figure 8.8. Mean cloud amount over the oceans (percentage cover), July

visibility when the clear horizon appears almost within hailing distance, to thick fog with visibility of the order of a metre.

When the visibility is good, its precise value is of largely academic interest. But when it is bad it may be a matter of life and death to know precisely how bad. Accordingly, although a whole scale of different visibility ranges is reported, when it comes to the statistical analysis of the observations, just two significant levels have been chosen by the British Meteorological Office. These are 10 km and 1 km which, in nautical terms, become 5·4 n. mile and 0·54 n. mile. These threshold values represent the upper limits of what may be called 'mediocre visibility' and fog.

Maps showing the distribution of visibility less than 5.4 n. mile (10 km) are available in the *Admiralty Routeing Charts* for the various oceans published by the Hydrographer of the Navy. These are of less vital concern to the mariner than are the corresponding maps of fog distribution, but they show some points of interest. For example they show the high frequency of haze surrounding the African coast in the vicinity of Cabo Verde, where vessels report visibility less than 5·4 n. mile on more than 30% of occasions from January to the end of May. This is in contrast with the low frequency of fog in this area.

Fog (visibility less than 1 km) is comparatively rare in the tropical and sub-tropical regions of the oceans, since conditions favourable for the advection of air sufficiently warm and moist relative to the temperature of the water surface do not exist in these regions. The distribution of fog over the oceans in January and July is shown in Figures 8.9 and 8.10, from which it can be seen that the frequency of fog may reach 40% in certain localities. Some of the regions of relatively high fog-frequency at sea are:

(*a*) THE NEWFOUNDLAND AREA. The reason for the high frequency of fog off the Grand Banks of Newfoundland is explained thus: the cold Labrador Current flowing southwards through the Davis Straits meets the warm Gulf Stream as the latter begins its journey across the Atlantic. When the winds in summer are from a southerly direction, they become heavily loaded with moisture as they pass over the warm waters of the Gulf Stream. They are chilled by this cold Labrador Current and the moisture they contain is condensed to form fog. If the wind shifts to the west or north-west the fog quickly clears. In the winter the winds are more often westerly or north-westerly; coming from the land, and having only a short journey over the sea, they are comparatively dry, and the fog frequency is consequently less at that season.

(*b*) THE NORTH-WEST PACIFIC. There is a similar explanation for the high frequency of fog in the north-west Pacific. The warm Kuro Shio, on its way north-eastwards across the Pacific, is met by the cold Kamchatka Current. The winds in summer are southerly and therefore warm and charged with moisture; cooled by the cold Kamchatka Current they cause frequent fogs. In winter the winds tend to be more westerly or north-westerly and therefore cool and dry, with only a short sea journey, and fogs are generally less frequent at that season.

(*c*) CALIFORNIA. Fog is most frequent between about latitudes 36°N and 39°N from June to the end of October when it is reported in 10–20% of all observations.

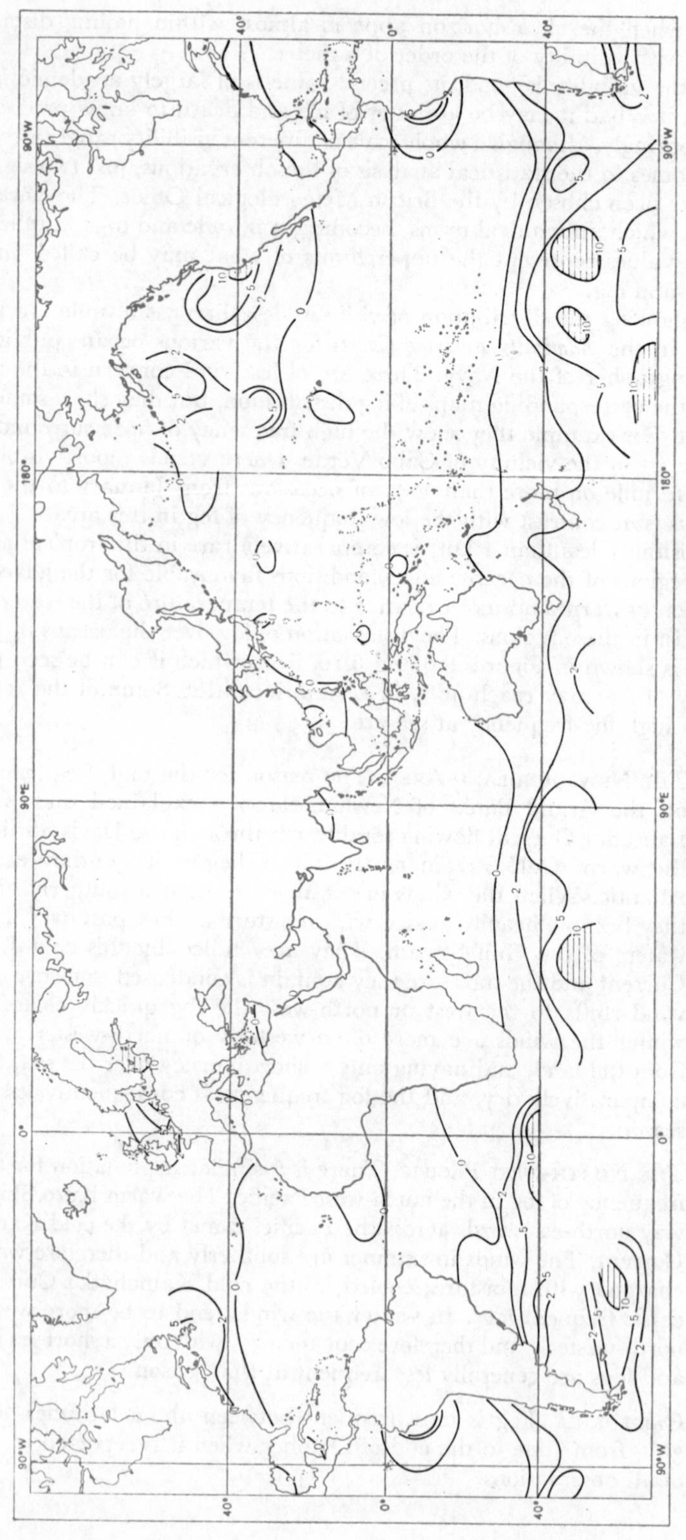

Figure 8.9. Percentage frequency of fog over the oceans, January

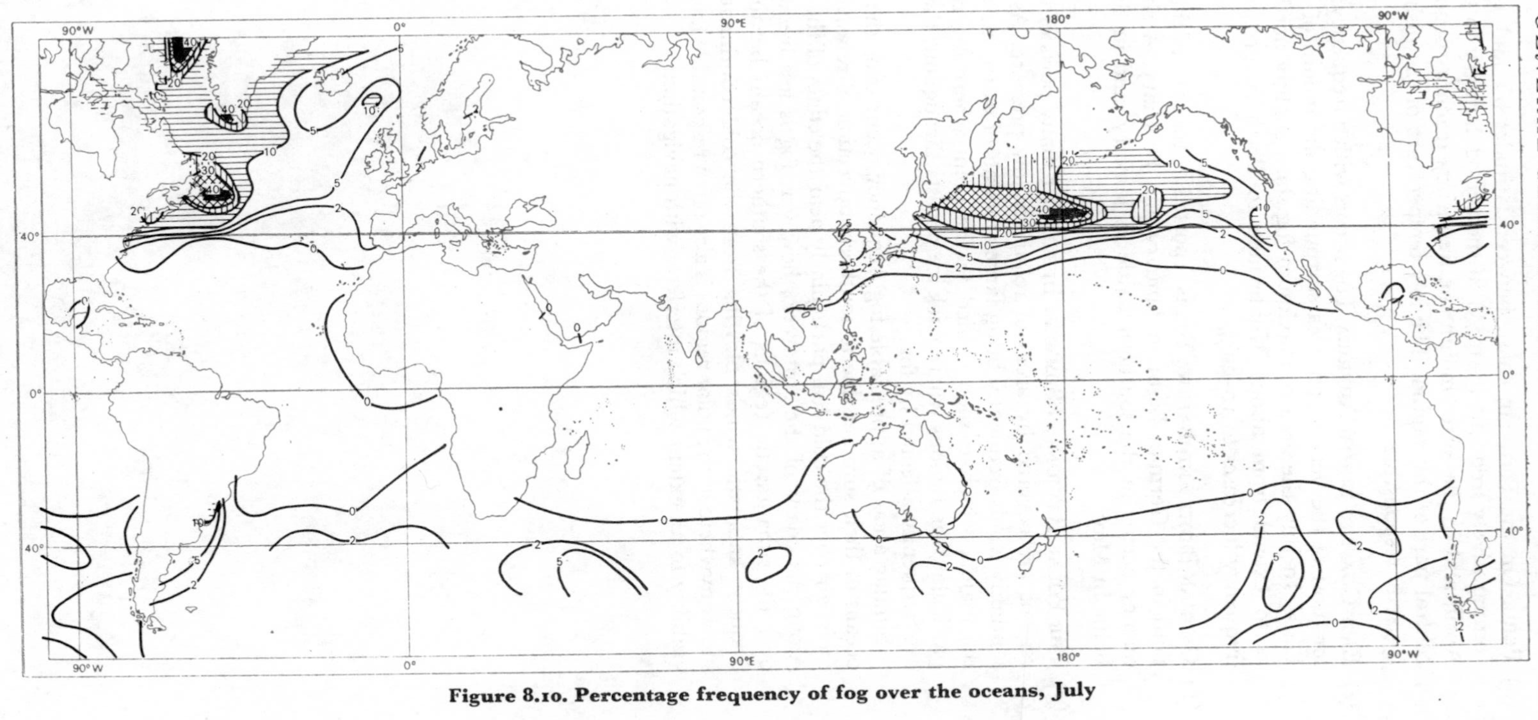

Figure 8.10. Percentage frequency of fog over the oceans, July

(*d*) PERU. On the Peruvian coast, between latitudes 6°s and 15°s, fog is comparatively frequent (5–10%) in the period January to the end of April, and in the latter month the frequency reaches 10–20% in the central part of this region. The low temperature of the inshore water is partly responsible.

(*e*) EAST COAST OF SOUTH AMERICA. Fog occurs with a frequency of 5–10% for most of the year in an area surrounding the mouth of Rio de la Plata. In July the area of maximum fog shifts slightly northwards so as to extend from about Montevideo to Porto Alegre, while the frequency increases to 10–20%.

(*f*) GERMAN BIGHT AND BALTIC. Fog is reported in 10–20% of the observations in the German Bight in January and February. A similar frequency occurs in the southern Baltic in January, and in the central Baltic in May.

(*g*) THE POLAR REGIONS IN SUMMER. In summer many parts of the Arctic regions are covered by areas of relatively low pressure. As a result, southerly winds frequently bring large amounts of warm moist air into the region, which is soon chilled to the point where fog results. In particular, the regions where pack ice persists throughout the summer are frequently affected by fog.

Similar areas of appreciable fog frequency occur over the southern oceans in their summer due to southerly advection of relatively warm moist air over the cold sea, particularly near the edges of the pack ice. Along the coasts of the Antarctic, however, fog is less frequent than over the subantarctic regions of the southern oceans because of the frequent katabatic winds blowing from the cold continent. Falling snow anywhere in the polar regions is also often responsible for reducing visibility to an extent which interferes with navigation.

PART III. WEATHER SYSTEMS

CHAPTER 9

STRUCTURE OF DEPRESSIONS

Introduction

In earlier chapters where the physical properties of the atmosphere and its general circulation have been described, the emphasis has chiefly been laid upon wind and pressure distribution. In the remaining chapters dealing with meteorology we shall be more concerned with the structure and physical characteristics of pressure systems such as the depression and the anticyclone, and with the factors controlling their movements, as the reader needs this preliminary knowledge before he can properly understand the principles of weather forecasting which are outlined in the later chapters.

We have already referred to depressions and anticyclones as the main features on a weather map, and have shown that they are associated with the movement of very large amounts of air between low and high latitudes, as well as with large-scale transfers of air at higher levels more nearly parallel to the equator, and mainly from west to east. As will be shown later, it is reasonable to assume that the behaviour of pressure systems is, to some extent, related to the physical characteristics of the air currents entering into and partaking in their circulation, and whenever these air currents have originated from widely separated latitudes, the degree of activity of the system is largely dependent upon the difference between the physical characteristics of these air currents. Before a more detailed study of depressions and anticyclones is made, closer attention must be paid to these large-scale movements of air.

Air Masses

A study of air movement on a large scale would be simplified if the air flowing over an area of several thousand square miles could be given a broadly descriptive title in accordance with one or more of its physical properties, assuming for this purpose that these properties had a nearly uniform value over the whole area. To be of any use, such a classification would have to be based on temperature and water-vapour content, since these are the physical properties of air of most concern to the meteorologist. At first sight the reader may well find it hard to believe that any properties of the atmosphere could be really uniform over such a wide area. He could reasonably point to the objection that temperature varies rapidly with height, and that the water-vapour content of the atmosphere decreases rapidly with height.

However, the temperature and water-vapour content of air usually change much more slowly in a horizontal direction than in a vertical. In fact study of synoptic charts has shown that sometimes air can acquire and retain substantially the same values of temperature and water-vapour content at any one level over areas comprising thousands of square miles. In any such area, the temperature and moisture characteristics of one sample vertical air column is very much like that of any other, since the horizontal homogeneity at the surface largely imposes a corresponding homogeneity at each level aloft. The whole amount of air overlying such a region is termed an AIR MASS, though the use of the term is usually restricted to air within the troposphere. The next steps are to see how air masses are formed and to look for the regions where their formation takes place.

Source regions

The essential characteristic of an air mass is that the distribution of temperature and humidity is broadly uniform throughout the air mass in a horizontal plane. Air masses are formed therefore over regions where the earth's surface temperature is nearly uniform and the winds are comparatively light. These factors ensure that air can remain in the region long enough to acquire the characteristic physical properties, which are largely determined by the nature of the underlying surface. For instance, the oceanic areas usually covered by the central regions of the subtropical anticyclones favour the formation of uniform air masses. Areas which produce this effect on the air above them are known as SOURCE REGIONS. Other typical source regions are snow-covered continents, the Arctic Ocean and extensive deserts such as the Sahara.

Figures 9.1 and 9.2 show very broadly the locations of the principal source regions of air masses for limited areas in January and July respectively. The mean flow patterns of the air masses from these source regions are indicated in a general way by continuous lines and arrows; the air masses thus indicated are of course those that occur *most frequently* in the areas concerned. The study of synoptic charts shows that the patterns of flow of these air masses on a given day will show marked differences from Figures 9.1 and 9.2, and that many areas will often be occupied by air masses different from those thus indicated. The largest variations in the positions of air masses are found in the temperate latitudes in both northern and southern hemispheres. Some indication of this variability is given for certain regions in Figures 9.1 and 9.2, where the paths most frequently followed by the other air masses which sometimes reach these regions is shown by dotted lines and arrows and the corresponding classification is enclosed in square brackets.

Classification of Air Masses

Air masses are classified as follows:

(*a*) An absolute classification based on the principal source regions, in which the following descriptive terms are used to describe the air masses:

(i) Arctic (A)
(ii) Maritime polar (mP)
(iii) Continental polar (cP)
(iv) Maritime tropical (mT)
(v) Continental tropical (cT)
(vi) Equatorial (E).

(*b*) A relative classification based upon the temperature of the air relative to the land or sea surface temperature in the area under consideration. According to this classification

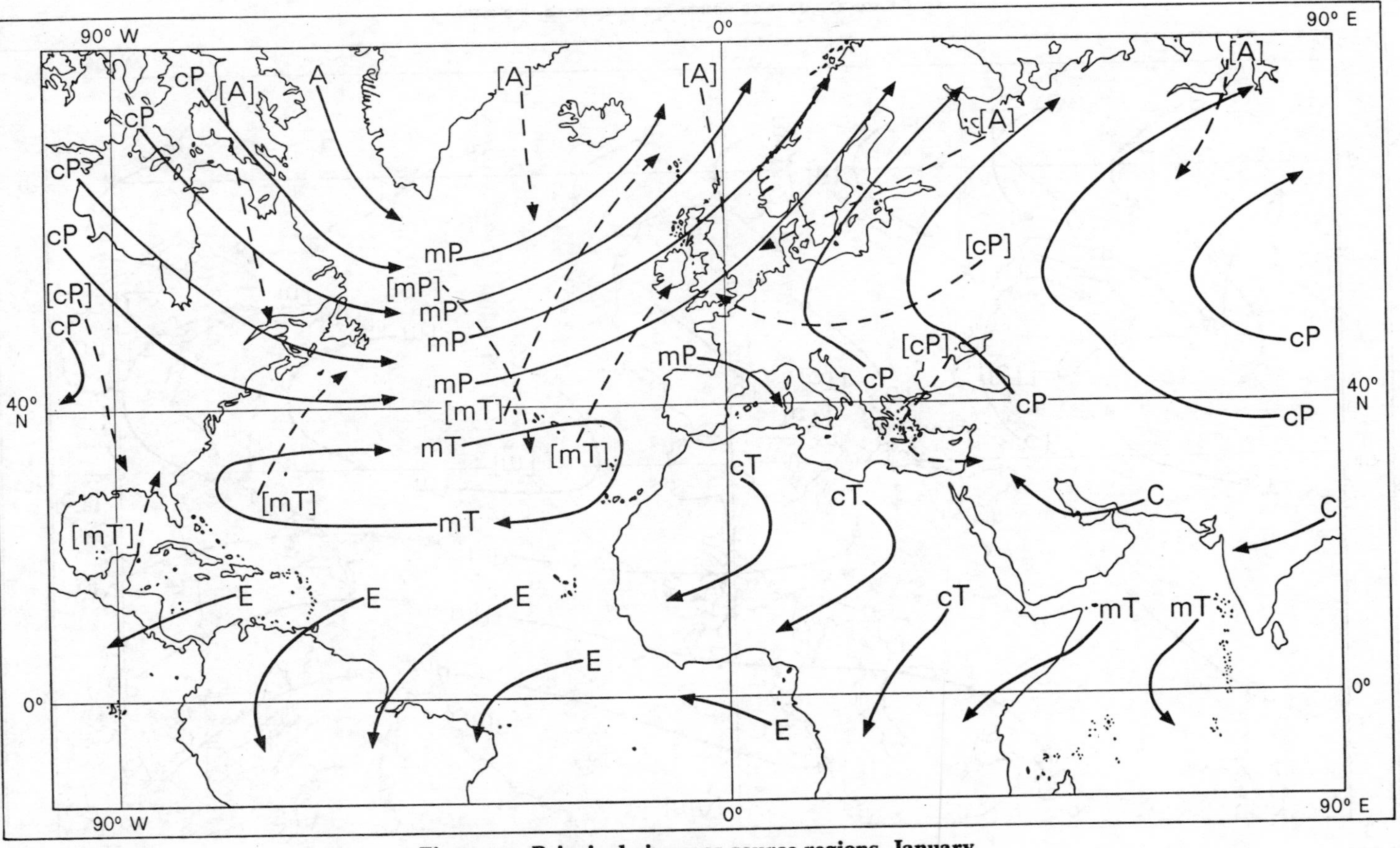

Figure 9.1. Principal air-mass source regions, January

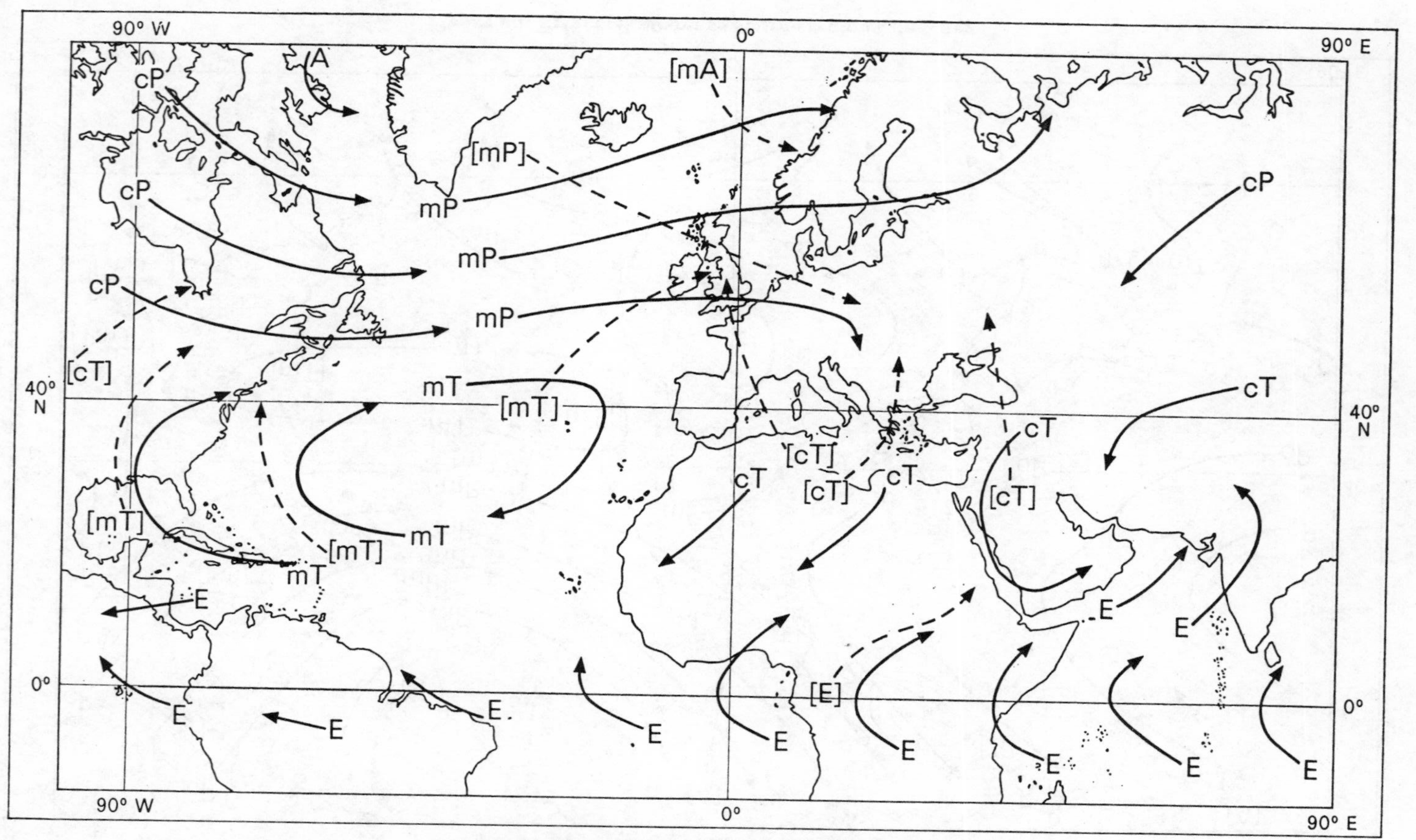

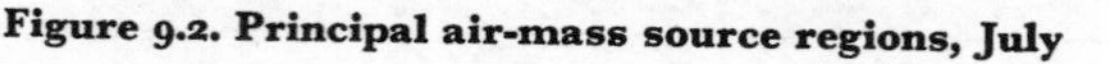

Figure 9.2. Principal air-mass source regions, July

(i) COLD AIR MASSES are those whose temperature near the surface is below the temperature of the underlying land or sea surface.

(ii) WARM AIR MASSES are those whose temperature near the surface is above the temperature of the underlying land or sea surface.

Since the temperature of the underlying surface will vary according to the weather recently experienced in the area considered, it will be understood that what is termed 'cold' on one occasion may be termed 'warm' on another. Nevertheless, the terms are useful since they enable one to make a number of inferences about the character of the associated weather. Used in this sense 'cold' air is being warmed from beneath and so tends to be thermally unstable, with cumulus-type clouds and good visibility except in showers. 'Warm' air is being cooled from beneath and so tends to be thermally stable, with stratified cloud and possibly fog.

Life History of Air Masses

As soon as it leaves a source region the properties of an air mass begin to be modified. These modifications mainly result from changes in the nature of the underlying surface, and changes in the radiation processes to which the air mass is subjected such as the length of the day and the mean solar elevation. The effect of these influences depends upon the time which has elapsed since the air mass left the source region; this time is known as the 'age' of the air mass.

The thermodynamical processes which produce modifications in the air mass include heating and cooling from below, and the addition or removal of water vapour by condensation or evaporation. Dynamical processes which produce modifications are convection, turbulence and subsidence.

Air-mass Properties

The properties of an air mass, while still in its source region, may be deduced from surface conditions. Evidently ARCTIC air masses and CONTINENTAL-POLAR air masses in winter are very cold since they originate over surfaces of land or sea which are covered with extensive ice or snowfields. The prevailing low temperatures result in an air mass, which originates from a large land area such as Greenland or Antarctica, having at first a very low moisture content before it has had time to pick up moisture from the sea surface. Since ice and snow surfaces are good radiators of heat the maximum cooling takes place in the lowest layers of the air. In such an air mass the lapse rate in the lower layers is thus small, and often shows a marked temperature inversion close to the ground.

In winter MARITIME-POLAR AIR MASSES are usually warmer than continental-polar air in the surface layers. Thus in maritime-polar air there is often a steep lapse rate in the lowest layers, the humidity also showing a rapid decrease with altitude since the moisture is mostly added to the surface layers from the sea.

Both CONTINENTAL-TROPICAL and MARITIME-TROPICAL source regions are warm so, unless the overlying air is already of an equal or higher temperature, it will be heated, though to a smaller extent in the air of maritime-tropical origin. In maritime-tropical air masses both relative and absolute humidity are high; however, in continental-tropical air in which very high dry-bulb temperatures are common, the relative humidity is generally less than in the maritime variety.

The main modification which an air mass undergoes while it travels is due to heating or cooling from below, that is, from the underlying surface. Heating from below creates a steep lapse rate of temperature, leading in turn to instability, convection and increased turbulence. The development of cumulus or cumulonimbus cloud, and showers, will follow when sufficient moisture is available in the air mass, or is added to the air mass by evaporation from the underlying surface. Strong convection generally results in good visibility, except in precipitation. Cooling from below is most effective in the layers nearest the surface and extends upwards only slowly due to the agency of turbulence, while in these circumstances convection is entirely suppressed. This is because an inversion of temperature is soon produced in the lowest layers of an air mass undergoing cooling from below. In very light winds, with little turbulence, the surface cooling of the air may be enough to cause condensation and fog. Stronger winds may create enough turbulence to prevent fog, but in such cases stratocumulus clouds often form just below the upper limit of the temperature inversion. Surface visibility remains poor on these occasions because the particles which constitute haze are not dispersed through a great depth of air but are confined within the lowest layer where they originated.

When an air mass subsides its temperature is raised at the dry adiabatic lapse rate. The relative humidity of the air mass also falls, partly due to adiabatic heating and partly because air is brought down from higher levels which has a lower moisture content than the air which it replaces. On the other hand, when an air mass is lifted, as in crossing a mountain range, its temperature falls adiabatically, often enough to produce condensation followed by cloud and precipitation.

Air-mass Boundaries—Fronts

Although each air mass is fairly homogeneous in a horizontal direction, the boundary zones of different air masses may be quite sharp. This is because mixing of air across the boundary between two distinct air masses having different temperature and humidity takes place rather slowly. On a weather map, the boundary zones of the different air masses are represented as lines, known as FRONTS. Thus a front may be regarded as a line at the earth's surface dividing two air masses. Boundaries between air masses are really 'zones of transition', but the small scale of the chart enables them to be regarded as lines. If we think in three dimensions, we can visualize two air masses separated by a surface whose intersection with the horizontal is the front as represented on the chart. This surface of separation is known as a FRONTAL SURFACE.

The name front was introduced during the First World War by analogy with the battlefronts. The analogy goes further, for most weather disturbances originate at fronts and the general picture, as successive disturbances move along the frontal zone, is one of 'war' between two or more conflicting air masses.

The principal frontal zones in the northern hemisphere are:

(*a*) The ARCTIC FRONT, in the Atlantic, separating Arctic air from maritime-polar air of the North Atlantic.

(*b*) The POLAR FRONT, in the Atlantic, which separates either continental-polar air of North America from maritime-tropical air of the North Atlantic, or maritime-polar air of the North Atlantic from maritime-tropical air of the North Atlantic.

(*c*) and (*d*) Similar Arctic and Polar fronts in the Pacific.

(*e*) The MEDITERRANEAN FRONT, separating the cold air over Europe in winter from the warm air over North Africa.

(*f*) The INTERTROPICAL CONVERGENCE ZONE. This was formerly known as the INTERTROPICAL FRONT but the name was changed because its character has little in common with the fronts of temperate latitudes. It is described more fully under the heading 'Doldrums, Intertropical Convergence Zone' in Chapter 7. What chiefly distinguishes the Intertropical Convergence Zone from the fronts of higher latitudes is that there is comparatively little difference between the temperatures on either side of the zone. Another difference is that it does not play a decisive role in the development of major depressions as do the fronts of higher latitudes.

Figures 9.3 and 9.4 show the mean positions of these frontal zones in January and July.

These frontal positions are very generalized. On any particular day positions vary very considerably from those shown on the maps, the greatest range in position and movement from day to day occurring in temperate latitudes. The latitudinal movement of air masses and fronts closely depends upon the distribution of pressure at any time. For instance, in certain synoptic situations over the North Atlantic it is not uncommon for maritime-tropical air to reach the Norwegian Sea area, or for maritime-polar air to penetrate south of the Azores and Madeira. Similarly continental-polar air from the Siberian anticyclone occasionally reaches Hong Kong just within the tropics, where it has been known to bring air temperatures down to freezing-point. Incursions of maritime-polar air to lower latitudes similarly bring cold spells to the interiors of South Africa and South America in winter. Even more extreme conditions occur in the prairie lands of western and north-western Canada, which are largely shielded by the Rockies from the moderating influence of the maritime air masses of the Pacific. At the same time these great plains are readily invaded by continental-arctic air in winter and by continental-tropical air in summer; so the range of temperature is very great over the year, and sometimes great extremes are recorded within a period of a few days. Very similar conditions occur in northern and north-eastern Siberia.

Frontal Surface as a Surface of Equilibrium

A frontal surface has been defined as the surface of separation between two air masses, but nothing has been said about the nature of this surface. At first sight it is difficult to see that such a surface will exist, and the question 'Why doesn't the air mix?' arises. In fact, although the air masses do mix to some extent through their surface of separation, the scale on which this mixing takes place is so small in comparison with the magnitude of a system such as a depression, and the air masses involved in its circulation, as to make it appear as though the air flow near the frontal surfaces takes place on both sides of an impenetrable boundary. In this respect the behaviour of air masses and that of the frontal surfaces separating them bears a close resemblance to that of two fluids, such as oil and water, which do not mix.

This analogy can be extended further. For instance, it can readily be shown that two fluids which do not mix and which have differing densities and

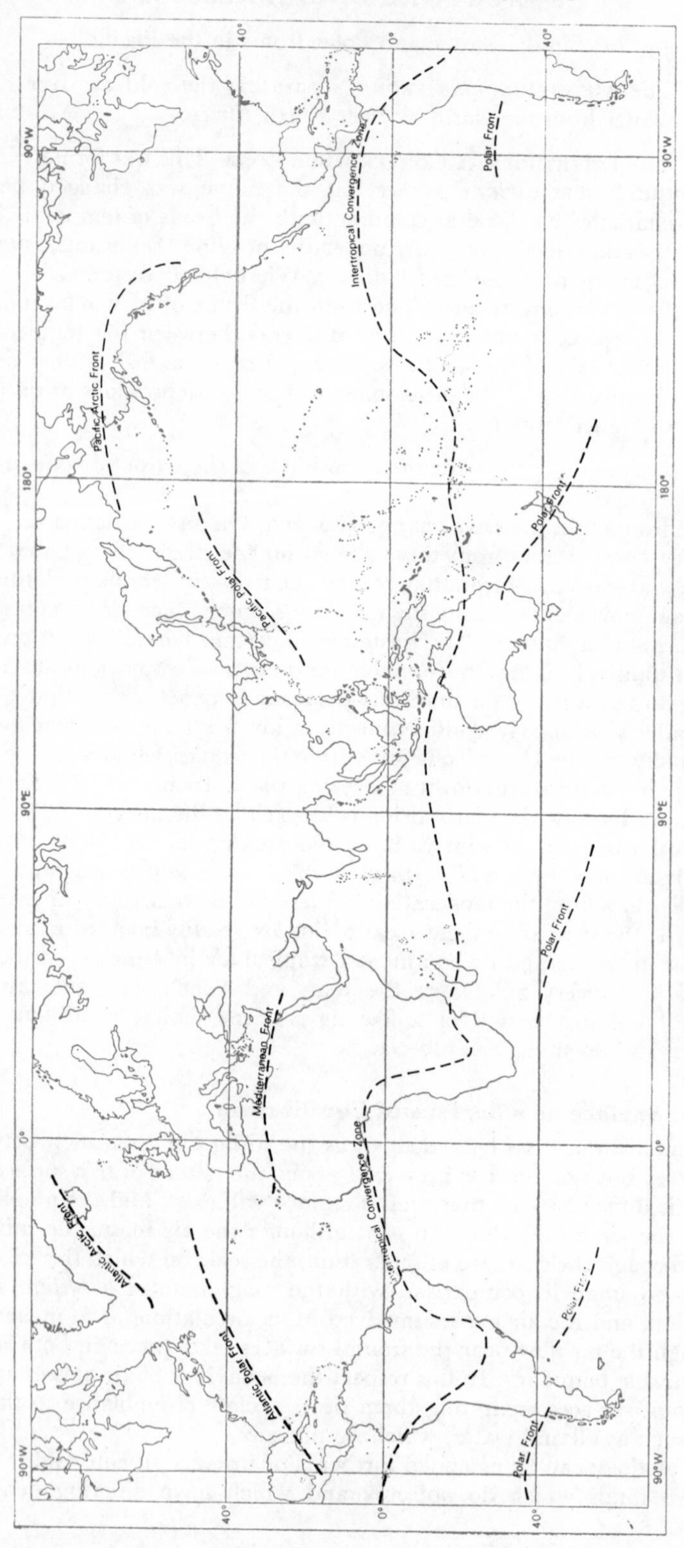

Figure 9.3. Mean position of frontal zones, January

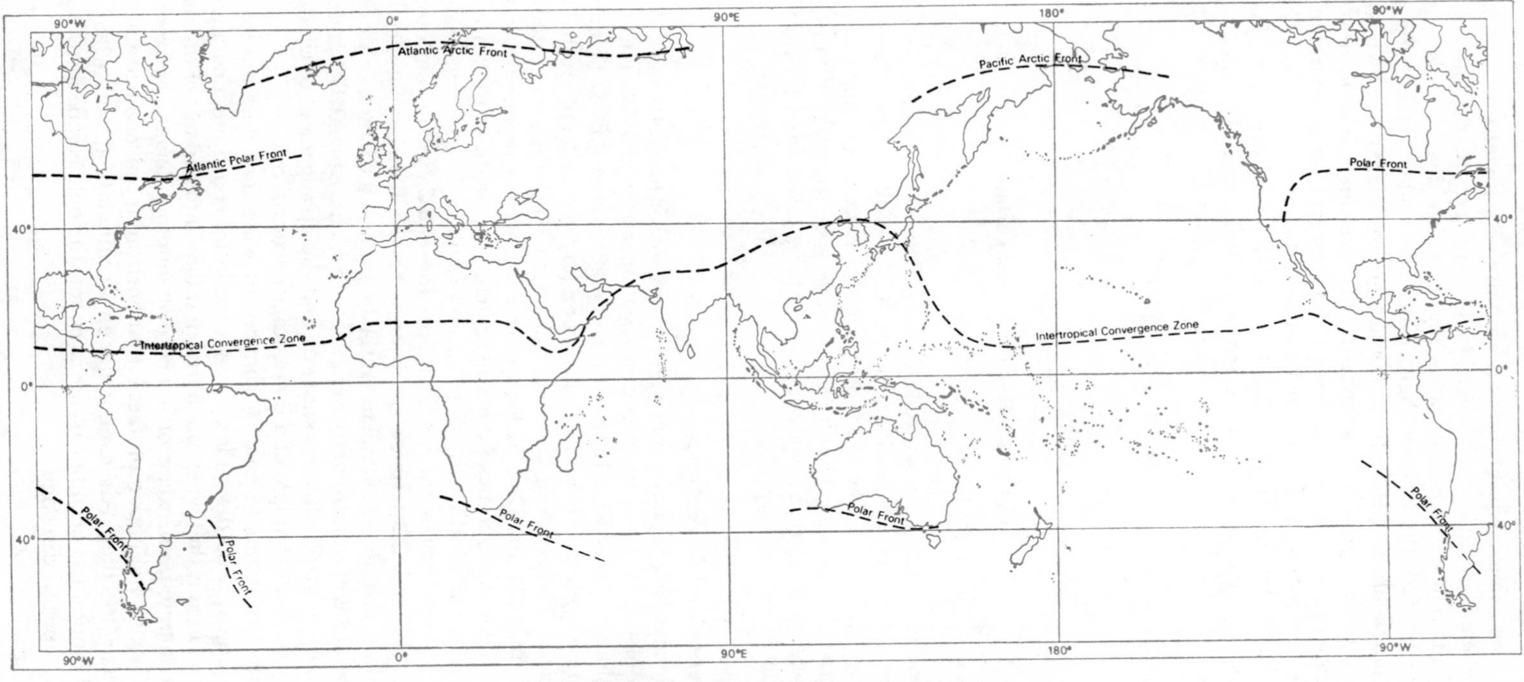

Figure 9.4. Mean position of frontal zones, July

velocities, can achieve equilibrium on a rotating earth along a plane surface of separation inclined at a small angle to the horizon. In the case of two air masses of different densities the colder, and therefore denser, air is found to lie as a wedge beneath the warmer, lighter air. The angle of inclination α is very small, its tangent being between 1/200 and 1/50. In Figure 9.5, which is a section and applies to the northern hemisphere, the colder air is imagined as blowing into the paper relative to the warmer air. **FS** then represents the frontal surface and **F** the line in which this surface intersects the horizontal plane **WC**. The fact that **FS** is not horizontal is due to the rotation of the earth. The actual magnitude of α depends also on the relative velocity and difference in density of the two types of air.

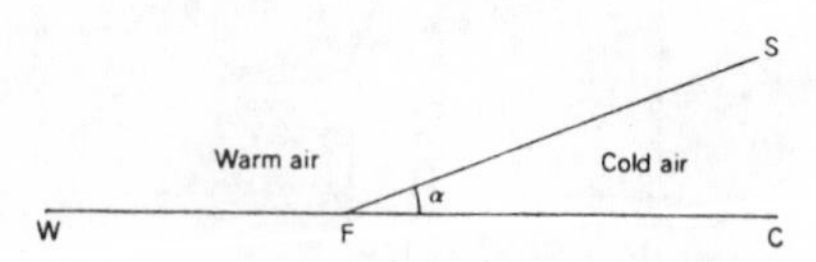

Figure 9.5. Slope of frontal surface

Fronts occur in sea water as well as in air. One of the finest examples, with which most mariners are familiar, is the 'Cold Wall' off the Newfoundland Banks, where the cold water of the Labrador Current meets the much warmer waters of the Gulf Stream. Here the front is very sharp, so sharp in places that ships athwart it have registered sea temperatures taken from forward and aft which differed appreciably.

Convergence and Divergence

An understanding of convergence and divergence is necessary in dealing with many problems in synoptic meteorology, especially those connected with pressure change. Consider a horizontal rectangular area **ABCD** (Figure 9.6a) and the space contained in an imaginary box (say 200 m deep) erected directly over this area. The box will have vertical sides whose edges touch the ground along **AB**, **BC**, **CD** and **DA**. In Figure 9.6a the arrows represent the average wind over the lowest 200 m of the atmosphere, the length of the arrows being proportional to the wind speed. More air crosses through the vertical side above **AD** than leaves through the vertical side above **BC**, yet no air crosses the boundaries **AB** and **CD** because the wind is always blowing parallel to these lines. Air must therefore accumulate in the box unless it can escape through the top, i.e. by means of upward vertical motion. In such a case **ABCD** is described as an AREA OF HORIZONTAL CONVERGENCE and the arrangement of winds shown in Figure 9.6a is an example of a CONVERGENT WIND FLOW.

There are two points to note. A convergent wind flow does not necessarily mean convergent wind directions, though the latter constitute a common case. The essential idea is that when the horizontal flow into and out of the restricted region is computed, an excess of inflow over outflow denotes horizontal convergence. The second important fact is that horizontal convergence and upward motion occur together. Air cannot go on accumulating, and when all the horizontal motion has been taken into account, the only escape for the air is in a vertically upward direction.

Figure 9.6b illustrates an example of depletion of air, in other words, HORIZONTAL DIVERGENCE. It is easily seen that horizontal divergence at the surface accompanies a downward vertical motion. This is the same as saying that the air is undergoing subsidence as described earlier in this chapter.

We have, therefore, the following results applying to motion of air near the earth's surface:

Upward motion of air is associated with areas of horizontal convergence in the lower atmosphere.

Downward motion of air is associated with areas of horizontal divergence in the lower atmosphere.

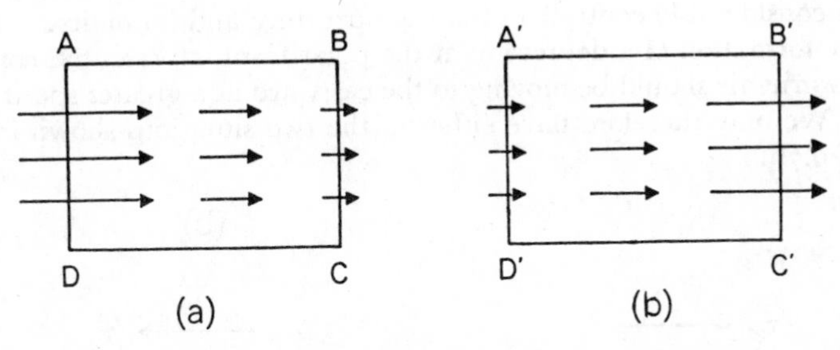

Figure 9.6. Convergence and divergence (arrow length represents wind speed)

Life History of a Depression

In Chapter 2 a depression was defined as a region of relatively low pressure with closed isobars.

Shortly after the First World War, Norwegian meteorologists, headed by Bjerknes, developed a theory of the formation, growth and decay of depressions in middle latitudes. More precisely, the achievement of the Norwegian meteorologists was to gather together the results of different workers and give a coherent account of the life history of each depression. The main advantage of the theory was that it related the life of depressions to the frontal zones and air masses separating them, and thus enabled the future behaviour of a depression to be forecast from a study of its previous history, as shown on a sequence of synoptic charts. Initially, meteorologists only recognized one frontal zone, that classified earlier as 'the polar front in the Atlantic', and as a result the name POLAR FRONT THEORY was given to the Norwegian work at an early stage. It has now been generally accepted among meteorologists and is outlined in the following paragraphs.

Cold surface air of polar origin has already been shown to be separated from the warmer air of lower latitudes along a surface of separation whose intersection with the earth's surface is known as a frontal zone, or more generally as a polar front. This polar front is not continuous around the earth but is extensive enough to justify the simple description given above. It is on this polar front that depressions of temperate latitudes form. (*See* also pages 114 and 115.)

The formation of a depression is assisted by a large temperature difference between the warm and cold air masses. It is for this reason, among others, that the regions where depressions most frequently form are in the western North Atlantic and western North Pacific and in the Southern Ocean, where the horizontal gradients of air and sea temperatures are greatest, particularly in winter.

In the first of these regions, the ocean area to the south of Newfoundland is particularly favourable for the development of depressions. Reference to Figure 7.1 will show that, in winter, there is a flow of cold air from the Canadian Archipelago towards this region. Moreover, in close proximity to this cold air, but slightly further south, the air flow is westerly. Although the mean pressure chart shows a continuous belt of high pressure lying east and west across the Atlantic in about lat. 30°N, this belt, when examined on a daily basis, is found to be made up of a series of separate anticyclones separated by minor cols, so that the westerly current of warm air on its northern flank is intermittently fed by still warmer air from its southern flank. This brings together air masses having a considerable contrast in both temperature and humidity.

For the formation of a depression at the polar front, an essential condition is that the warm air should be moving to the eastward at a greater speed than the cold air. We may therefore have either of the two situations shown in Figures 9.7a and 9.7b.

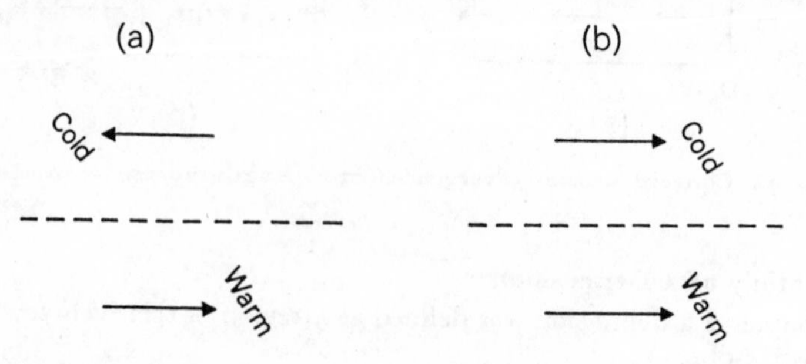

Figure 9.7. Formation of a depression

In general, such a depression begins as a small wave-like disturbance on a frontal surface. As it develops a circulation it becomes a larger system and moves away to the E or NE in the northern hemisphere. (Figures 9.8a–d represent successive stages in the development of a depression.) The warm air overrides the cold air at the warm front; the cold air undercuts the warm air at the cold front. Simultaneously a fall of pressure occurs over the centre; in other words, the depression deepens. Along with the process of development, there is a general motion of the system as a whole with the approximate direction and speed of the warm air. Thus we see that the polar front zone is the 'breeding ground' for temperate-zone depressions. Each of these depressions has its own warm and cold fronts which it retains for at least part of its journey across the ocean.

Warm Fronts

When a warm air mass replaces a cold one, the line on which the frontal surface meets the ground is known as a WARM FRONT. The warm air overlies the cold air, which remains as a narrow wedge in contact with the ground. Earlier in this chapter it was explained that upward motion in the atmosphere is associated with horizontal convergence. At a warm front the warm air is flowing up the frontal surface over a wide area, and it is this extensive upsliding of air which is associated with the convergence of the warm air leading to the sequence of clouds and precipitation shown in Figure 9.9.

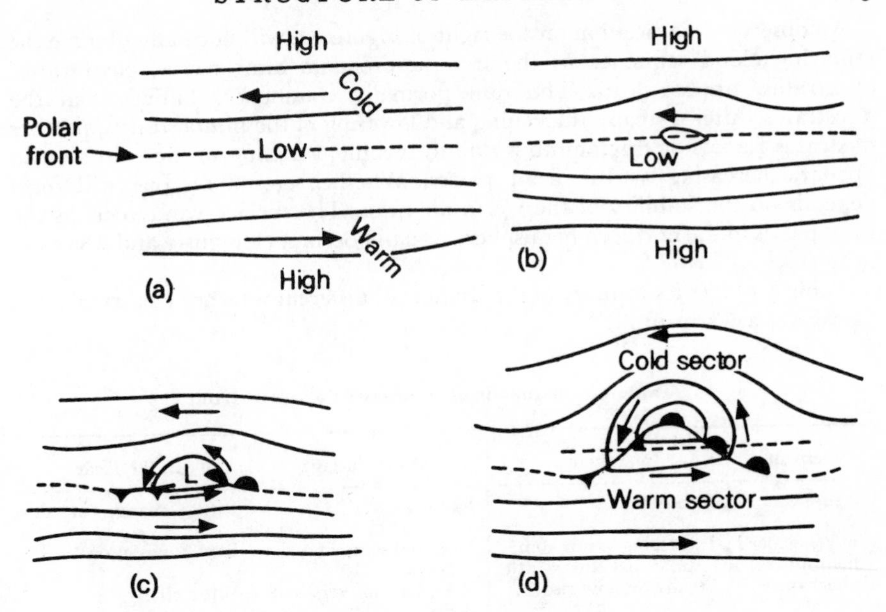

Figure 9.8. Four stages in the development of a depression

Sequence of Clouds and Weather at a Warm Front

Figure 9.9 gives the weather sequence at a typical warm front in the northern hemisphere. The vertical scale is much exaggerated, since in practice the angle of slope is only about 1/100 to 1/200.

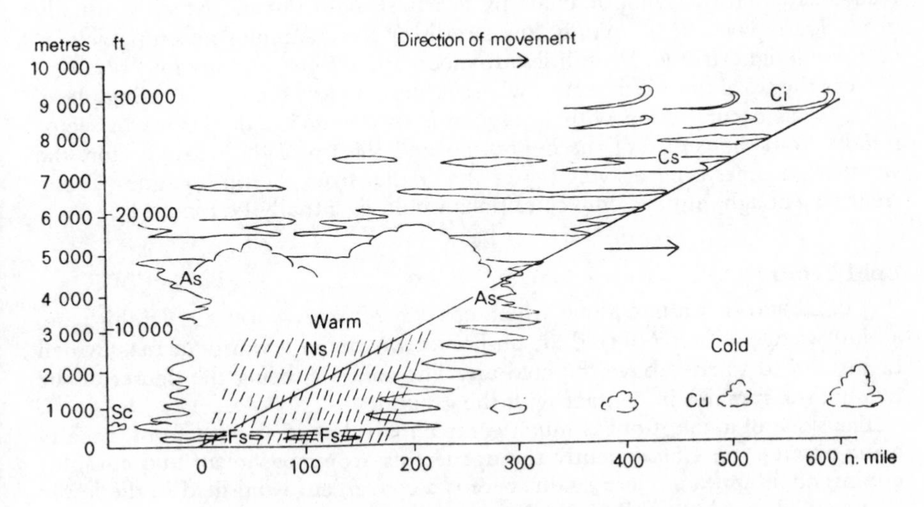

Figure 9.9. Vertical section through a warm front

An observer in a position on the right of Figure 9.9 will normally observe the following cloud sequence as the front approaches him: cirrus, cirrostratus, altostratus, nimbostratus. The rain normally commences falling from the altostratus. After a steady thickening and lowering of the nimbostratus, patches of stratus fractus or fractonimbus due to turbulent mixing in the cold air will appear, increasing as the front passes. Whether cumulonimbus will form depends on the stability of the warm air mass. The surface wind veers as the front passes in the northern hemisphere, usually becomes less gusty and decreases in force.

Table 9.1 gives a summary of the changes of different weather elements at the passage of a warm front.

Table 9.1. Sequence of weather at a warm front

Element	In Advance	At the Passage	In the Rear
Pressure	Steady fall	Fall ceases	Little change or slow fall
Wind (northern hemisphere)	Increasing and sometimes backing a little	Veer and sometimes decrease	Steady direction
Temperature	Steady or slow rise	Rise, but not very sudden	Little change
Cloud	Ci, Cs, As, Ns in succession; scud below As and Ns	Low Ns and scud	St or Sc
Weather	Continuous rain or snow	Precipitation almost or completely stops	Mainly cloudy, otherwise drizzle, or intermittent slight rain
Visibility	Very good except in precipitation	Poor, often mist or fog	Usually poor; mist or fog may persist

Warm Sector

After the passage of the warm front comes the WARM SECTOR, the portion of the depression where warm air is in contact with the earth's surface, which is recognizable on the synoptic chart by nearly straight isobars. At sea it usually gives cloudy conditions, mainly low stratus or stratocumulus accompanied by occasional light drizzle. Visibilities are moderate or poor, and fog may occur.

Near the tip of the warm sector, where occlusion (*see* page 119) is taking place, thick cloud occurs, often with heavy rain. In the region of the warm sector remote from the centre of the depression and the tip of the warm sector, the weather becomes progressively better the further from the tip one goes; if one went far enough, anticyclonic conditions would eventually be reached.

Cold Fronts

A COLD FRONT is a line along which cold air replaces warm air. In this case, a blunt wedge or 'nose' of cold air pushes its way under a warm air mass which is thus forced to rise above the cold air. The cold air, being the denser of the two masses, remains in contact with the ground.

The slope of a cold front is much greater than that of a warm front, usually being about 1/50. Consequently the upcurrents are more violent and cumulonimbus often appears. Here again there is a convergent wind field in the lower layers which produces this vertical motion.

Sequence of Clouds and Weather at a Cold Front

Figure 9.10 shows the typical sequence of clouds and weather at a cold front. The vertical scale is much exaggerated. Provided gaps in the low cloud layers (stratus or stratocumulus) ahead of the front are large enough, an observer in a position on the right of Figure 9.10 will observe cirrocumulus or patchy cirrus followed by altocumulus, slowly thickening into altostratus and then cumulonimbus. At the front, cumulus or cumulonimbus predominates and heavy rain for a relatively short period is typical; the surface wind veers quite quickly in the northern hemisphere and usually becomes squally and increases somewhat in force.

Once the frontal weather has passed, the kind of weather that will be experienced will depend entirely on the character of the cold air mass. If this is unstable it will be characterized by cumulus or cumulonimbus cloud and occasional showers.

Table 9.2 gives a summary of the changes of different weather elements at the passage of a cold front.

Table 9.2. Sequence of weather at a cold front

Element	In Advance	At the Passage	In the Rear
Pressure	Fall	Sudden rise	Rise continues more slowly
Wind (northern hemisphere)	Increasing and backing a little, often becoming squally	Sudden veer and sometimes heavy squall	Backing a little after squall, then often strengthens and may steady or veer further in a later squall
Temperature	Steady, but fall in pre-frontal rain	Sudden fall	Little change or perhaps steady fall; variable in showers
Cloud	Ac or As, then heavy Cb	Cb with low scud	Lifting rapidly, followed by As or Ac; later further Cu or Cb
Weather	Usually some rain; perhaps thunder	Rain, often heavy, with perhaps thunder and hail	Heavy rain for short period but sometimes more persistent, then mainly fair with occasional showers
Visibility	Usually poor	Temporary deterioration followed by rapid improvement	Usually very good except in showers

Occlusions

As the depression progresses on its journey, the upsliding of the warm air at the warm front and the undercutting by the cold air at the cold front gradually diminish the extent of warm air at the surface, the latter being ultimately lifted from the ground and raised to greater altitudes. This shutting-off of the warm air from the ground is known as OCCLUSION. When the process has finished, the depression is said to be OCCLUDED. On the weather map this is shown by the cold front moving faster than the warm front and catching up with it, first near the centre where the fronts are close together and then at successively greater distances from the centre. Figure 9.11a–c show successive stages in the occlusion of a depression. The line at the surface dividing the cold air which was previously ahead of the warm front from that which was previously behind the cold front is called an occlusion.

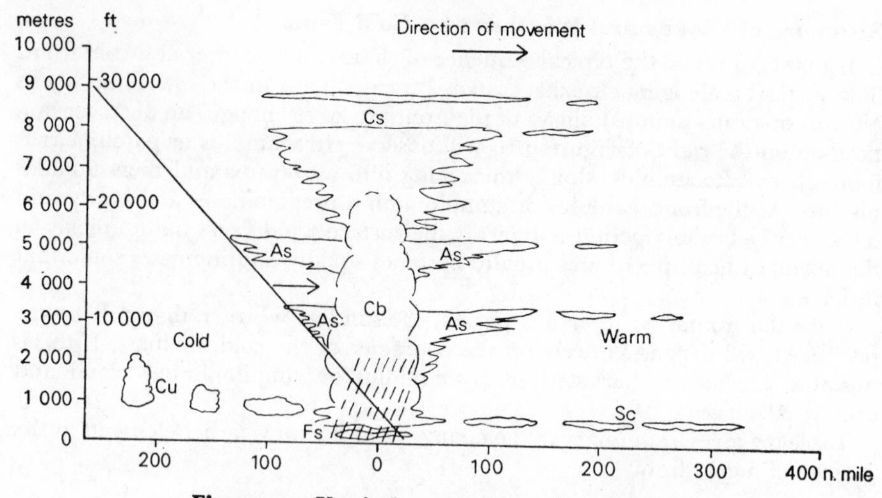

Figure 9.10. Vertical section through a cold front

Most of the depressions which reach north-west Europe from the Atlantic are already occluded, and thus a large proportion of the fronts arriving over this area are occlusions. This was one reason why the acceptance of the Bjerknes idea on the development of temperate zone depressions was so long delayed, since the existence of warm sectors was not at first evident.

It is unlikely that the air masses on each side of the occlusion will have identical properties in view of their different history, so that the air following the occlusion may be either warmer or colder than the air preceding it. In the first event, the occlusion is said to be of the warm-front type, or simply a WARM OCCLUSION; in the second case, the occlusion is of the cold-front type (COLD OCCLUSION).

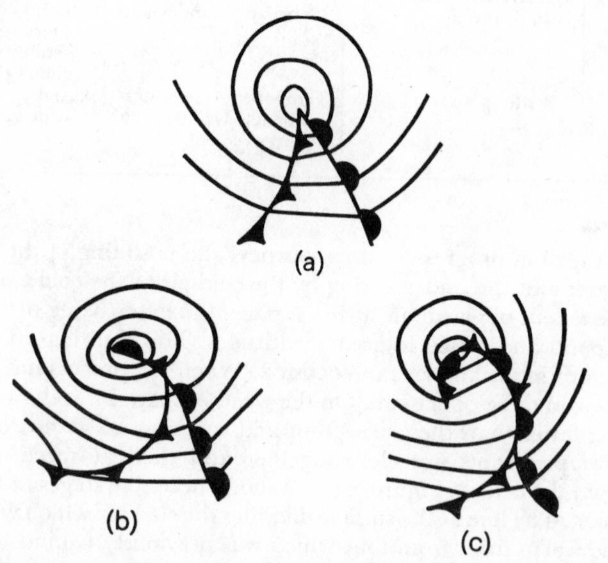

Figure 9.11. Three stages in the development of an occlusion

Figures 9.12a and b show examples of warm and cold occlusions respectively in sections. Figure 9.12a shows what happens when the cold air in the rear of the occlusion is less cold, and therefore less dense, than the cold air ahead of it. Both the warm air and the less-cold air override the preceding cold wedge. The discontinuity between the warm air and the less-cold air is still to be found aloft, and the line in which this surface meets the original warm-front surface is referred to as an UPPER COLD FRONT. The upper cold front is marked by cumulonimbus clouds and rain of a showery type, which are superimposed on the normal warm-front distribution of weather. The wind veers at an occlusion in the northern hemisphere; usually the veer is more marked at a cold occlusion than at a warm occlusion.

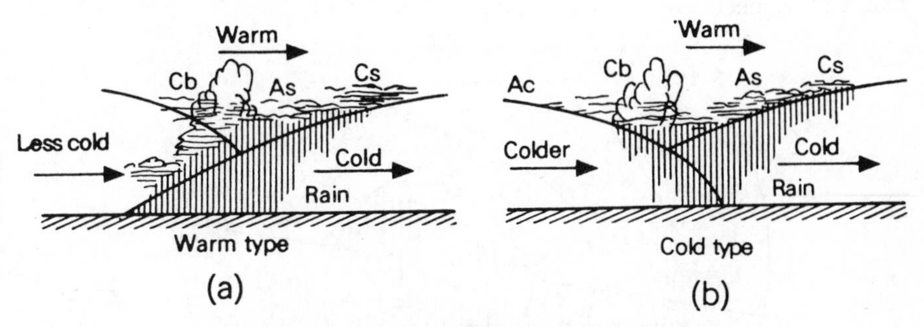

Figure 9.12. Vertical sections through occlusions

Referring to Figure 9.12b, the cold air in the rear of the occlusion, being colder and denser than the cold air ahead of it, necessarily remains in contact with the ground and constitutes the under-cutting wedge, obliging the less-cold air ahead of it to rise. The discontinuities between the two types of cold air and the warm air still exist aloft but are no longer apparent at the ground.

Ahead of the occlusion the weather characteristics are those of a warm front, since the warm air continues to rise over the cold wedge of air. At the occlusion the wind veers in the northern hemisphere, as it does at the passage of all fronts, but there is no rapid clearance; there is only a belt of more intense rainfall with heavy cumulonimbus cloud formed by the upthrusting of both warm and cold air masses by the colder following wedge. Squally conditions with an increase of wind in the colder air also occur frequently.

Occlusions which penetrate into the cold continents from the oceans in winter are generally of the warm type, because maritime-polar air is warmer than continental-polar air. In the summer, cold occlusions bring cold maritime air into the relatively warm continents.

Filling up and Dissolution of Depressions

Once a depression is fully occluded, it will usually fill up and within a few days will no longer be recognizable on a synoptic chart. In the typical case of an eastward-moving depression in the northern hemisphere, by the time the depression is fully occluded it will have moved well to the north of the surface boundary between the original warm and cold air, and so will have become remote from the temperature contrast which originally provided the energy for its development.

During the later stages of occlusion and filling the weather usually tends to improve slowly. The frontal cloud and rain gradually become less extensive, but convective cloud and showers may increase depending upon the character of the cold air and the other factors to which it is subjected. Weather in old depressions may thus range from conditions of little cloud to widespread cloud with showers and sometimes thunderstorms.

General Distribution of Weather in a Warm-sector Depression

Various aspects of the depression have, so far, been discussed separately. Figure 9.13 shows a plan of the distribution of cloud and weather in a typical warm-sector depression, considered as a whole. Figures 9.14a and b show sections taken to the north and south of the centre, taken along the lines **XX′** and **YY′**, respectively.

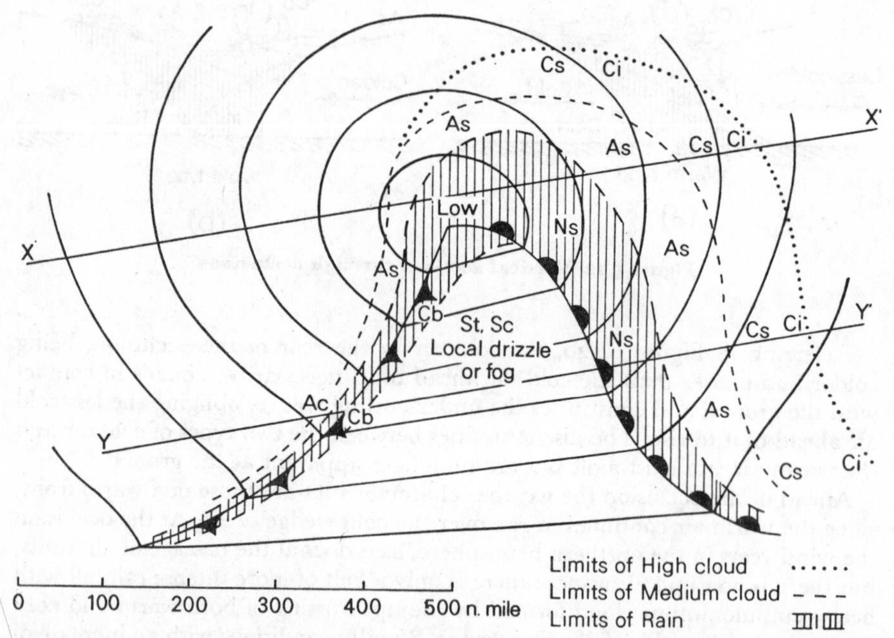

Figure 9.13. Distribution of cloud and weather in an idealized warm-sector depression

Relationship between Fronts and Isobars

It will be noticed that on any synoptic chart the isobars change direction sharply when they encounter a front. The change is such that the front lies in a trough of relatively low pressure. Let us consider Figure 9.13 and the changes occurring at the fronts on moving from left to right along the line **YY′**. At the cold front the isobars bend sharply towards the centre (or to the left relative to the direction of motion). Again on passing the warm front the isobars bend in the same sense. While in the warm sector and proceeding in the direction of the warm-sector isobars, the pressure remains constant. However, after passing the warm front, cold air is encountered. This is shallow at first but becomes progressively deeper as one continues in the same direction. Since cold air is denser than warm, this means that the pressure rises on passing the warm front and continues to rise as one continues to move further into the cold air following the direction of the warm-sector isobars. This implies that the isobar in question must be deflected to the left after passing the warm front. Similar reasoning shows that on crossing the cold front in the opposite direction, the line of constant pressure must be deflected to the right in the cold air.

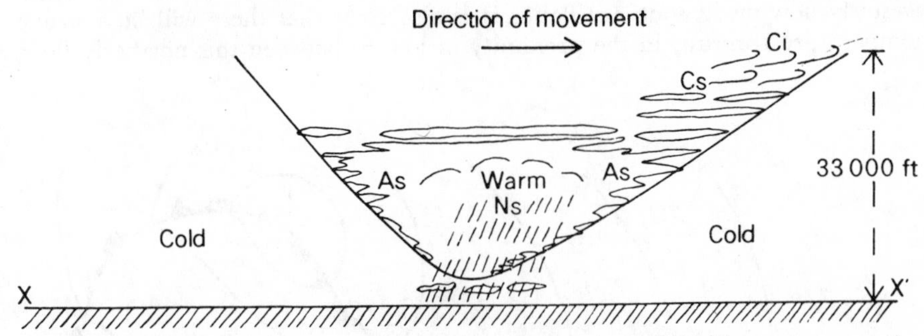

Figure 9.14a. Section of typical depression along XX′ in Figure 9.13

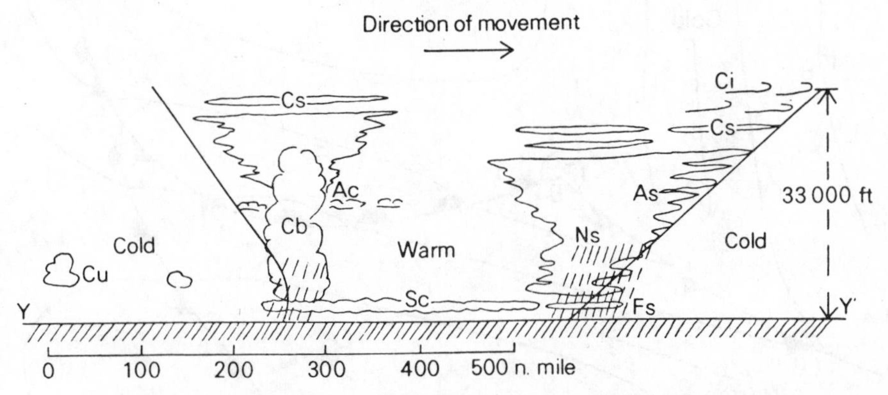

Figure 9.14b. Section of typical depression along YY′ in Figure 9.13

Because fronts are associated with pressure troughs, the pressure can be expected to fall as a front approaches, and to rise, or fall less rapidly, after its passage. The latter can usually be timed by noting the pressure discontinuity on a barograph.

A frontal passage can also be detected by noting the associated wind veer. If one faces the wind, the latter will shift towards the right (in the northern hemisphere) as the front passes. In the vicinity of the British Isles the most common wind sequence experienced in warm-sector depressions is south-westerly ahead of the warm front, veering westerly in the warm sector and finally veering north-westerly at the cold front. This assumes that the centre passes to the north of the observer.

Families of Depressions

We have seen how a depression originates as a disturbance on the polar front and dies after its warm sector has been occluded and the centre has become remote from the main polar front. The conditions which favoured the development of the original depression tend to be reproduced on the relatively slow-moving part of the cold front which, in the typical Atlantic case, lies well to the south and west of the position reached by the parent depression.

In Figure 9.15, low A is the original (parent) depression, now filling up. It can be seen that in the vicinity of the incipient low B there is a large wind shear between the northerly flow on the western flank of low B, and the south-westerly flow on its southern flank. It is also clear that there will be a strong temperature contrast, in the proximity of low B, between this northerly flow,

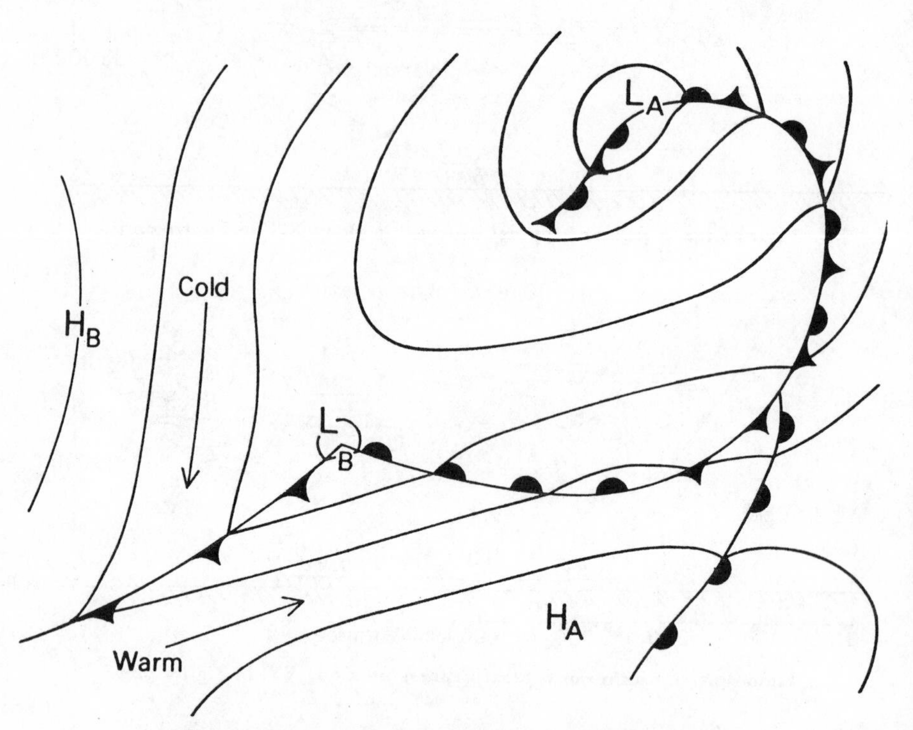

Figure 9.15. Two members of a family of depressions

and the converging south-westerly flow which derives from the southern side of the more southerly anticyclone. These conditions favour the development of a depression in the position indicated by low B. This accordingly deepens and intensifies while the original depression fills up. As the new depression moves away to the north-east and becomes occluded, the conditions favourable for cyclogenesis tend to be re-established on the slow-moving portion of its trailing cold front. This process may be repeated several times and so gives rise to a series or family of depressions in which each member tends to originate in something like the same longitude, but with the latitude displaced successively towards the south. Eventually the cold air behind one of the depressions sweeps through to the trade winds and the series is broken, the next depression forming much further north on a regenerated polar front.

The number of depressions in a family varies, but averages about four. The depressions are normally separated by ridges of high pressure which give brief, fair intervals between rainy periods. Usually the time interval between such periods in a series of depressions is between 24 and 48 hours.

Secondary Depressions

By 'secondary' is meant a depression embedded in the circulation of a larger or more vigorous depression, known as the primary. (*See* Figure 2.2.) In general, the secondary moves around the primary in a cyclonic direction. When the primary is weak and the secondary strong, as often occurs during the 'filling-up' stage of the primary depression after it has become well occluded, the primary may be absorbed in the circulation of the secondary as described earlier, or the two depressions tend to rotate around each other in a cyclonic sense. When the depressions have formed on the polar front, the secondary may be simply the next member of the family. This will play a secondary role in its early stages but will often intensify and may ultimately become dominant.

Non-frontal Depressions

Most depressions of temperate latitudes form on the polar front, and for this reason emphasis has been placed in the foregoing paragraphs on the formation of these 'frontal' depressions. Other types of depression exist, however, which are not connected with frontal zones. They are as follows:

(*a*) Thermal depressions.

(*b*) Depressions due to vertical instability.

(*c*) Depressions due to topography ('Lee depressions').

Thermal Depressions

The formation of thermal depressions is due to unequal heating of adjacent surface areas, and land and sea distribution plays a big part in determining their location. In winter the cooling of the continents induces higher pressure over the land than over the sea. In summer pressure tends to be lower over the land than over the sea. This effect may be seen by comparing Figures 7.1 and 7.2 showing world pressure distribution for January and July. The low pressure over Asia in July, known as the South Asiatic Monsoon Low, is pronounced enough to control the atmospheric circulation over a vast area.

Examples on a smaller scale are shown by the occurrence of low pressure over inland seas in winter, e.g. Mediterranean, Black Sea, Caspian Sea.

Depressions due to Vertical Instability

Thermal instability in the vertical plays an important part in the development of many types of depression. In the simple thermal depression just described, the low pressure is accounted for by the difference in temperature between the heated air, and its surroundings. Assuming a stable temperature lapse rate, these depressions are usually rather shallow and innocuous. If, however, the lapse rate should increase to a point where it exceeds the limit of stability (*see* 'Stability of the Atmosphere'—Chapter 1), a deeper and more serious depression may result. Especially if the air becomes saturated, the liberation of latent heat will contribute to the process of convection which, if it becomes widespread, may lead to the development of a vigorous depression.

The temperature of inland waters in early winter is relatively high compared with the air temperature over the surrounding land. In consequence, when a fresh outbreak of polar air spreads over such waters, the air often becomes unstable and a depression results. This occurs frequently over the Great Lakes of North America, and over the northern Mediterranean.

Depressions which show no frontal structure sometimes develop within a polar air mass which is moving over progressively higher sea temperatures. The first indication is usually just a bulge in the isobars, but this may later develop into an extensive system. Sometimes, such a depression, which originated well away from any fronts, may later approach a front and draw it into its developing circulation. The depression then becomes indistinguishable from those which originated on a frontal boundary.

Vertical instability is important in the formation of tropical cyclones (*see* Chapter 11) and the much smaller rotating systems such as tornadoes, dust devils and waterspouts (*see* Chapter 7, 'Local Winds').

Depressions due to Topography ('Lee depressions')

In a few localities where the winds sometimes blow across a mountain range which is sufficiently high and continuous to act as a barrier, the resulting distortion of the wind flow leads to the formation of a depression in the lee of the mountain range. (*See* Figure 9.16.) A depression formed in this way is known as a LEE DEPRESSION.

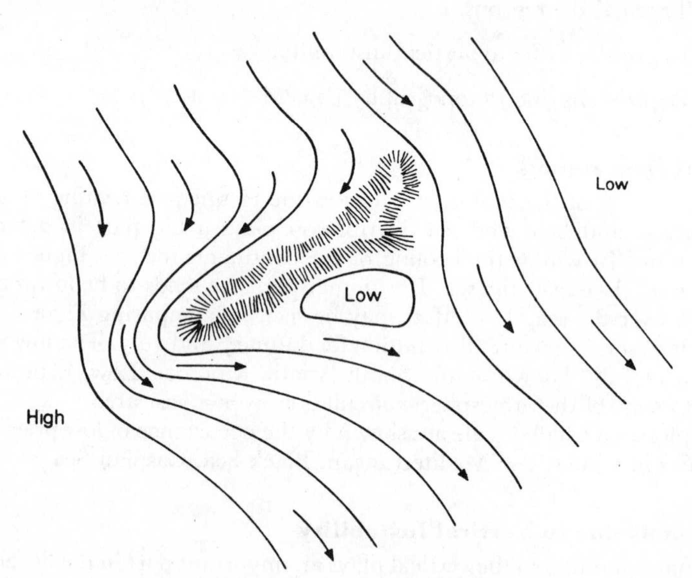

Figure 9.16. A lee depression

While it is probable that a depression of a sort would arise from purely dynamic causes, in a manner similar to the formation of an eddy in a current of water, in many cases the causes are complex. Vertical instability again may play an important part. With a fresh outbreak of cold air, it is possible for relatively warm air at low levels on the sheltered southern side of the mountains to be overrun, at height, by appreciably colder air. The resulting instability and convection may produce an active depression whereas without this factor the lee depression would normally be relatively insignificant.

Lows frequently form in this way in the Golfe du Lion following a north-westerly outbreak. The Alps tend to produce a lee depression but the instability produced by cold north-westerlies overriding warmer, moister air in contact with the Mediterranean contributes largely to the activity of such lows. Lee depressions also tend to form on the southern side of the Atlas mountains in North Africa in north-westerly situations. Depressions arising in this way sometimes travel east and may intensify on reaching the central Mediterranean.

The formation of shallow, mostly slow-moving depressions in central China, which are of great importance there on account of the rainfall they produce, is attributed to the distortion in the flow of westerly winds when crossing mountain ranges in their path.

Winds aloft above a Depression, Thermal Winds

To understand fully the structure of a depression one must consider not only the wind and temperature field at the earth's surface but also the corresponding fields at higher levels in the atmosphere so as to obtain a three-dimensional view of the phenomenon. The familiar pattern of wind and temperature associated with a warm-sector depression as shown on a surface chart changes progressively as one ascends to higher levels. Although the mariner does not have so vital an interest in the conditions at these levels as does the aviator, he should have some general knowledge of the flow pattern aloft in order to understand the movement of depressions. For some purposes depressions can be regarded as vortices, having a considerable vertical extent, which are carried along in the general flow of the surrounding air. Precisely what the 'general flow' is can be judged more easily from the flow pattern at higher levels than at the surface, because the pattern is normally most complex near the earth's surface and becomes progressively more simple as one ascends to higher levels.

Charts showing winds and temperatures aloft are commonly drawn at meteorological centres for various pressure levels which include 1000, 700, 500, 300, 200 and 100 mb. These levels are roughly equivalent, in terms of height, to the earth's surface, 3, 6, 9, 12 and 16 km. One of the most convenient of these upper levels for the study of the movement of depression is the 500-mb (approx. 6 km) level.

Figure 9.17 shows the flow pattern associated with a warm-sector depression both at the surface (full lines) and at 500 mb (dotted lines).

It will be seen that, in contrast with the closed isobars associated with the depression near the surface, the only indication of the depression at the higher level is a sinuosity in the generally west-to-east flow pattern. The latter exhibits a ridge slightly in advance of the surface warm front, and a trough somewhat in the rear of the surface cold front.

Let us consider why the flow pattern changes with height.

First consider a region wherein the surface isobars are straight and parallel and the pressure gradient is uniform. The surface geostrophic wind, being directly proportional to the pressure gradient, will have the same value over the whole region. In addition, if the lapse rate is the same everywhere over this pressure field, it follows that the horizontal differences of temperature will be the same at all levels over the area as they are at the surface. In practice this

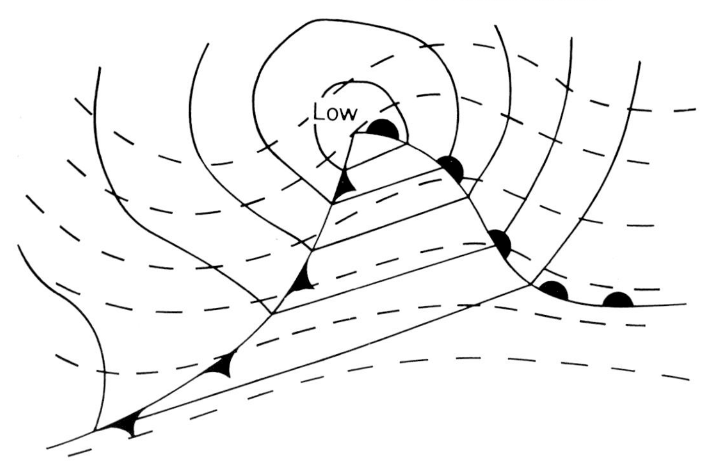

Figure 9.17. Flow patterns associated with depression at surface level and at 500 millibars

state of affairs is never found to exist in the atmosphere; on the contrary, at higher levels in the troposphere the air is often found to be colder over a depression than it is at the same levels outside the depression, and similarly the air at the centre of an anticyclone is often warmer than at the same levels over surrounding areas. We shall now consider in more detail what effect is produced upon the pressure and wind distribution aloft by the presence of colder air at higher levels in the middle of a depression, and correspondingly by the presence of warmer air at higher levels in an anticyclone.

The effect can be illustrated very simply. Suppose we consider three columns of air **A**, **B** and **C** (as in Figure 9.18), where the bases of all three columns have the same surface pressures and surface temperatures. The lapse rates in columns **A** and **C** are also the same, so that the temperatures and pressures in these two columns have the same value at corresponding levels above the surface. Now suppose that the lapse rate in column **B** is steeper than in columns **A** and **C**; that is to say, at higher levels the air in this column is colder than the air at corresponding levels in columns **A** and **C**. It is readily inferred that the pressure in the colder column **B** will decrease more rapidly with height than the pressure in the other two columns. As a result we have the situation shown in Figure 9.18. Here, for the sake of simplicity, the pressure at the ground level is everywhere shown as 1000 mb. At the level **PP′**—since the lapse rates in columns **A** and **C** are equal—the pressures in columns **A** and **C** will have the same value (say 500 mb), but will have some value *below* 500 mb in column **B** at this level. From these considerations it can be understood why a synoptic chart drawn for a higher level (for instance, 6 km) will show an area of low pressure where the air is relatively cold. Another and more convenient method which makes use of the same result, is to draw synoptic charts which show the contours of standard pressure surfaces by means of isopleths joining points at which the given pressure surface is at equal heights above sea level. These isopleths are equivalent to isobars on a surface at a given height, and are normally drawn at height intervals of 60 m.

Buys Ballot's law relating wind and pressure gradient is independent of pressure, hence it follows that a cyclonic circulation is set up at higher levels around a cold column of air, and an anticyclonic circulation at higher levels

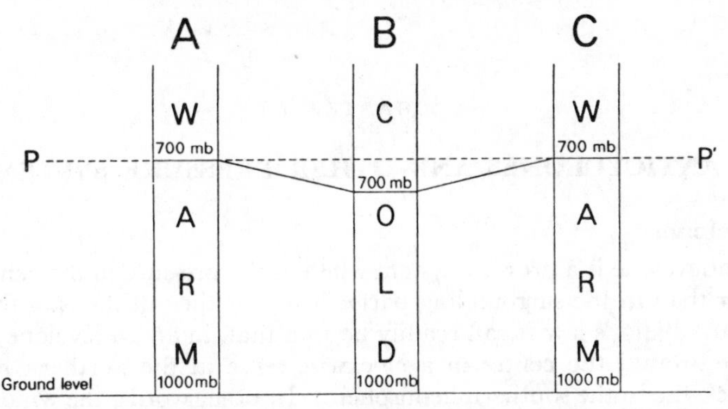

Figure 9.18. Relation between fall of pressure with height, and the lapse rate

around a warm column of air in the troposphere, and that these circulations are independent of the surface pressure gradients. The component of wind resulting from this cause alone is described as the THERMAL WIND for the layer in question. The upper wind at any level may conveniently be regarded as consisting of two vector components, namely the 'surface geostrophic' component and the 'thermal wind' component determined by the gradient of mean temperature between the surface and the level in question. In consequence the variation of wind direction and speed with height depends on the distribution of temperature at the higher levels. Alternatively it may be regarded as being dependent upon the pattern of distribution of areas of warm and cold air (commonly called warm and cold pools) at these higher levels.

When the pressure and temperature gradients are parallel and the colder air above coincides with the low pressure, the wind increases with height; similarly when the pressure and temperature gradients are parallel but the warmer air lies over the low pressure the wind decreases with height; in both these cases the wind direction does not alter with height so long as the pressure and temperature gradients remain parallel.

In most cases the temperature pattern aloft is not the same as that at the surface, and in consequence the wind changes both in direction and speed as one ascends to higher levels in the vicinity of a depression, as can be seen from Figure 9.17.

A knowledge of the flow pattern at these higher levels (e.g. 500 mb) aids in forecasting the movement of depressions since the latter commonly move approximately in the direction indicated by the 500-mb wind above the centre. In theory the 1000–500-mb thermal wind is the more appropriate guide but in practice there is often little difference between this direction and that of the actual wind at 500 mb.

CHAPTER 10

ANTICYCLONES AND OTHER PRESSURE SYSTEMS

Anticyclones

An anticyclone is a pressure system wherein the pressure in the central region is higher than in the surrounding parts. It is thus the antithesis of the cyclone. From Buys Ballot's law it can readily be seen that, in an anticyclone, the winds circulate around the centre in a clockwise sense in the northern hemisphere (anticlockwise in the southern hemisphere). In other words, the winds circulate around an anticyclone in the opposite sense to that in which they circulate around a cyclone or depression.

The weather typically associated with an anticyclone is, in many respects, in direct contrast with that which characterizes the depression. Thus, in contrast with the strong winds, dense cloud, and rain or snow which is typical of the central region of a depression, the central region of most anticyclones is one of light winds and fair weather. This is true for the subtropical high-pressure areas and for anticyclones in temperate latitudes in summer. Although cloudy skies are somewhat frequent in anticyclones over the ocean in temperate latitudes, precipitation even as drizzle is rather uncommon near the middle of an anticyclone because active fronts do not penetrate such regions. In winter over the continents in temperature latitudes, and particularly near industrial areas, the weather in an anticyclone can seldom be described as 'fair'. The reason is that in these circumstances fog is common and often persists for some time in the lowest layers of the atmosphere, with serious effects upon the movement of shipping in estuaries and coastal waters.

We have already seen that, at the earth's surface, the air does not flow exactly along the isobars but is deviated slightly from high to low pressure. If this is applied to a closed system of isobars with high pressure on the inside, it is apparent that air is continually being lost from the area; in other words, the anticyclone is an area of horizontal divergence in its lowest levels. As explained earlier, horizontal divergence at the earth's surface implies a downward motion of the air which is accompanied by adiabatic heating. Hence the relative humidity of air in the lower levels of anticyclones is lowered, clouds tend to be evaporated, and the frequent occurrence of fair weather in anticyclones is thus explained. Also, in an anticyclone which has persisted for several days, subsidence often leads to the formation of a temperature inversion in the lowest few hundred metres. Anticyclones can be divided into two main classes—'cold' or 'warm'.

Cold Anticyclones

A cold anticyclone is one in which the air at the surface, and in the lower layers of the troposphere, is colder than the air in adjacent regions. The air in the anticyclone is thus denser than the surrounding air, level for level. The high pressure of a cold anticyclone is therefore due, primarily, to the density of the lower layers of the troposphere being greater than the density of the same layers in the area surrounding the anticyclone.

Examples are the 'semi-permanent' anticyclones over the continents in winter.

These anticyclones most often occur over Siberia and less frequently over North America. They are not strictly permanent because these areas are occasionally invaded by travelling depressions; but after each period of cyclonic activity, high pressure tends to be re-established, and may then exist for weeks with little change.

These seasonal anticyclones are predominant enough to show on the winter chart of mean pressure (Figure 7.1), and they control the atmospheric circulation over wide areas. They are the 'source regions' of continental-polar air. Countries which normally enjoy mild maritime conditions, occasionally experience the rigours of a continental winter during an invasion of continental-polar or arctic air. In the British Isles, this sometimes occurs in winter months when a separate anticyclone develops over northern Europe, resulting in persistent easterly winds across the North Sea and the British Isles.

On the east coasts of North America and Asia the outflow of continental-polar air maintains a sharp temperature contrast off the coast, where a warm ocean current runs. These temperature contrasts favour the formation and development of depressions, which then travel eastwards.

Although cold anticyclones are of limited vertical extent (i.e. the height to which an anticyclonic circulation of air extends does not exceed 3000 m), they play an important part in the low-level atmospheric circulation in winter.

In addition to these seasonal cold anticyclones, other more transitory ones exist. The frontal depressions of temperate latitudes travel eastwards in families, the depressions being separated from each other by ridges of high pressure travelling with them and bringing the weather clearances that occur after the passage of each depression. After the last member of the family has formed, the cold polar air in its rear begins to sweep towards the equator and an anticyclone builds up in the cold air. Its history is then a matter of circumstance. If formed over land, in winter, the transitory anticyclone will be maintained and perhaps intensified, and eventually become an extension of the main semi-permanent anticyclone. If formed over the sea, or over the land in summer, the transitory anticyclone either collapses rapidly or becomes transformed into a warm anticyclone, the cold air being heated adiabatically by subsidence.

Warm Anticyclones

In a warm anticyclone the air throughout the greater part of the troposphere is warmer, level for level, than that in the surrounding regions. As this fact, alone, would favour low, rather than high, surface pressure it is clear that the explanation of the high pressure must involve factors other than temperature.

Typical examples of warm anticyclones are the oceanic subtropical belts of high pressure which are a major feature of the average pressure distribution of both hemispheres. Because they constitute such a major feature it is natural to seek an explanation in terms of the general atmospheric circulation (*see* Chapter 7).

Neglecting, for a moment, the earth's rotation, the excess of heating in equatorial regions, relative to that at the poles would favour low pressure in the equatorial regions and high pressure at the poles. Low-level convergence and ascending air in the former region would lead to a poleward flow of air aloft which would feed high-level convergence, descending airflow and low-level divergence in the polar regions.

Considering the northern hemisphere, the northward flow of air aloft is progressively diverted by the geostrophic force (*see* Chapter 3) towards the right, making the wind progressively more westerly. Ultimately equilibrium is reached when the geostrophic force just balances the force due to the pressure gradient and the corresponding winds are approximately westerly. This occurs between about latitudes 35° to 40°N. Thus the northward flow of air aloft is, as it were, blocked by the girdle of westerly winds, leading to an accumulation of air and resulting high pressure in a latitudinal belt somewhat to the south of the southern boundary of the westerly winds. Similar considerations apply in the southern hemisphere.

The warm anticyclones of the oceanic subtropical belt are characterized by subsiding air. Consequently the weather there is generally fine, with little or no cloud and good visibility. The subtropical anticyclones are the source regions of maritime-tropical air masses, providing the warm air that feeds travelling depressions of temperate latitudes. The charts of mean pressure show the centres of these anticyclones as being located between latitudes 30° to 35°N and S; in fact, in addition to showing a regular variation of mean position between seasons, their positions vary between corresponding seasons in different years. The causes of these movements are not yet known, variations in the strength of the equatorial ocean currents and of the amount of ice in the North Atlantic are among many other factors which have been suggested to account for them, but convincing proof of any such relation remains to be found. However, there is no doubt that these variations have an important effect on the seasonal weather in many localities. For example, in summer the Azores anticyclone lies farther north than in winter, so that fewer depressions affect the British Isles in summer than in winter, and therefore summer is less windy and rainy on the whole than winter in these islands. Most spells of dry warm weather in the British Isles result from occasional NE extensions or offshoots of the Azores anticyclone taking up a position over the British Isles; and from time to time such spells may last with little interruption for weeks on end.

Upper-air soundings made above anticyclones show that the air is relatively dry, which is to be expected in view of the subsidence taking place. They may also show an inversion of temperature, known as a 'subsidence inversion' because it is formed as a result of subsidence, which is most effective in a layer a short distance above the ground. As a result the highest temperature in an anticyclone is often at a level of about 300–600 m above the surface; this is particularly true over the sea. To some extent this inversion aids fog formation. At sea this happens because the moisture evaporated from the sea is all kept in the lowest layers, particularly if the airstream is moving towards higher latitudes, that is towards progressively colder sea surfaces. Over land other factors, such as radiation to clear skies and high moisture content, are assisted by the absence of wind associated with the inversion to produce a fall of temperature enough to cause fog by condensation.

Clouds forming in the lower layers of an anticyclone, as, for example, when moist air from the sea flows over land that is being heated by the sun, tend to spread out at the temperature inversion, giving a layer of stratocumulus. This is more typical of the boundaries of an anticyclone than the centre, where the subsidence frequently prevents cloud formation.

Trough of Low Pressure

A trough of low pressure is a pressure system in which the isobars resemble the contours around a river valley. Pressure is lowest with the V-shaped, or U-shaped, isobar of lowest value. Frequently a front lies in the trough, and the isobars may show a marked kink or bend at the front. A number of troughs do not contain fronts; these are known as non-frontal troughs. They are usually well rounded and do not show any sharp bend in the isobars such as occurs at a front. Figures 10.1a and b illustrate both types of trough. A non-frontal trough

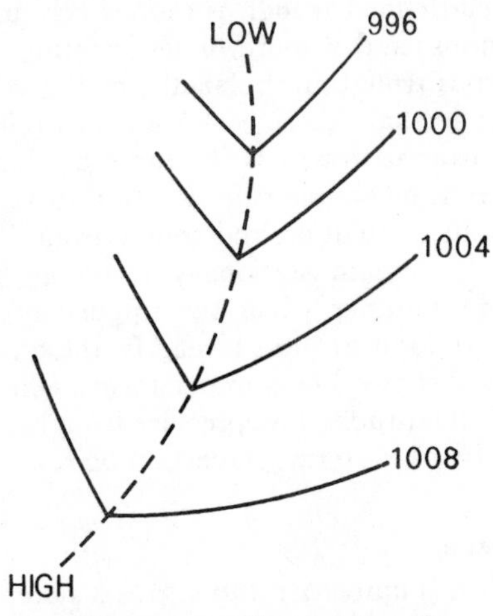

Figure 10.1a. Frontal trough

typically occurs in the cold air mass behind an occluded depression, and the resulting increase of pressure gradient often leads to a renewal of strong winds or gales after the depression centre has passed. The line joining the points in the trough where the lowest pressures occur at each point on the ground during the passage of the trough is known as the 'trough line'. Bad weather generally occurs on the advance side of troughs and a clearance appears after the passage of the trough line.

To some extent a trough can be regarded as an extension to, or offshoot from, a depression. Like the depression, the trough is a region of convergence and tends to be associated with dense cloud and precipitation. The cloud and

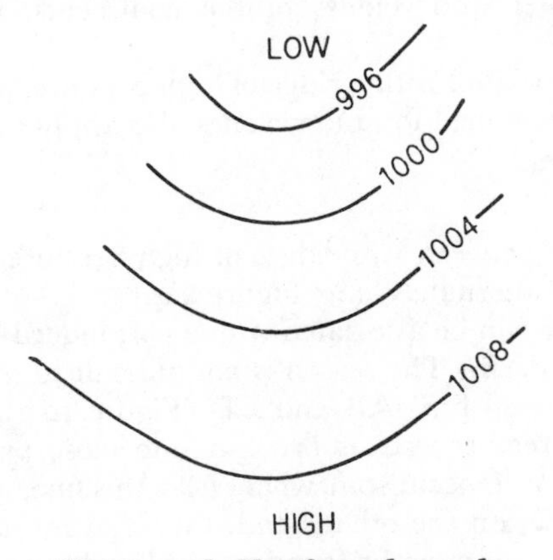

Figure 10.1b. Non-frontal trough

weather associated with frontal troughs has already been described in connection with warm and cold fronts and occlusions, in Chapter 9.

The weather associated with non-frontal troughs covers a wider range of variation. It depends, among other things, upon how well marked the trough is in terms of isobaric configuration. If the trough is barely discernible, the associated weather may be little worse than it is on either side. On the other hand, a steep-sided, well-defined trough is more likely to be marked by large cumulus or cumulonimbus, and a more or less continuous line of showers or thunderstorms. A vigorous trough, in polar air, may give a weather sequence, as it passes, very similar to that described for a cold front.

The movement of a frontal trough is the same as that of its front, but for non-frontal troughs there is no simple rule. When associated with a depression, they are swept along in its circulation and rotate around its centre.

Some troughs tend to remain stationary for long periods in temperate latitudes, particularly in summer when the upper-air circulation pattern is weak, i.e. when winds are light at most levels. In these circumstances weather is mostly cloudy with local showers or thunderstorms, which may be heavy near an active front. In tropical latitudes, low-pressure troughs are regions favourable for the development of tropical storms in certain ocean areas.

Ridge of High Pressure

A ridge of high pressure (Figures 2.1 and 2.3) is a system of curved isobars in which pressure is higher on the inside than on the outside. It therefore has anticyclonic properties. The isobars in a ridge are not always closed. Ridges are often located in the outer regions of anticyclones, and weather conditions in a ridge are often very like those in an anticyclone.

Fast-moving ridges occur between individual members of a depression family, moving along the polar front in temperate latitudes. They bring intervals of fair weather between the periods of rain or showers associated with each passing depression and its fronts. With a moving ridge a progressive deterioration of weather is usual after the passage of the axis of the ridge. When depressions follow one another a short distance apart, as often occurs in the Southern Ocean and in winter in the North Atlantic, the duration of the fair interval may be a very few hours.

Ridges formed as offshoots from the subtropical anticyclones, or from the semi-permanent winter anticyclones of the continents, usually move fairly slowly.

Fine weather is associated with a ridge of high pressure, although the qualifying circumstances mentioned for anticyclones also apply with ridges.

Col

A col is the region between two ridges of high pressure and two troughs of low pressure situated alternately. (*See* Figure 2.4.)

No definite weather can be associated with a col; indeed it is usually a region where sharp changes occur. The reason is not difficult to see. Suppose we cross the col with two diagonal lines **AB** and **CD** (Figure 10.2), **AB** is a line along which air from different sources is brought into close proximity. We might therefore expect a front to occur somewhere near this line, and this often occurs in practice. Along **CD**, on the other hand, the adjacent air flow is not front-forming. Thus a front can often be found in a col and it must lie along or close to the line **AB**. A front is not always found in a col, however, because its formation requires favourable conditions over a period of time.

The centre of a col is associated with very light and variable winds, and thus it is an area in which fog may occur, particularly in autumn and winter over land. When a col lies over a land area in summer, thunderstorms commonly occur when the necessary moisture and instability are present at higher levels.

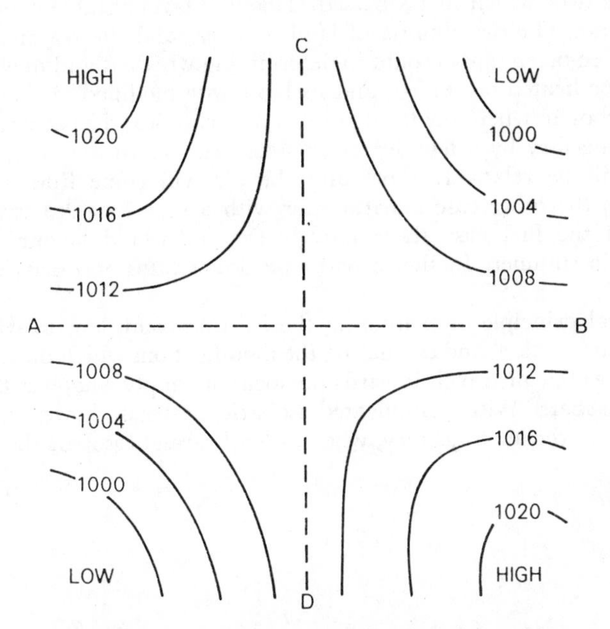

Figure 10.2. Pressure distributions in a col

Straight Isobars

No special significance attaches to straight isobars in themselves, but in some circumstances useful conclusions can be reached concerning the associated weather. The special case of the weather associated with the comparatively small region of straight isobars within a warm sector has already been discussed in Chapter 9.

In other cases straight isobars may extend over a considerable area around a station for which a forecast is required. This situation implies that the station is not directly under the influence of either low pressure or high pressure. In this case the weather will be determined, to a large extent, by the conditions obtaining upwind of the station and the progressive modification of these conditions by the character (especially temperature) of the intervening surface over which the air must travel.

Straight isobars have the advantage that one can make a reasonably good guess concerning the likely air trajectory over a considerable distance. To take an extreme case, let us suppose that the focus of attention is the Irish Sea, and the isobars run from N to S with low pressure over Russia and high pressure well to the west of Ireland. Suppose also a gradient for fresh to strong winds. Under these conditions the wind direction will be about NNW and the air will be flowing rapidly to the specified area from polar regions far away to the north.

This will imply very cold air aloft which is being progressively warmed from below by the warmer seas. Accordingly one would expect unstable conditions with large cumulus or cumulonimbus, and showers.

With the same orientation of isobars, but with pressure low to the west and high to the east, the air will have its origin far to the south and in consequence will have been progressively chilled from beneath by the decreasing surface temperatures over which it has passed. This will favour stable conditions and stratified cloud. The distribution of land and sea, and the season of the year both need to be taken into account in determining whether, and to what degree, the air will be heated or chilled, and its lapse rate modified, during its travel. In the case chosen, where the focal point is the Irish Sea, the backing or veering of the air trajectory by a few degrees can be critical. In the one case the sea trajectory will be relatively short and the air will come from the African continent. In the other case maritime air with a very long sea track is to be expected. In the first case the relatively dry air would favour clear skies, particularly in summer. In the second case, low stratus and drizzle might be expected.

The general principle of examining the existing conditions to windward of the location in question, and estimating the modifications which these conditions will undergo as the air travels towards the location, apply whatever the orientation of the isobars. With complicated isobaric patterns, however, it is very difficult to trace the air trajectory, whereas, with straight isobars the problem is simplified.

CHAPTER II

TROPICAL REVOLVING STORMS

General

Tropical revolving storms which when fully developed are known as hurricanes, typhoons or cyclones, depending on their area of occurrence, constitute one of the most destructive of atmospheric phenomena. The most violent winds experienced are those associated with these storms. Wind speeds of 147 knots (Riehl, 1972) have been reported but often the wind speed is not known, and there have been several cases in which the anemometer has been destroyed after recording a wind of over 100 knots. Such winds can cause widespread damage. It has been estimated that, in an average year, the Atlantic and Gulf Coast States of America suffer over $100m damage and 50 to 100 fatalities from hurricanes. This is only one of the areas affected. Torrential rain occurs in the inner parts of the storm (1170 mm (46 inches) in 24 hours has been reported) and can cause serious flooding. At sea and on the coasts damage can be due to high waves (one of 22 m was reported in hurricane 'Camille', 1969). Finally, very serious flooding may be caused in coastal regions by what is known as the 'storm surge'. This is an increase in the normal water level which occurs in advance of a storm due to the effect of the violent winds on the abnormally rough sea surface.

When a storm approaches the coast, abnormally high water level may result, particularly if the coastline is concave. In the storm of 12 November 1970 in the Bay of Bengal approximately 300 000 deaths were reported, largely due to the coincidence of the storm surge with a high astronomical tide.

Definitions

A tropical revolving storm may be defined as a roughly circular atmospheric vortex, originating in the tropics or subtropics, wherein the winds which blow in converging spiral tracks (anticlockwise in the northern hemisphere and clockwise in the southern hemisphere) reach or exceed gale force (Beaufort force 8).

In the tropics, as elsewhere, depressions of varying degrees of intensity occur.

The smallest tropical disturbance normally takes the form of a shallow trough in the easterly current which is more marked in the upper level flow than at the surface, and moves towards the west. These disturbances are referred to as 'easterly waves'. A reduction of pressure which is large enough to be marked by a closed circulation of winds is termed a 'tropical depression'. This is indicated on a synoptic chart by one or more closed isobars.

The term 'tropical depression' can be used in a general sense to cover any depression in the tropics whatever its intensity.

Since it is obviously of great importance to be precise in the definition of the various degrees of intensity of tropical depressions the following nomenclature has been laid down by the World Meteorological Organization (WMO):

1. 'Tropical Depression'—when the associated winds do not exceed Beaufort force 7.

2. 'Moderate Tropical Storm'—when the associated winds are of forces 8 and 9.

3. 'Severe Tropical Storm'—when the associated winds are of forces 10 and 11.

4. 'Hurricane' (or local synonym)—when the associated winds reach force 12.

The foregoing nomenclature for the various intensities of tropical disturbances, while accepted by many countries, including Britain and the United States of America, is not yet universally accepted, so that the mariner is liable to encounter some variation in terminology. For example, the Indian Meteorological Department use the term 'moderate cyclonic storm' and 'severe cyclonic storm' instead of the 'moderate tropical storm' and 'severe tropical storm' as defined above. Such variations in terminology, however, are not likely to worry the mariner because warnings will normally include a statement of the maximum associated wind speed. From this the intensity of the disturbance can be assessed, irrespective of the nomenclature used. The important thing to be understood is that whether it be called hurricane, typhoon, cyclone or any other local synonym, the same physical phenomenon is involved. What varies is the degree of intensity. Since the latter is liable to increase rapidly without much warning, it is important not to take a complacent attitude when given a warning of, say, a 'tropical depression', since within a comparatively short period (perhaps 12 hours) this may intensify to produce hurricane-force winds.

Description and Structure of Tropical Revolving Storms

These storms have some features in common with the depressions of higher latitudes. Both are atmospheric vortices in which the airflow in the lowest layers of the atmosphere follows nearly circular spiral tracks towards a centre of low pressure. Both are associated with strong winds and low pressures.

If we confine our attention to intense, mature systems, the differences between tropical and higher-latitude storms include those of size, wind profile and structure.

As regards size, a typical depression of temperate latitudes will have a diameter of some 1000 to 1200 n. mile. The tropical storm is much smaller, typically about 400 to 500 n. mile in diameter. Since, in general, the central pressure in a tropical storm is not much different from that in a middle-latitude storm (960 mb in a typical case), the pressure gradient in the former is generally considerably greater than in the latter. The winds are also generally stronger than in a middle-latitude storm and the variation of wind speed with distance from the centre of the storm is different. In the average storm of higher latitudes, there is a gradual, almost uniform increase from speeds such as 15 knots on the periphery to, say, 50 knots near the centre. In the tropical storm, with its much smaller radius, the wind increases relatively slowly at first and then progressively more and more rapidly to a peak of, say, 90 knots near the inner margin of an annulus which, typically, extends from about 15 to 30 n. mile from the centre of the storm. Within this annulus is the 'eye' of the storm where the wind suddenly drops to speeds below 15 knots. Figure 11.1 shows how the wind speed near sea level varied with distance from the centre, in the case of hurricane 'Hilda', 1964.

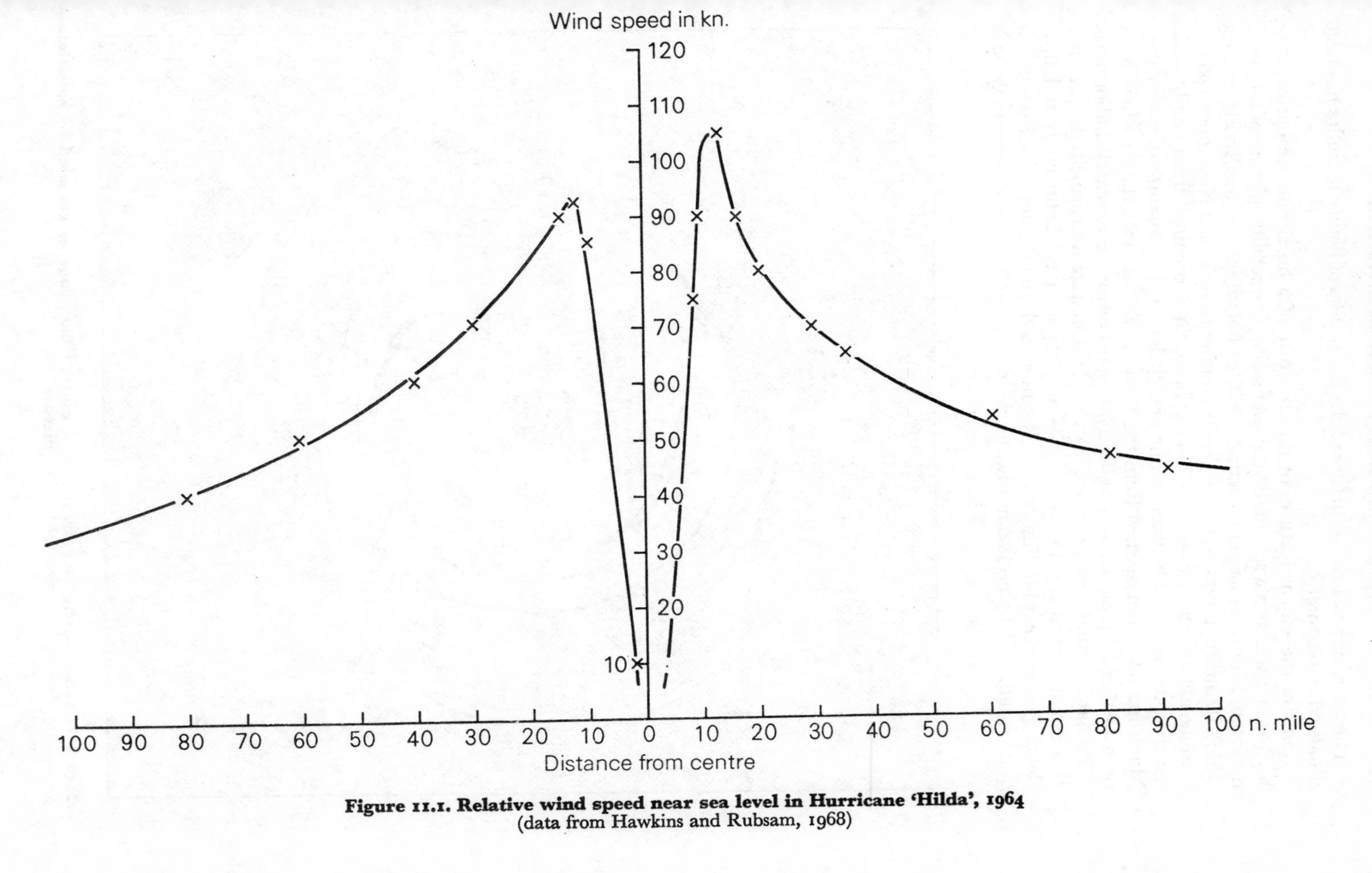

Figure 11.1. Relative wind speed near sea level in Hurricane 'Hilda', 1964
(data from Hawkins and Rubsam, 1968)

The three-dimensional structure of the depression of higher latitudes has been discussed in Chapter 9.

Some of the salient features of the structure of a hurricane are indicated in Figure 11.2 which is a schematic representation of a satellite picture of a mature hurricane. For comparison an actual satellite photograph is shown in Figure 11.3 which illustrates a typhoon near the Philippine Islands on 2 November 1967.

Since about 1960 the study of cloud photographs obtained from satellites has led to considerable advances in our knowledge of the structure of hurricanes. Much has also been learned from the many reconnaissance flights which have been analysed by the United States National Hurricane Research Laboratory.

Perhaps the most characteristic feature of the mature hurricane is the presence of a small central region—the 'eye'—of comparatively light winds and little cloud. It is indicated in Figure 11.2. Relative to the total extent of the storm it is quite small. It is by no means always cloud free but any cloud there may be is

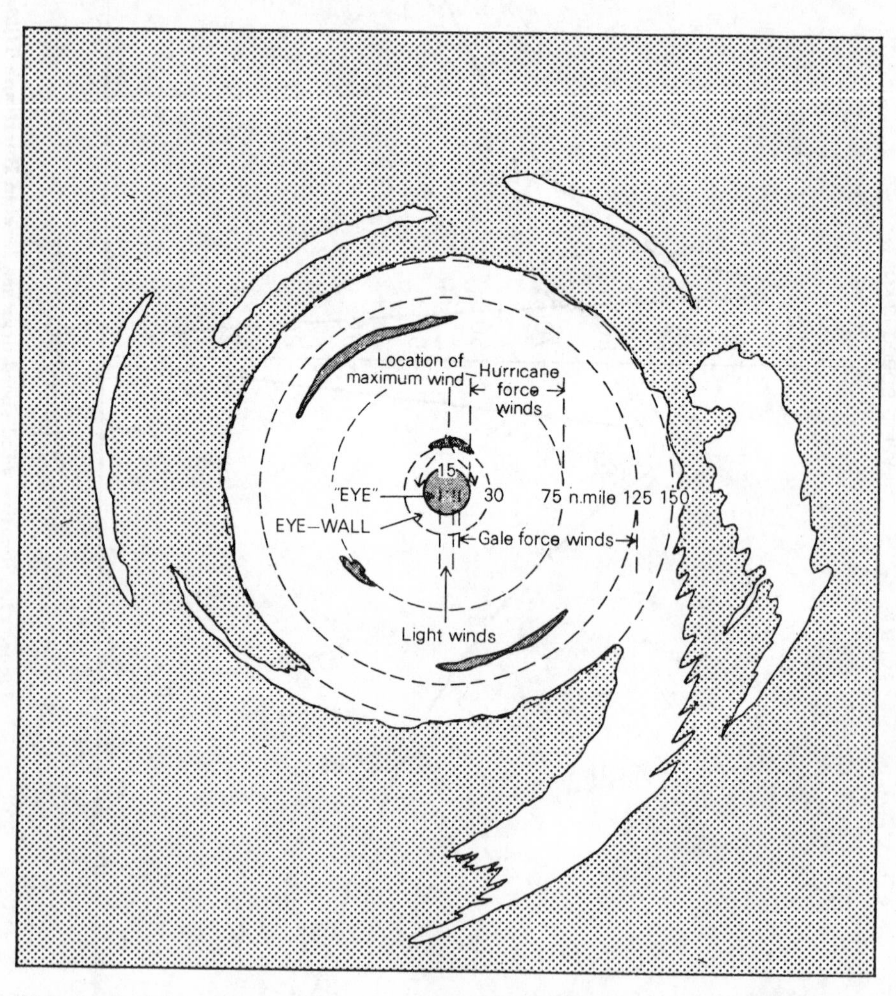

Figure 11.2. Schematic representation of mature hurricane as revealed by satellite picture

thin and insignificant in comparison with the very dense mass of cloud which immediately surrounds the 'eye' and extends vertically to great heights. This is known as the 'eye-wall' which forms an annulus, some 15 n. mile wide, on the average, surrounding the 'eye'. This is a region of very dense cloud extending from near the sea surface to the tropopause. In satellite pictures it is sometimes possible to distinguish the 'eye-wall', as a narrow ring of very white (i.e. dense) cloud from the slightly less-white surrounding mass of cloud which is less dense.

Figure 11.3. Satellite photograph of typhoon near the Philippine Islands on 2 November 1967

Outside the 'eye-wall' two areas may be differentiated. These are an inner area wherein the cloud is mainly continuous, and an outer area where there are long spiral cloud bands or 'feeders' converging on the central cloud mass, but wherein the cloud is generally broken.

Over the whole area of the hurricane, with the exception of the inner part of the 'eye', the low-level wind directions conform to spiral tracks converging towards the centre. From the light winds on the outside edge of the storm, typically about 250 n. mile from the centre, the winds gradually increase with decreasing distance from the centre. In the typical mature hurricane the wind reaches gale force (Beaufort force 8) at about 125 n. mile from the centre, and hurricane force (Beaufort force 12) at about 75 n. mile from the centre. The maximum wind is reached near the inner margin of the 'eye-wall', whereafter the speed drops very rapidly to practically zero in the centre of the 'eye'. This sudden decrease is clearly indicated in Figure 11.1.

The structure of the inner part of the hurricane, including the 'eye' and the 'eye-wall', is most easily understood in terms of the three-dimensional motions involved. In the outer part of the vortex the spiral inflow is largely horizontal. In the inner part the flow, while continuing to spiral towards the centre, acquires a progressively increasing vertical component so that finally, in the inner part of the 'eye-wall' the inflow ceases and the motion may be likened to that of a person ascending a spiral staircase. At great heights (approaching the tropopause) the vertical motion decreases and the horizontal motion increases, this time diverging from the centre of the column. Thus in the vertical plane there is inflow in the lower layers of the atmosphere, upflow in the inner regions (concentrated in the eye-wall region) and outflow near the tropopause. In contrast with the violent upflow in the eye-wall, several authorities agree that within the eye there is slow downward motion.

Another peculiarity of the tropical storm which differentiates it from the higher-latitude storm is its thermal structure. In the tropical storm the temperatures in the central eye are appreciably higher than elsewhere over most of its vertical extent. In the lowest levels there is relatively little variation across the storm but at all levels between 3000 m and 13 000 m temperatures are considerably higher in the centre than further out. In one case the temperature was about 7°C higher in the central region than at a distance of 80 n. mile from the centre at all these levels. It should be noted that the thermal structure is symmetrical with respect to the storm centre. There are no fronts or warm sectors as in the higher-latitude depression.

Associated Weather

Heavy rain, sometimes with thunder and lightning, accompanies all tropical revolving storms. Rain tends to occur in spiral bands in the outer region and to become more intense and more widespread in the inner region, reaching a maximum in the eye-wall. With the illumination reduced by very dense cloud, the effect of torrential rain is to reduce the visibility here to fog levels. Inside the eye the rain ceases, visibility improves, the clouds break and the wind falls off rapidly. Very high seas persist, however. After the centre has passed the sky again becomes overcast, rain returns with renewed violence, visibility drops and the wind suddenly increases to all its former fury, but now blowing from an almost opposite direction. After the second crossing of the eye-wall conditions gradually improve as the winds decline.

Irregular spells of heavy rain are likely to continue intermittently, corresponding with the crossing of the spiral feeder-bands of dense cloud.

Causes

While a great deal is known about the structure, mechanism and behaviour of tropical storms it has to be admitted that our knowledge is still incomplete. This can be judged from the fact that, while we can trace the development of a severe hurricane from its initial detection as a weak tropical depression, many such depressions occur which do not develop into hurricanes and, as yet, there are no agreed criteria for determining whether a depression will intensify or not.

Most authorities are agreed about the necessity of at least two pre-requisites for the development of a tropical storm:

1. A sufficiently large sea area wherein the sea surface temperature is in excess of 26°C (some would say 27°C).
2. The location should be sufficiently removed from the equator for the Coriolis force (*see* Chapter 3 p. 28) to be effective—in practice this means that the latitude must not be less than 5° north or south.

Other conditions which have been claimed to be necessary are:

3. A region of small vertical wind shear in the troposphere. This means that the development must be remote from the subtropical jet stream.
4. A pre-existing depression of whatever origin.

Although there is still much argument about the detailed physics, there is little doubt that the main source of energy in the hurricane is the latent heat released by the condensation of the moisture in the ascending tropical air mass. One can imagine relatively stagnant tropical air masses becoming gradually hotter and moister in the lower layers as a result of prolonged insolation and evaporation until a point is reached when the air becomes convectively unstable and there is, as it were, an ebullition. It is significant that, after moving inland, a hurricane gradually decreases in intensity, whereas, if its track is such that it later returns to the sea it may well increase again in intensity after reaching the sea. Further information concerning the causes and character of tropical storms may be found in 'Tropical Cyclones' by F. E. Fendell, *Advances in Geophysics*, Volume 17, 1974.

Regional Distribution (in general)

Two of the necessary conditions for development—a minimum distance from the equator, and remoteness from the subtropical jet stream, mean that tropical storms are limited to a band of latitude on each side of the equator. Figure 11.4 shows some of the tracks followed by tropical storms from which an idea can be obtained of the regions where they are liable to occur.

Storms are more liable to originate in higher latitudes (up to about latitude 30°N) in the North Atlantic and western North Pacific than elsewhere. Once having developed, the tropical storm or hurricane may persist for many days (one persisted as a hurricane for 20 days). During its life it tends to drift to higher latitudes and if it persists until poleward of about latitude 35° it gradually changes its characteristics, losing its tropical character and acquiring that of a middle-latitude depression. Usually, by the time a depression of tropical origin has reached latitude 45°, it has ceased to have the attributes of a tropical storm and is treated as a middle-latitude depression.

A glance at Figure 11.4 will show that there is a complete absence of tropical storms in the South Atlantic and in the east of the South Pacific. The reason for this anomaly is thought to be the absence, in these parts, of sufficiently high sea surface temperatures.

Movement (in general)

Tropical storms, in general, move with a velocity which approximates to the tropospheric average of the velocities in their immediate surroundings. As most of them originate on the poleward side of the equatorial trough, i.e. in the trade-wind belt, they mostly, in their early stages, move from east to west with a small additional poleward component.

While moving on a westward track the storms move relatively slowly, averaging about 8 to 10 knots. As they drift to higher latitudes in many cases the poleward component progressively increases at the expense of the westward component until the storm is moving towards the pole. At about this time it begins to be influenced by westerly winds and gradually turns towards the north-east (northern hemisphere) or south-east (southern hemisphere) and accelerates, typically to 20 to 25 knots. There is thus a typical parabolic path which hurricanes tend to follow. This is illustrated in Figure 11.5. The point of the track where the curvature is greatest, i.e. the 'nose' of the parabola, is often referred to as 'the point of recurvature'.

While the parabolic track is certainly a characteristic feature of tropical storm trajectories it is estimated that not many more than half of all storms follow such tracks. Of the remainder, some continue on some sort of westward track without recurvature as indicated, and others can best be summarized as 'erratic', describing complicated loops and, at times, moving in the opposite direction from that generally expected for the position and season.

For an idea of the probable and possible storm tracks in a particular area, the mariner should refer to the appropriate *Pilot* or *Marine Atlas*.

Seasonal Variation (in general)

Tropical storms are generally most frequent in the autumn months of the hemisphere concerned. This is in accord with the requirement, previously mentioned, for the sea temperature to be sufficiently high. Thus, in most of the northern hemisphere, the months when storms are most frequent are August and September. In the southern hemisphere the frequency is highest in January, February or March. The Bay of Bengal and Arabian Sea areas constitute an exception to this simple pattern of autumn maximum. In these areas there is a double maximum, one in late spring and the other in late autumn. This may be explained by the complicating effect, in this region, of the Asian monsoon. Firstly, the high cloud amounts associated with the summer monsoon interfere with the normal seasonal insolation so that the sea temperature has its maximum in May instead of August; secondly, the criterion of small vertical wind shear is not fulfilled in the early autumn months.

There is also a seasonal trend in the most likely path to be followed by a storm. While remembering that many storms do not conform to the typical parabolic track described in the preceding section, it is worth noting that, on average, the geographical location of this track shifts northwards and southwards following the sun. For example, in the North Atlantic, the mean latitude of recurvature (i.e. the latitude of maximum curvature of the parabolic track) varies from latitude 13°N in March, through latitude 21°N in June, to latitude 30°N in August, and returns to latitude 17°N in December. Further details about tracks are given later under regional headings.

Brief Details and Behaviour in Individual Areas

It is recommended that the appropriate *Pilot* volume should be consulted for tracks and additional information within a particular area.

West Indies and North Atlantic. The mean monthly and annual frequencies of tropical storms, including hurricanes (force 8 and above), in this area over the period 1941 to 1968 (Atkinson, US Air Weather Service, *Technical Report* 240) is as follows:

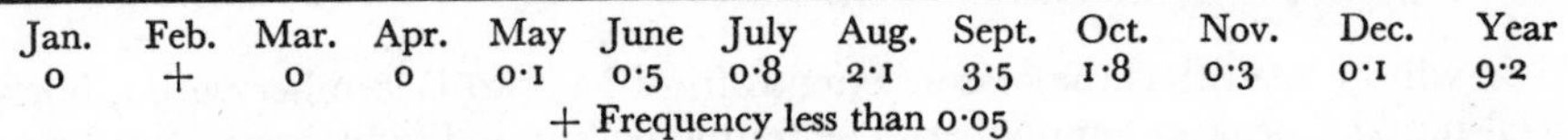

Jan.	Feb.	Mar.	Apr.	May	June	July	Aug.	Sept.	Oct.	Nov.	Dec.	Year
0	+	0	0	0·1	0·5	0·8	2·1	3·5	1·8	0·3	0·1	9·2

+ Frequency less than 0·05

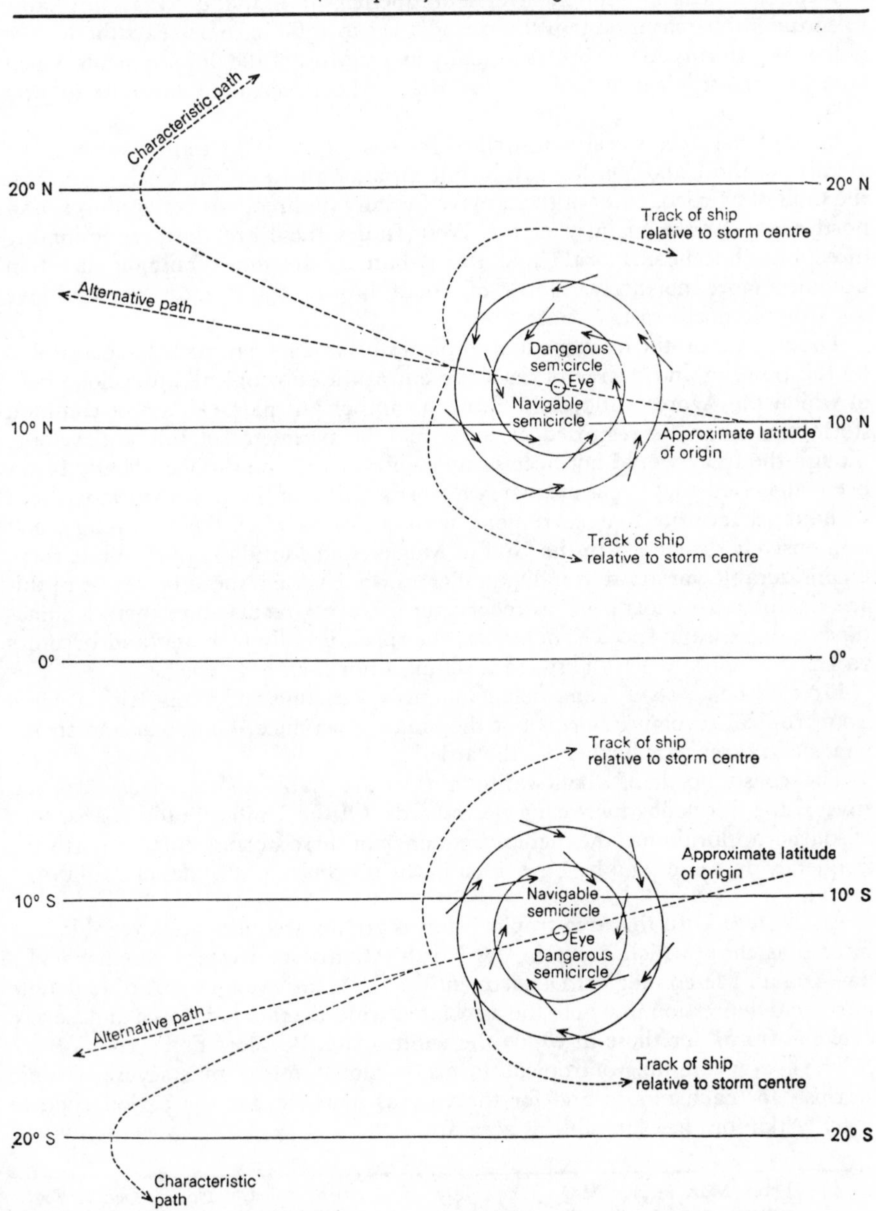

Figure 11.5. Characteristic paths and wind circulation of tropical storms in northern and southern hemispheres

The corresponding frequencies for storms which reached hurricane force are:

Jan.	Feb.	Mar.	Apr.	May	June	July	Aug.	Sept.	Oct.	Nov.	Dec.	Year
0	0	0	0	+	0·2	0·5	1·5	2·3	1·1	0·1	+	5·7

+ Frequency less than 0·05

It will be seen that the season extends from May to December with a fairly well-marked peak in September. The frequencies in individual years are liable to considerable variations. In the period 1941 to 1968 in some years there were as few as 5 storms and in others as many as 14. Most of the developments which later became tropical storms were first detected between about latitudes 10° and 27°N.

Most of the storms originate well to the east of the West Indies, some in the vicinity of the Cabo Verde Islands, but some originate in the Caribbean or in the Gulf of Mexico. Most of them move initially in directions between west and north-west. In the vicinity of the West Indies these are the predominating directions, but near Florida and the Bahamas, the most common direction becomes more northerly. North of about latitude 30°N most storms move towards the north-east.

To some extent the movement of tropical storms can be regarded as controlled by the position and degree of development of the subtropical anticyclone belt, of which the Azores anticyclone forms an important part. The most common storm track can be regarded as following the perimeter of this anticyclone, though the track would not necessarily conform with the surface isobars. It has been suggested that occasions when a storm continues its westward movement without recurvature may correspond with occasions when the anticyclone belt is intensively developed in the Gulf of Mexico and Florida region. While there is considerable variation in individual cases, the average speed of storms in this area is about 10 knots prior to recurvature. During recurvature there is sometimes a decrease in speed. Thereafter, the speed usually increases and becomes 20 knots or more when the storm is set on a north-easterly course.

Most of the coasts of Venezuela, Columbia, Panama and Costa Rica are free from tropical revolving storms but they may experience rough seas and strong winds as storms pass further northwards.

The coasts north of about latitude 15°N are liable to experience tropical storms, the frequency increasing northwards. Of the United States coasts, that of southern Florida has the highest frequency of these storms. Further north the frequency decreases and becomes least in the northernmost States of the USA.

ARABIAN SEA. In this area tropical storms are not frequent and are seldom as intense as those which occur in the North Atlantic or western North Pacific. The Indian Meteorological Department use the term 'cyclonic storm' to denote a tropical depression in which the associated wind reaches gale force and 'severe cyclonic storm' for those in which the winds reach Beaufort Force 10.

The average number of tropical storms ('cyclonic storms' plus 'severe cyclonic storms') for each month and for the year as a whole, for the period 1890 to 1967 (Atkinson, loc. cit.) in this area are:

Jan.	Feb.	Mar.	Apr.	May	June	July	Aug.	Sept.	Oct.	Nov.	Dec.	Year
+	0	0	0·1	0·2	0·2	+	+	0·1	0·2	0·2	0·1	1·1

+ Frequency less than 0·05

Of the total number, roughly half were 'severe' (i.e. force 10 or more). It will be seen that storms are most frequent in two distinct periods, namely in May and June, and in October and November. These periods roughly correspond with the transition periods between the south-west and north-east monsoons. No doubt it is the effect of the south-west monsoon which prevents the peak frequency from occurring in early autumn as in other parts of the northern hemisphere.

Of the storms affecting the Arabian Sea some originate within the area but others originate over the Bay of Bengal and move westwards to the Arabian Sea. In the earlier months, April to the end of June, the storms originate in the Arabian Sea and initially move, for the most part, towards directions between north and north-west. Although some tracks conform to the characteristic parabolic shape, and recurve towards the north-east, a peculiarity in this region is that some storms, after following a northerly or north-westerly track, gradually curve towards the left and ultimately move west-north-westwards. The positions where the storms were first reported were well to the south (mostly south of latitude 10°N) in April but shifted northwards in later months and were mostly north of latitude 10°N by June.

In the later period, September to the end of December, most of the storms originate in the Bay of Bengal and move over, or to the south of, southern India, to the Arabian Sea. Some continue moving westwards or west-north-westwards, while others recurve towards the north-east.

Their rate of advance varies considerably but averages about 7 knots.

BAY OF BENGAL. The average number (per month and year) of tropical storms (or cyclonic storms) having winds of Beaufort force 8 or more, during the period 1948 to 1967 (Atkinson, loc. cit.), was as follows:

Jan.	Feb.	Mar.	Apr.	May	June	July	Aug.	Sept.	Oct.	Nov.	Dec.	Year
0·1	0	0	0·1	0·7	0·1	0·1	0·1	0·4	0·8	0·7	0·5	3·6

The figures indicate that February and March were without storms during the period though they have been reported on rare occasions in earlier periods. The maximum frequency occurs in October (nearly one per year on the average) as part of a period extending from September to the end of December when storms are relatively frequent (about 1 in 2 years). A secondary maximum in May is more well marked in that the frequency in that month is almost as great as in October although in April and June storms are very infrequent.

Most storms have their point of origin in the Bay itself. The season starts in late April with storms originating in the vicinity of the Andaman Sea near the Nicobar Islands (latitude 6–9°N, longitude 92–94°E) and as it progresses they originate more and more to the north. By June most storms lie north of 15°N and by August north of 18°N. In succeeding months, however, the storms also break out southwards so that from October to the end of December most of the Bay is liable to be affected.

In addition to those which originate within the Bay, some storms originate as typhoons in the Philippine or South China Sea and re-intensify after entering the Bay from the east. This happens most frequently in September and October.

Most of the storms in the early stages head in directions between west and north. Some continue in these directions while others recurve towards the north-east. Most of the storms encountered south of latitude 15°N may be expected to be moving in directions between west and north-west whereas north of latitude 15°N they may be moving in any direction between north-west and north-east.

Average speeds of advance of fully developed storms are about 10 knots but after recurving towards the north-east they may accelerate. Storms in the Bay of Bengal are liable to affect any part of the coasts of India, West Pakistan, Burma and Sri Lanka.

In addition to the normal destruction caused by tropical storms, the coasts of this region are particularly liable to flooding by the associated storm surge. This is due to the high astronomical tides in this area which may coincide with the storm surge, and to the peculiarity of the bottom topography which tends to confine the accumulation of water. It is claimed that approximately 300 000 lives were lost in the cyclone of 12 November 1970 when the water level rose about 6 metres above mean sea level.

Western North Pacific, including the South China Sea. The average monthly and annual number of tropical storms and typhoons (as hurricanes are called in this region) over the period 1959 to 1968 (Atkinson, loc. cit.) were as follows:

Jan.	Feb.	Mar.	Apr.	May	June	July	Aug.	Sept.	Oct.	Nov.	Dec.	Year
0·4	0·6	0·4	0·9	1·5	1·6	5·0	6·8	5·3	4·3	2·4	1·3	30·5

January, February and March are the months of lowest frequency, and the months July to October are those with the highest frequency (between 4 and 7 storms per month on the average).

The majority of these storms originate somewhere in the vast region between the Philippine Islands and longitude 170°E, between latitudes 5° and 20°N. Although typhoons may originate anywhere in the above region, the most prolific source is near the Caroline Islands (latitudes 5° to 10°N, longitudes 137° to 160° E). A few typhoons originate in the middle of the South China Sea.

By far the greater number of the western North Pacific typhoons follow the accepted pattern of movement, namely to the westward from their source and recurving to the north-east on reaching higher latitudes. Others travel westwards and dissipate over the Philippines or mainland coast. A few may find their way into the Bay of Bengal (*see* under Bay of Bengal). There is a case on record of a typhoon which approached the northern end of Taiwan in a west-north-westerly direction and then turned to the southward, passing through the Taiwan Strait from north to south. The course and speed of each typhoon is influenced by the existing general meteorological situation. Secondary typhoon centres, as violent as the primary, have been known to develop within the typhoon area. As with all tropical revolving storms, the speed of advance varies in each storm, but in general will be about 10 knots before recurvature and 20 knots thereafter.

In addition to the general indications of the approach of a tropical revolving storm outlined later in this chapter, the direction of the wind over the seas adjacent to the northern Philippines may be a guide to the existence of a typhoon. In these waters a steady south-west or north-west wind should be suspected, since there is no season in the year when these winds predominate. (In this area the so-called south-west monsoon more commonly blows from a direction between south and east.) Should the wind in the northern part of the South China Sea or Philippine Archipelago blow steadily from the south-west during the months of June, July, August or September, the mariner may be fairly confident that there is a typhoon to the northward of him. Average monthly and annual frequencies of storms (including typhoons) for the South China Sea alone (period 1961 to 1968, Atkinson, loc. cit.) reveal a different seasonal trend from that shown by the inclusive figures given above and are quoted separately as follows:

Jan.	Feb.	Mar.	Apr.	May	June	July	Aug.	Sept.	Oct.	Nov.	Dec.	Year
0	0·1	0	0·1	0·5	0·2	0·5	0·6	0·9	0·5	0·5	0·4	4·3

Here the main maximum is in September, with all the months July to December having a moderate frequency, and there is a secondary maximum in May reminiscent of the regime in the Bay of Bengal.

Eastern North Pacific. In this region tropical storms are sometimes referred to as 'cordonazos' from a Mexican word meaning 'a lash of the whip'. They affect the western coasts of the North American continent from Costa Rica to lower California. The average monthly and annual frequencies of storms (force 8 and over) during the period 1965 to 1969 (Atkinson, loc. cit.) were:

Jan.	Feb.	Mar.	Apr.	May	June	July	Aug.	Sept.	Oct.	Nov.	Dec.	Year
0	0	0	0	0	1·8	2·2	4·2	4·0	1·8	0	0	14·0

These figures show a well-marked season from June to October with a maximum of about 4 storms per month in August and September and roughly 2 per month in the other months of the season. It should be noted that these figures may need to be revised when a longer period of satellite surveillance of the area becomes available.

The area liable to be affected by the storms lies between about latitudes 10° and 40°N and between about longitudes 87 and 140°W. Most of the storms originate in the south-eastern parts of this area and move towards the west-north-west or north-west, roughly parallel with the coast. Some of them recurve to follow the typical parabolic track and may strike the coast moving northwards or north-eastwards.

Generally, in the early part of the season, these storms occur at some distance from the coast, but from September onwards their tracks are much closer to it. In this region it is estimated that about one in three of the tropical storms intensifies to reach hurricane force.

The storms occurring near the coasts of Mexico tend to be smaller in size than either those of the Caribbean region or those more remote from the coast, e.g. in the vicinity of Islas Revilla Gigedo. They are no less frequent, however. In September 1918 a hurricane struck La Paz and every craft in the bay was driven ashore, many being destroyed. The town was partly wrecked and several people were killed. In October 1959 a storm struck the west coast of Mexico and caused over 1300 deaths.

South-west Indian Ocean (African Coast to Longitude 100°E). The average monthly and annual frequency of tropical storms or cyclones (force 8 and over) in this area during the period 1931 to 1960 (Atkinson, loc. cit.) were as follows:

Jan.	Feb.	Mar.	Apr.	May	June	July	Aug.	Sept.	Oct.	Nov.	Dec.	Year
2·3	2·0	1·5	0·7	+	+	+	0	+	0·1	0·3	0·9	7·8

+ Frequency less than 0·05

The season may be said to extend from October to the end of April with the peak frequency in January and February. Storms occur most frequently between Madagascar and about longitude 70°E, and between about latitudes 10° and 35°S. However, they also occur between Madagascar and the African coast, and may affect any part of the South Indian Ocean between longitude 70°E and Australia within the latitude band 5° to 35°S.

The storms which occur between the African coast and longitude 100°E can be divided into two groups along the meridian of longitude 80°E. Inspection of storm tracks in individual months shows a relatively high frequency of storms between Madagascar and longitude 70°E. There is then a region between longitudes 70° and 80°E where storms are relatively few. Then between longitudes 80° and 100°E there is an increase in frequency in some months, although in this eastern area the frequencies are not so great as in the western region.

In the western region (west of longitude 80°E), the area chiefly affected lies between about latitudes 10° and 35°S, although storms have been reported as far north as latitude 6°S. While there is considerable variation in individual cases, the average latitude of recurvature is about 20°S. In the eastern region (east of longitude 80°E), the average track would seem to lie further north, since the northern limit of the area affected is about latitude 5°S and the average latitude of recurvature is about 15°S. Few of the recorded tracks in this region extend beyond about 30°S but, in this area where ship reports are scanty, there must remain some doubt about the southern limit of the area affected.

A fairly high proportion of the recorded tracks conform to the characteristic parabolic shape with the storm heading west-south-west in its early stages and later recurving and finally heading south-eastwards. Recurvature of the storms normally takes place well to the east of Rodriguez in October and again in April and May, and between Rodriguez and Madagascar in November and December. In January and February they usually pass near to Mauritius or Réunion during recurvature. In the months December to April, inclusive, some storms have passed through the Moçambique Channel.

The speed of advance of the storms varies but averages about 8 knots prior to recurvature, after which it increases.

SOUTH-EAST INDIAN OCEAN (LONGITUDES 100° TO 135°E). Since the advent of satellite pictures more tropical storms have been detected in this region than previously, presumably because some of the storms of earlier years were not reported.

The following figures giving average monthly and annual frequencies of tropical storms (for 1962 to 1967, Atkinson, loc. cit.) may need to be revised when larger periods of data based on satellite coverage become available.

Jan.	Feb.	Mar.	Apr.	May	June	July	Aug.	Sept.	Oct.	Nov.	Dec.	Year
1·8	1·4	2·0	0·2	0	0	0	0	0	0	0·4	1·2	7·0

Although earlier books refer to the name 'Willy-Willy' as the local synonym for tropical storm, it is understood that the term is now largely obsolete.

The season lasts from November through April with the months January, February and March having the greatest frequency. In previous years storms have sometimes occurred in all months except June. The storms originate over the sea to the north of Australia and in most months are confined to latitudes south of latitude 8°s. In April, however, they occur in the Arafura Sea as far north as latitude 5°s. Some originate in the Gulf of Carpentaria and some move westwards into the area from the Pacific.

Most of the storms follow the characteristic parabolic path, moving initially west-south-westwards and later recurving towards the south and finally towards the south-east. Many follow a track roughly conforming to the general lie of the coasts of northern and western Australia. A number of storms recurve in longitudes east of longitude 115°E and so strike the north-west facing coast of Australia while heading south or south-eastwards. These gradually decrease in intensity after moving inland but some have retained some of their circulation until they reached the Australian Bight where they experienced a partial rejuvenation.

South-west Pacific. Average monthly and annual frequencies of tropical storms (force 8 and over) between longitudes 135°E and 150°W over the period 1947 to 1961 (Atkinson, loc. cit.) are as follows:

Jan.	Feb.	Mar.	Apr.	May	June	July	Aug.	Sept.	Oct.	Nov.	Dec.	Year
1·9	1·4	1·6	0·7	0·1	0·1	0	0	0	0	0·1	0·7	6·6

Earlier records claim that storms have occurred in all months but there is no doubt that they are very rare from July through September. The main season is December to April, inclusive, and January, February and March are the months when storms are most likely.

The majority of these storms originate in an area between latitudes 10° and 20°s and between longitudes 150°E and 180° but some originate between latitudes 5° and 10°s, and in any longitude west of 150°W.

Examination of recorded tracks shows a certain number which conform to the characteristic of the southern hemisphere path, initially moving towards west-south-west and later recurving to the south and finally south-east. In so far as we can establish a mean latitude of recurvature, this is between about latitudes 15° and 20°s. North of 15°s the majority of storms may be expected to be travelling in directions between west and south but, for the rest of the region, south-easterly tracks predominate. Most of the recorded tracks are south-easterly throughout their existence, but this may mean that the storm escaped detection in its earlier stages. In this area a number of storms follow irregular paths, for example, moving initially north-westwards and later recurving to the north-east!

Some storms strike the Queensland coast and are said to be most likely between Cooktown and Rockhampton. A peculiarity in this region is that warning signs (*see* later section) in the shape of swell may be suppressed due to the presence of reefs and shoals which may prevent the swell from reaching a vessel when close to the shore.

Average speeds of advance of storms are reported as around 8 knots prior to recurvature, rather slower during recurvature and increasing to 15 knots or more thereafter.

Warning Signs

Warnings of the position, intensity and probable movement of a tropical storm may be received at any time by radio from a Meteorological Service or from another ship (*see Admiralty List of Radio Signals*, Volume 3). In most tropical storm areas the responsible Meteorological Services take considerable care to issue comprehensive warnings to shipping by radio when such a storm is known to be developing, and issue frequent bulletins concerning the storm's progress. In the hurricane season it is very important, therefore, that careful watch be kept aboard ship for radio storm warnings. However, the shipmaster should not rely on such warnings to the extent of ignoring his own observations. There is always the possibility that the ship might be close to a rapidly developing, or erratically moving, storm and could be aware of the danger well before a warning could be issued from a Meteorological Service in view of the inevitable delays in receiving information, analysing it, drafting warnings and broadcasting them. Since, as we have seen, some storms move capriciously, there is also the possibility of a storm, for which a warning has been issued, not behaving as forecast. It is therefore very important that the master of every ship in waters liable to tropical storms should be constantly on the alert for any sign of such a storm and should if necessary take evasive action and send a radio message to the nearest coastal radio station and to other ships as soon as possible, in accordance with the *International Convention for the Safety of Life at Sea.*

The following paragraphs summarize the various signs that give warnings of a tropical storm:

(*a*) In the open sea, when there is no land between the ship and the storm centre, swell from the direction of the storm will probably give the earliest indication, since it travels at greater speed than the storm itself.

(*b*) If in a tropical storm area the barometer reading, corrected for height, latitude, temperature, index error and diurnal variation* is reading 3 mb or more below the mean pressure for the time of year, as shown in the *Admiralty Pilots* or in an appropriate *Climatological Atlas*, the mariner should be on his guard. If the reading thus corrected is 5 mb or more below the normal, there can be little doubt that a tropical depression is in the vicinity, probably not more than 200 nautical miles away. It is desirable to read the barometer at frequent intervals; the safest plan is to read it hourly in these areas during the storm season, even when conditions appear to be normal. Any cessation of the diurnal range of pressure in the tropics should be regarded with suspicion. When a tropical storm passes fairly near to the ship there are usually three distinct phases in the fall of the barometer:

(i) A slow fall, with the diurnal variation still in evidence, usually occurring 500 to 120 n. mile from storm's centre.

(ii) A more marked fall, during which the diurnal variation is almost completely masked. This usually occurs at from 120 to 60 n. mile from the centre.

The barometer may be very unsteady throughout this phase.

(iii) A rapid fall occurring at from 60 to 10 n. mile from the centre.

(*c*) An appreciable change in the direction and strength of the wind during the cyclone season should be viewed with suspicion.

---

* *See* the tables for diurnal variation, 11.1 and 11.2 on page 164.

(*d*) These storms are frequently preceded by a day of unusual clearness and remarkable visibility. The atmosphere at such times is oppressive. These conditions are followed by extensive cirrus cloud, often in convergent bands which point towards the storm's centre. This cirrus shows no disposition to clear away at sunset but reflects lurid colouring then and also at sunrise. It later becomes reinforced by a thick layer of altostratus and ultimately by cumulus fractus and stratus fractus (scud). The progress of the scud is marked by rain squalls of increasing frequency and violence. Rain is one of the most prominent features of a tropical revolving storm; in the outer portions it is intermittent and showery, whilst in the neighbourhood of the centre it falls in torrents. The rain area extends further in advance of the centre than in the rear.

Satellite Pictures and Land-based Radar

The taking of 'photographs' of cloud cover by satellite started in about 1960 and the process is now established as a matter of routine. The satellite cameras and radiometers scan and record continuously while the earth rotates beneath them. By this means an almost continuous surveillance of the cloud structure over the ocean is maintained. Because the cloud structure associated with a tropical storm has a characteristic form and appearance (*see* Figure 11.2), the occurrence of such a storm can usually be detected by the examination of these pictures.

This has led to a big advance in our knowledge of the structure and behaviour of these storms. Many storms are now detected by this means which would formerly have been undetected owing to the lack of shipping in their immediate vicinity. Techniques have been evolved, particularly in the United States of America, whereby estimates of the size of the storm, its duration and speed of advance, and even the associated wind speeds can be made from measurements of various parameters in these pictures.

The information is used to supplement reports from shipping and other sources in the preparation of storm warnings.

Radar Pictures

Although the normal ship-borne radar is of limited use because of its short range, land-based and airborne radar can achieve ranges of about 200 n. mile and can play an important part in the precise location and monitoring of storms. Some coasts which are particularly liable to storms have a series of radars mounted in elevated positions and these give precise information about the size, shape and speed of approaching storms. Radar is also used by reconnaissance aircraft and enables the inner structure to be seen through the general veil of cloud.

Warnings broadcast by Meteorological Services

The responsible Meteorological Services in areas liable to experience tropical storms have elaborate organizations for detecting tropical storms and issuing warnings, generally by radio, from various coastal stations. The stations which issue these warnings are set out in the *Admiralty List of Radio Signals*—Vol. 3 which gives details of frequencies and times of broadcast, and the regions covered. Maps on a world scale showing the location of the warning centres in the various geographical regions are to be found in Vol. 3a—'Weather Reporting and Forecast Areas', of the same publication.

Action to be taken when in the vicinity of a Tropical Revolving Storm

When the master of a vessel decides that he is in the vicinity of a tropical storm his first care should be to get as far away as possible from the vortex and to warn other shipping of the danger. It is realized that at such times masters are preoccupied with the safety of their own ships, but a concise weather report from a ship, in plain language or by means of the International Code of Signals will almost certainly result in timely advice being given to other ships who may be in the path and also to the inhabitants of islands and coastal communities where life and property may be threatened. The master should therefore lose no time in transmitting a priority message by radio addressed to the nearest coastal radio station and repeated to all ships. This is required by Regulations 2 and 3 of Chapter V of the *International Convention for the Safety of Life at Sea* (1960). The text of such a message might read as follows:

> TTT Storm. Appearances indicate approach of hurricane. 1300 GMT. September 14. 2200N, 7236W. Barometer corrected 1003·7 mb. Tendency down 6 mb. Wind NE, force 8. Frequent rain squalls. Course 035°, 9 knots.

As long as the ship is under the influence of the storm similar messages should be transmitted at least every three hours. Vessels of the Voluntary Observing Fleet should also provide observations in the meteorological code to the appropriate coastal radio station.

In order that he may place his ship in a position of comparative safety the mariner should:

(*a*) determine the bearing of the centre of the storm and endeavour to estimate its distance from the ship,

(*b*) determine the semicircle in which the ship is situated,

(*c*) plot the probable path.

It should be remembered that it is quite impossible to estimate the distance of the centre by the height of the barometer, or by its rate of fall, alone.

The bearing may be ascertained by the application of Buys Ballot's law, namely, face the wind and the low-pressure area will be on your right if you are in the northern hemisphere, and on your left if you are in the southern hemisphere. In northern latitudes the centre will bear about 12 points (135°) to the right at the beginning of the storm, i.e. when the barometer begins to fall. When the barometer has fallen 10 mb the centre will bear about 10 points to the right, and when it has fallen 20 mb the centre will be about 8 points to the right (to the left in the southern hemisphere). The nearer the observer is to the centre, the more nearly does the angular displacement approach 8 points, i.e. the direction of the wind more closely follows the isobars.

The distance of the centre from the ship will depend on so many factors that it is almost impossible to estimate without the aid of information from other sources. From the barometric pressure and force of the wind one can, however, arrive at certain broad conclusions. For instance, if the corrected barometer is 5 mb below the normal for the time of the year, the centre of the storm is probably not more than 200 n. mile away. At this distance the wind will probably have increased to force 6. If the wind is force 8, the centre is probably within about 125 n. mile. The semicircle of the storm, whether right-hand or left-hand, is reckoned from the point of view of an observer heading in the direction of progress of the storm. In the northern hemisphere the right-hand semicircle is known as the 'dangerous semicircle', and the left-hand semicircle as the 'navigable semicircle'. In the southern hemisphere the dangerous semicircle is the left-hand one, and the navigable semicircle the right-hand one. *See* Figure 11.6 for an example for the northern hemisphere.

Determination of the semicircle in which the ship is situated depends on a true appreciation of whether the wind is veering, backing or remaining steady in direction. Unless the relative speeds and directions of the ship and the storm field are known (and, in the early stages, it is unlikely that they will be), the prudent shipmaster will reduce speed, or heave to, and watch for a shift of wind, carefully watching the barometer at the same time. If he does not heave to, the master must work out a relative motion problem to determine exactly the real wind shift. If the wind veers he is in the right-hand semicircle, if it backs he is in the left-hand semicircle, if it remains steady in direction he is in the path of the storm. The movement of the barometer will enable him to subdivide his semicircles into quadrants. The barometer falls before the trough line and rises after it. The rules for determining the semicircle hold good in either hemisphere.

The path of the storm may be approximated by taking two such bearings of the wind direction with an interval of from two to three hours between them, provided that allowance is made for the ship's movement.

If the vessel is between latitudes 5° and 20°N, the most likely direction of motion of the storm is roughly west-north-westwards but the chance of the heading being more northerly increases with increasing latitude. Similarly between latitudes 5° and 20°S, the most likely direction of movement is roughly west-south-westwards, with the chance of the heading being more southerly increasing with increasing latitude. (See also the earlier regional headings for details of the average latitude of recurvature and its seasonal variation.)

A piece of tracing paper on which is drawn a diagram representing the average winds and tendency of the barometer in a tropical storm (*see* Figure 11.6), appropriately adapted to the scale of the chart currently in use, will be found useful for studying the behaviour of the wind in the storm field. One can lay off the ship's course and speed on a chart and manipulate the tracing paper as necessary to indicate the relative motion of the ship and the storm. Many examples which may be experienced according to the ship's position, course and speed relative to that of the storm can thus be illustrated, but it must be remembered that the forces and directions of the wind shown on the tracing paper are average or approximate.

Practical Rules for avoiding the Centre of a Tropical Revolving Storm

In whatever situation a ship may find herself, the matter of vital importance is to avoid passing within 80 n. miles or so of the centre of the storm; it is preferable to keep outside a radius of 250 n. mile of more, because at this distance the wind is generally not more than force 6 and freedom of manoeuvre is maintained. If a ship has about 20 knots at her disposal and shapes a course that will take her most rapidly away form the storm before the wind has increased above the point at which her movement becomes restricted, it is seldom that she will come to any harm.

Sometimes a tropical storm moves so slowly that a vessel, if ahead of it, can easily outpace it or, if astern of it, can overtake it. Since, however, she is unlikely

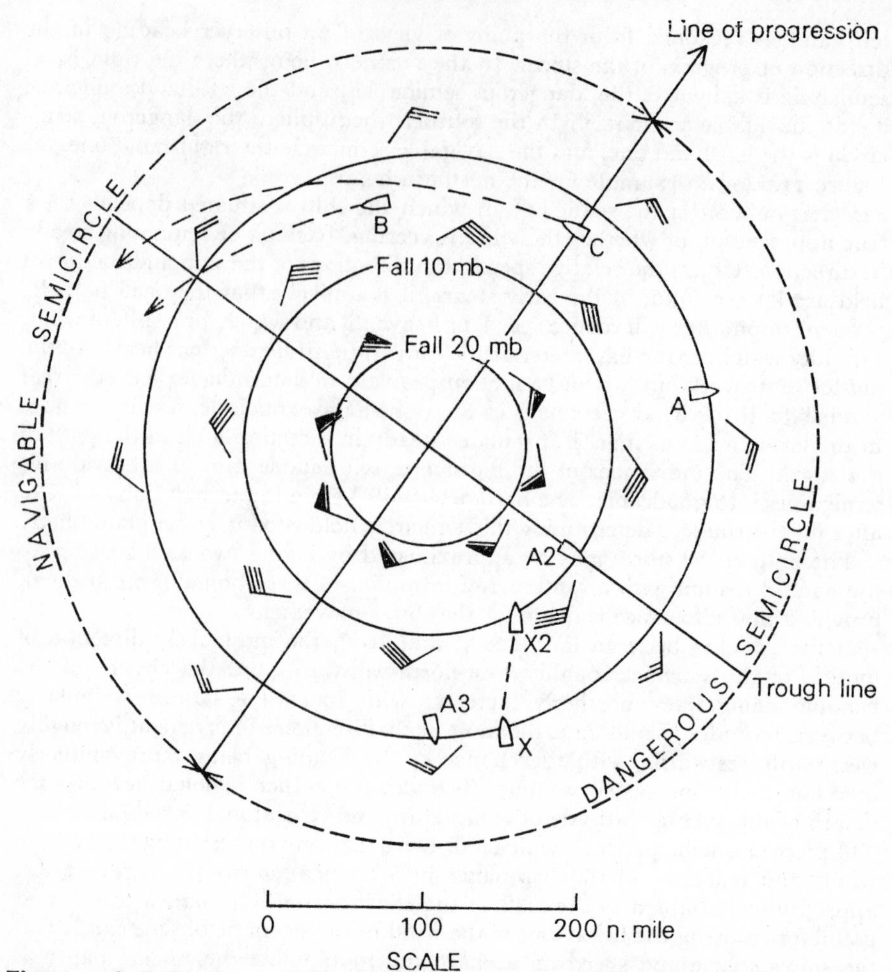

Figure 11.6. Rules for avoiding centre of a tropical revolving storm (northern hemisphere)

(A detailed description of these rules for the northern and southern hemispheres is given on page 159.)

Vessel A is in the right-hand semicircle with a falling barometer. Power-driven vessels should proceed with the wind ahead or on the starboard bow so as to be heading away from the storm field. Vessels under sail heave to on the starboard tack. Such vessels, and very low-powered vessels, may be driven nearer to the centre (position **A2**) by the wind. As the trough-line passes the barometer will start to rise and, when in position **A3**, the vessel will be able to resume course.

Vessel B is in the left-hand semicircle with a falling barometer, runs with the wind on the starboard quarter whether power-driven or under sail, altering course to port as the wind backs, and traces a course relative to the storm as shown by the pecked line.

Vessel C is in the direct path of the storm and acts as vessel B.

Vessel X is overtaking the storm and converging on its centre. If the vessel heaves to, the master will find that the wind is veering and the barometer rising. He will then know that he is in the rear quadrant of the right-hand semicircle. If the vessel is hove to, keeping the wind on the starboard bow for a few hours, the storm will draw away and the vessel can eventually resume her original course. If, however, the master does not heave to, he will possibly find the wind backing and, with a falling barometer, he might incorrectly assume that he is in the advance quadrant of a left-hand semicircle. If he acts on this assumption and puts the wind on his starboard quarter he may eventually find himself in the dangerous quadrant of the storm.

to feel seriously the effects of a storm so long as the barometer does not fall more than 5 mb (corrected for diurnal variation) below the normal, it is recommended that frequent readings should be made if the presence of a storm in the vicinity is suspected or known, and that the vessel should continue on her course until the barometer has fallen 5 mb, or the wind has increased to force 6 when the barometer has fallen at least 3 mb. If and when either of these events occurs, she should act as recommended in the following paragraphs until the barometer has risen above the limit just given and the wind has decreased below force 6. Should it be certain, however, that the vessel is behind the storm, or in the navigable semicircle, it will evidently be sufficient to alter course away from the centre.

In the northern hemisphere. If the wind is veering, the ship must be in the dangerous semicircle. A power-driven vessel should proceed with all available speed with the wind 1 to 4 points (depending upon her speed) on the starboard bow and should subsequently make alterations of course to starboard as the wind veers. If the vessel has insufficient room to make much headway when in the dangerous semicircle, she should heave to in the most comfortable position relative to the wind, preferably with the wind on her starboard bow so that she is heading away from the centre of the storm. A sailing-vessel in these circumstances should heave to on the starboard tack and make alterations of course to starboard as the wind veers. If the wind remains steady in direction, or if it backs so that the ship seems to be nearly in the path or in the navigable semicircle, a power-driven vessel should bring the wind well on the starboard quarter and proceed with all available speed, subsequently altering course to port as the wind backs. A sailing-vessel under these circumstances should run with the wind on the starboard quarter, altering course to port as the wind backs. Figure 11.5 illustrates the rules to be followed in the northern hemisphere.

In the southern hemisphere. If the wind is backing, the ship must be in the dangerous semicircle. A power-driven vessel should proceed with all available speed with the wind 1 to 4 points (depending upon her speed) on the port bow and should subsequently make alterations of course to port as the wind backs. If the vessel has insufficient room to make much headway when in the dangerous semicircle, she should heave to in the most comfortable position relative to the wind, preferably with the wind on her port bow so that she is heading away from the centre of the storm. A sailing-vessel in these circumstances should heave to on the port tack and make alterations of course to port as the wind backs. If the wind remains steady in direction, or if it veers, so that the ship seems to be nearly in the path or in the navigable semicircle respectively, a power-driven vessel should bring the wind well on the port quarter and proceed with all available speed, subsequently altering course to starboard as the wind veers. A sailing-vessel in these circumstances should run with the wind on the port quarter, altering course to starboard as the wind veers.

In either hemisphere, if there is insufficient room to run when in the navigable semicircle and it is not practicable to seek a safe and effective shelter before the storm begins to be felt, a power-driven vessel should heave to in the most comfortable position relative to the wind and sea, bearing in mind the proximity of land, and a sailing-vessel should heave to on the port tack in the northern hemisphere and on the starboard tack in the southern hemisphere.

If a ship finds herself in the direct path of the storm and has no room to run into the navigable semicircle, she should consider, bearing in mind possible recurvature, whether she should endeavour to make her way into the dangerous

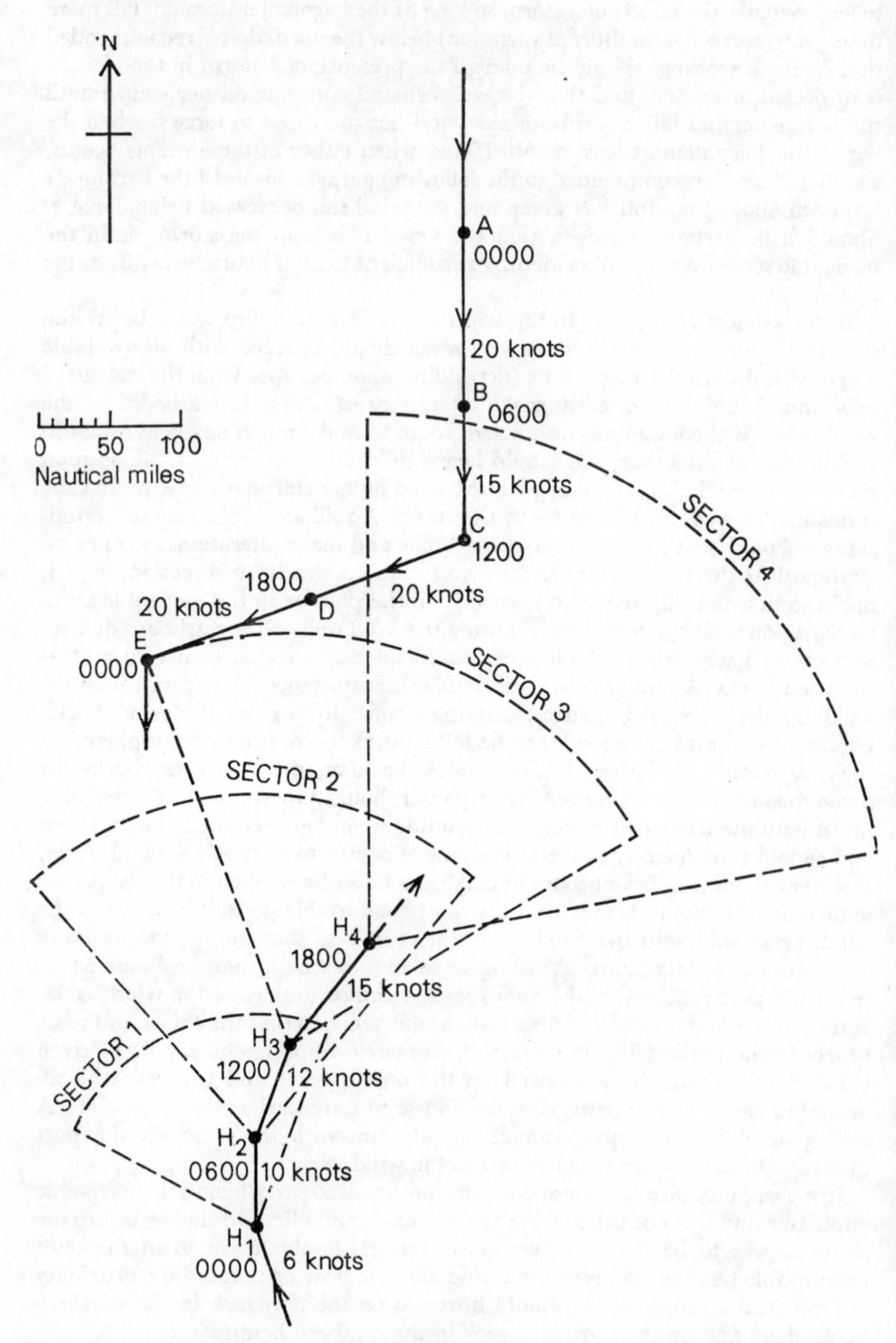

Figure 11.7. Use of safety sector for keeping a ship clear of a tropical storm (northern hemisphere)

A ship in a position **A** at midnight steaming 180° T at 20 kn receives a report of a tropical storm to the south of her with centre at **H1** moving north-north-westwards at 6 kn. Sector 1 is drawn but no action is taken at this time since if the storm continues on its course the ship will pass more than 200 n. mile away from the centre.

Six hours later, when the ship is at **B**, the storm is reported to be centred at **H2** and moving northwards at 10 kn. Sector 2 is drawn and it is apparent that if the storm continues on this path, the closest approach could be 150 n. mile or considerably less. Speed is therefore reduced to 15 kn and the plot maintained.

At 1200, with ship at **C**, the storm is reported at **H3** now moving north-north-eastwards and having accelerated to 12 kn. Sector 3 is drawn and from the plot it is now apparent that if the ship continues on her southerly course she will steam into dangerous proximity to the storm. Heaving to at this stage will only allow the storm to draw closer to the ship; therefore a bold alteration of course to 250° T is made and speed increased to 20 kn to clear the storm field.

At 1800 with ship at **D**, the storm is reported at **H4** moving north-eastwards at 15 kn and Sector 4 is drawn. Even if the path of the storm should change to a northerly direction the closest approach now is not likely to be less than 200 n. mile. To ensure an adequate margin of safety the ship maintains a course of 250° T until midnight and then reverts to her original course of 180° T or an amended southerly course to make her destination.

It will be seen from the diagram that the safety sector is merely a rule-of-thumb method of keeping clear of the storm field. Its effectiveness depends on the reception of radio reports giving the position of the storm centre and its progress, and its accuracy on the assumption that the storm will not alter course more than 40° without being detected. If no reports of the position and progress of the storm centre are received, it will be impossible to plot a sector and the mariner must be guided by his own observations and those received from other ships in the vicinity, and by careful attention to the 'Practical rules for avoiding tropical storms'.

semicircle (where she may at least be better off than remaining in the direct path of the storm) and continue to proceed to windward as fast as she can, so as to get as far as possible from the centre, or to heave to.

In the South Indian Ocean on the south side of a tropical storm a strong SE trade wind associated with a falling barometer may be experienced. It is difficult to know exactly when this wind becomes part of the circulation of an oncoming storm, though the falling barometer may give an indication. If in this area during the storm season the SE trade wind increases to a gale, the prudent shipmaster will reduce speed or heave to and obey the above rules. If the wind remains steady in direction and the barometer falls, he will assume that he is in the direct path of a tropical revolving storm. An additional indication of whether the ship is in a tropical revolving storm field or merely in the belt of trades may sometimes be had from the behaviour of the clouds. If their movement is persistently more from the south than the surface SE wind, the wind being felt is probably part of a storm circulation.

In order to be on guard for an erratic movement in the path of a tropical revolving storm, it is as well to plot a 'danger area' on the chart as an added precaution (*see* Figure 11.7). From the reported position of the centre of the storm, lay off its track and the distance it is expected to progress in 24 hours. From the reported centre, lay off two lines 40° on either side of the track. With the centre of the storm as centre and the estimated progress in 24 hours as radius, describe an arc to cut the two lines on either side of the track. This will embrace the sector into which the storm centre may be expected to move within the next 24 hours. Although it is impossible to say with certainty how a particular storm will behave, the sector does at least provide a margin of safety and it can be reasonably expected that the storm's track will be somewhere inside the sector during the next 24 hours. In taking avoiding action, provided there is sufficient sea room, the mariner would do well to endeavour to get his ship outside this sector as early as possible. If, after a few hours, the direction of the storm is found to have changed, another sector should be drawn with reference to the new estimated path of the storm and action taken to get out of that sector.

The most difficult situation is encountered when the ship finds herself at or near the point of curvature of the storm. This situation is shown in Figure 11.8 a–d, and it will readily be seen how necessary is a constant watch on the weather even when it has been decided in which semicircle the ship is situated.

Although the term 'navigable semicircle' has been used, very violent winds will be experienced in this area and a vessel should clear it as soon as possible.

If in harbour, whether alongside or at anchor, the mariner should, in the tropical storm season, be just as careful as at sea in watching his barometer and the weather and, if a storm is threatened, in watching the shifting of the wind and estimating the movement of the storm relative to himself, so that he can take timely and seamanlike precautions well before the storm reaches his area. If the storm is known to be heading towards the harbour, it is often preferable to put to sea, provided there is plenty of sea room outside, rather than encounter the storm in harbour.

Riding out a tropical storm, the centre of which passes within 50 n. mile or so, in a harbour or anchorage, even if some shelter is offered, is an extremely unpleasant and hazardous experience, especially if there are other ships in company. The extreme violence and gustiness of the wind and its sudden shifts of direction involve great risk of the anchors dragging. The torrential rain and driving spray may impair visibility to such an extent as to make it very difficult to see if such is in fact the case.*

Discretion must, of course, be used. In the case of a low-powered or small vessel having insufficient warning to enable her to gain a reasonable distance from the storm or from the shore by putting to sea, it may be preferable to remain in a reasonably sheltered harbour. If a vessel of this type receives warning of an approaching storm when at sea, and there is considered to be insufficient time or sea room to avoid the dangerous part of the storm area, it may be advisable to seek shelter. In the China Seas for example, there are so called typhoon harbours which are listed in the *Admiralty Pilot*. In all cases, however, the mariner must use seamanship and initiative. It would, for instance, be imprudent to make for harbour and attempt to pick up hurricane moorings if the storm was already being felt. It would be equally imprudent to remain at anchor, even in a sheltered harbour, without making preparations to sail as soon as the first warning of an approaching tropical storm is given.

---

* For some 10 hours during the worst of the hurricane which devastated the town of Kingston, Jamaica, in August 1951, a British ship, having previously buoyed both her anchors with brightly painted drums and with plenty of cable out, maintained her anchorage by steaming so as to keep the buoys in sight ahead. The master reported that he averaged three-quarter speed on his engines throughout the period. (The *Marine Observer*, Vol. 23 (1953), page 112).

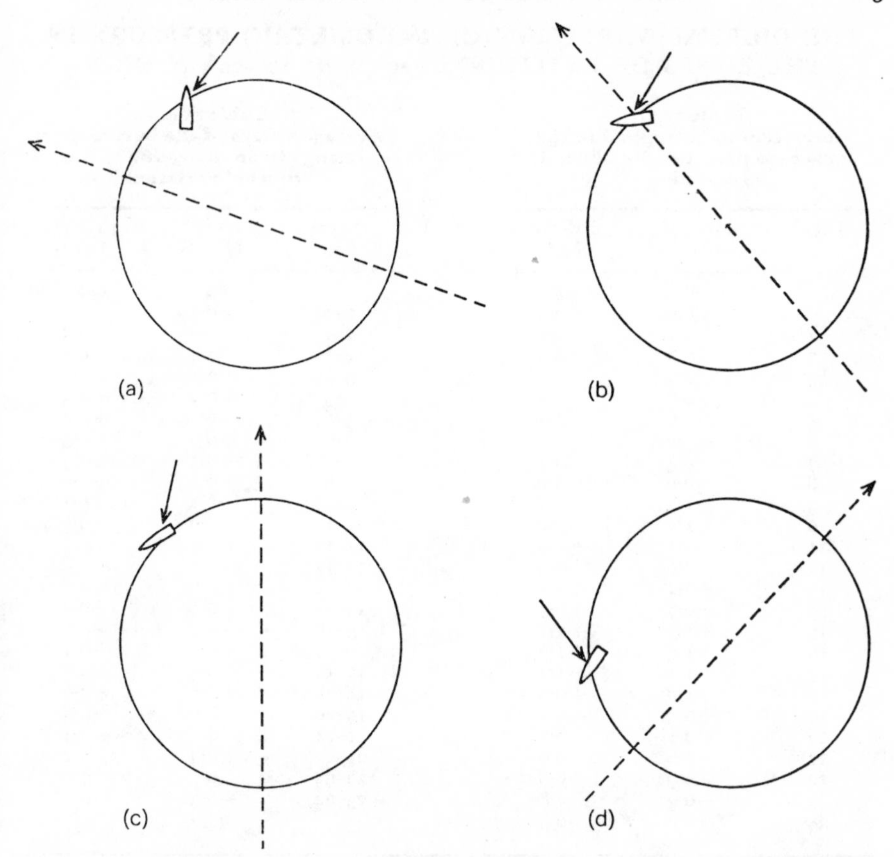

Figure 11.8a–d. Movement of a ship in a tropical storm near the point of recurvature (northern hemisphere)

This illustrates the necessity of watching the wind direction carefully when in the vicinity of a tropical revolving storm. It is assumed that a vessel is in the path of an advancing storm in the northern hemisphere and that, unknown to the meteorological authorities or the master of the ship, the storm has reached the point of recurvature. From a succession of observations made when the ship was hove to, head on to wind and sea, it was found that the wind was shifting to the right and it was, accordingly, assumed that the vessel was in the right-hand semicircle of a storm moving to the NW.

During the interval between Figures 11.8a and 11.8b the wind was observed to change very slightly, a little to the right at first, then back to the previous direction again but gaining in force. From this observation, it was correctly assumed that the vessel was now in, or nearly in, the direct path of the storm and accordingly course was altered to bring the wind on the starboard quarter as in Figure 11.8b.

From that time onwards the wind was observed to shift progressively to the left as in Figure 11.8c, suggesting that the ship was now in the navigable semicircle. The wind continued to shift to the left more rapidly, with the barometer rising, which showed that the storm was clearing to the eastward (*see* Figures 11.8d).

Had the manoeuvre shown in Figure 11.8a been continued without paying careful and frequent attention to the shifts of wind being experienced, the ship might have been seriously involved in the direct path and actually in the centre of the storm and suffered severely in consequence.

THE DIURNAL VARIATION OF BAROMETRIC PRESSURE IN THE ZONES OF LATITUDE 0°–10° AND 10°–20°, N OR S

Table 11.1.
Correction to be applied to the observed pressure for diurnal variation

Local time	0°–10° N or S	10°–20° N or S
	mb	*mb*
0	−0·6	−0·5
1	−0·1	−0·1
2	+0·3	+0·3
3	+0·7	+0·7
4	+0·8	+0·8
5	+0·6	+0·6
6	+0·2	+0·2
7	−0·4	−0·3
8	−0·9	−0·8
9	−1·3	−1·1
10	−1·4	−1·2
11	−1·1	−1·0
12	−0·6	−0·5
13	+0·1	+0·1
14	+0·7	+0·7
15	+1·3	+1·1
16	+1·5	+1·3
17	+1·4	+1·2
18	+1·0	+0·9
19	+0·5	+0·3
20	−0·1	−0·2
21	−0·6	−0·6
22	−0·9	−0·8
23	−0·9	−0·8
24	−0·6	−0·5

Table 11.2.
Average values of the barometric change in an hour, due to the diurnal variation

Local time	0°–10° N or S	10°–20° N or S
	mb	*mb*
0–1	−0·5	−0·4
1–2	−0·4	−0·4
2–3	−0·4	−0·4
3–4	−0·1	−0·1
4–5	+0·2	+0·2
5–6	+0·4	+0·4
6–7	+0·6	+0·5
7–8	+0·5	+0·5
8–9	+0·4	+0·3
9–10	+0·1	+0·1
10–11	−0·3	−0·2
11–12	−0·5	−0·5
12–13	−0·7	−0·6
13–14	−0·6	−0·6
14–15	−0·6	−0·4
15–16	−0·2	−0·2
16–17	+0·1	+0·1
17–18	+0·4	+0·3
18–19	+0·5	+0·6
19–20	+0·6	+0·5
20–21	+0·5	+0·4
21–22	+0·3	+0·2
22–23	0·0	0·0
23–24	−0·3	−0·3

These tables are based on observations made in British Ships, at the hours 0000, 0400, 0800 1200, 1600, 2000 local time, between 1919–38.

In the tropics, should the barometer, after correction for diurnal variation (Table 11.1) be as much as 3 millibars below the monthly normal for the locality, as shown on meteorological charts, the mariner should be on the alert, as there is a distinct possibility that a tropical storm has formed, or is forming. A comparison of subsequent hourly changes in his barometer with the corresponding figures in Table 11.2 will show whether these changes indicate a real further fall in pressure, and, if so, its amount.

Caution: When entering a barometric pressure in the log, or when including it in a wireless weather report, the correction for diurnal variation must not be applied.

PART IV. WEATHER FORECASTING

CHAPTER 12

THE WEATHER MAP, ANALYSIS, AND FORECASTING

General Remarks

In this chapter the main considerations underlying the preparation of the weather map are described, together with a brief account of the way in which the meteorologist uses weather maps in order to identify air masses and fronts and to prepare weather analyses and forecasts.

The weather map is the basic tool of the forecaster. On it are plotted observations of pressure, temperature, wind and other weather elements received from a large number of stations. These observing stations form a close network over the land. At sea, observations are provided voluntarily by merchant ships and are also received from ocean weather stations. The observations are made at the same time (GMT) so that the map gives a bird's-eye view, or synopsis, of the weather over a large area at a particular time; hence the term synoptic chart.

The advantage of a synoptic chart for studying the weather was apparent long before the invention of radio. Admiral FitzRoy began collecting reports daily by telegraph in London in 1860 from which weather maps were drawn, and he shortly afterwards organized a system of storm warnings. In 1868, Alexander Buchan, using observations taken from ships' meteorological log-books, drew up a series of synoptic charts to show the travel of depressions over the Atlantic. An example of these charts is shown in Figure 12.1. Later the Meteorological Council published a series of synoptic charts of the Atlantic for the year 1882–83. The observations on these charts were extracted from ships' logs long after the actual date, and hence the charts were only of value for investigation (Figure 12.2). It is interesting to compare these with the modern synoptic charts as drawn up by selected ships making use of the Atlantic Weather Bulletin, which are shown at Figures 12.3 and 12.4.

From the synoptic viewpoint, some inherent differences between ship and shore stations should be considered. Land stations form a fixed network whose density and distribution may be varied at will, according to requirements. If an additional reporting station is required in a certain locality, there is usually no great difficulty, in principle, in establishing one, except in isolated areas. At sea, however, there are few reporting stations in fixed positions. The forecaster must rely upon ships' reports and hope to have a good enough selection to complete his chart over the ocean. On some occasions an additional ship's report in a key position makes a great difference in drawing up a synoptic chart. A case in point occurred in the North Atlantic on 14 February 1975. At 1200 GMT on the previous day there was a low, 974 mb at 45°N, 52½°W, with a cold front trailing south-westwards. The centre moved east-north-east on successive charts while the cold front moved east and south-east across the ocean, with winds veering north-westerly behind it. Although there was some earlier evidence of troughing behind the cold front, the first clear indication of a development was the report from m.v. *Cargo Vigour* in position 38°N, 53°W at 1200 GMT on 14 February. This

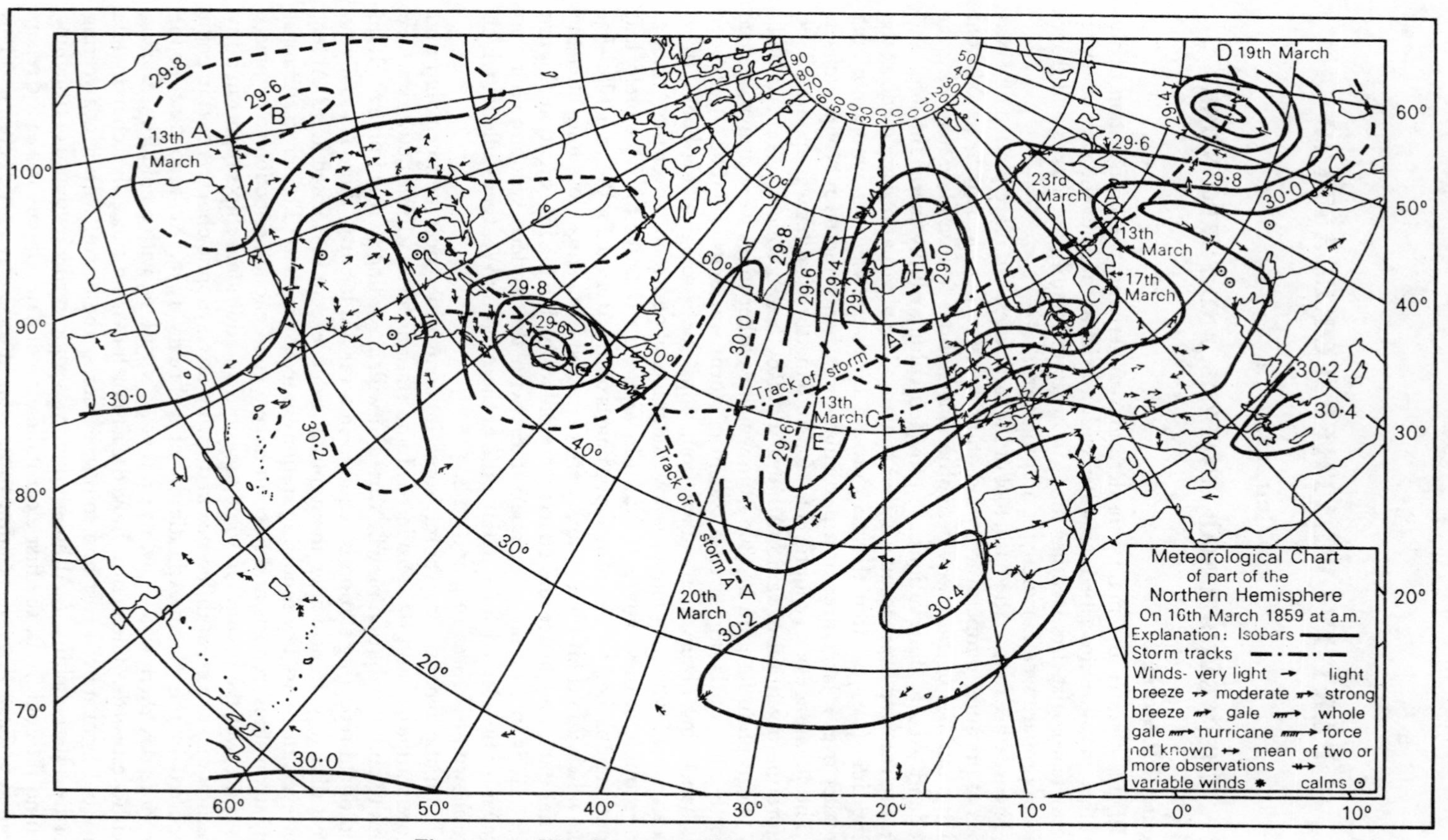

Figure 12.1. Weather map of North Atlantic, 16 March 1859

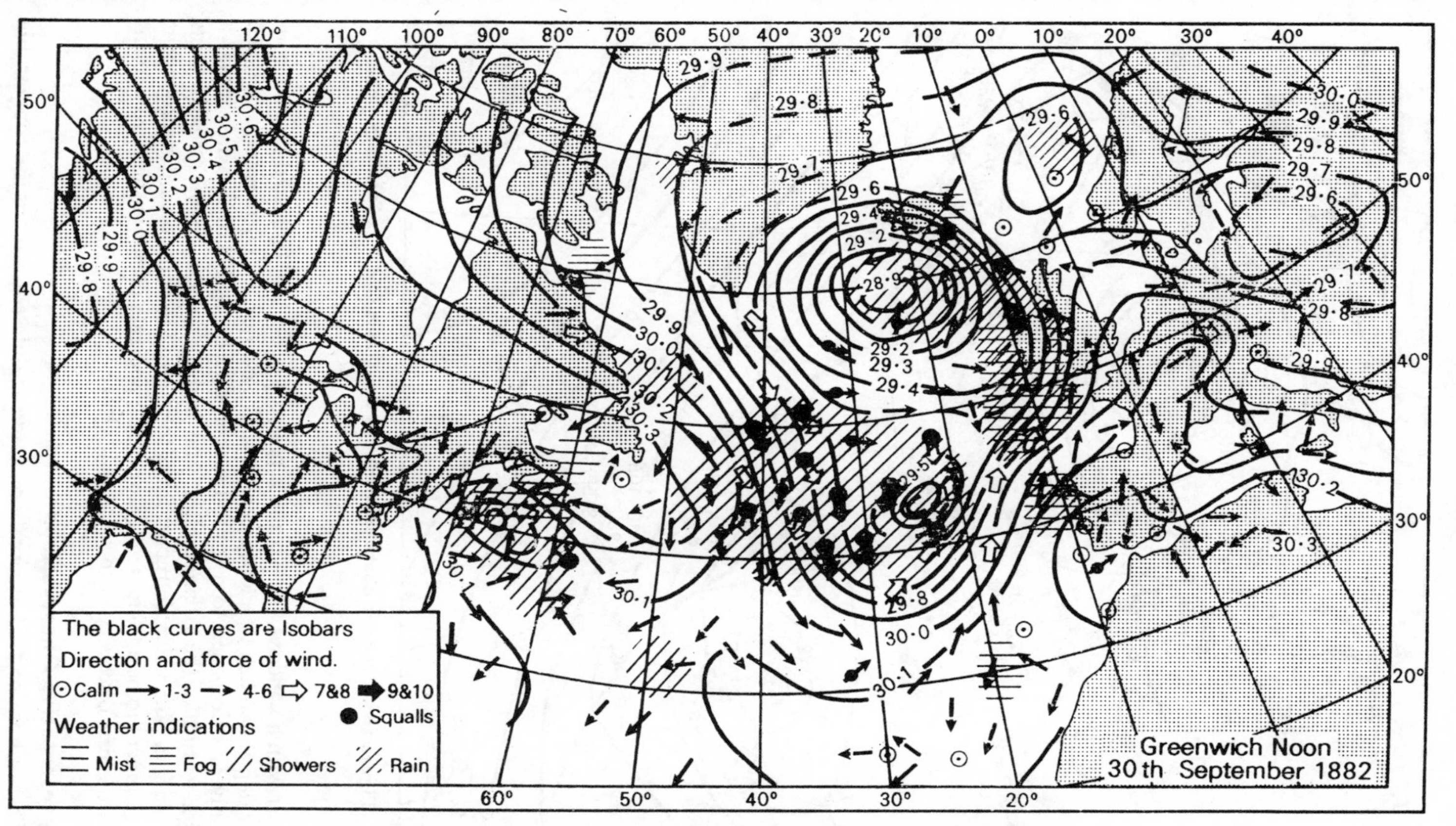

Figure 12.2. Weather map of North Atlantic, 30 September 1882

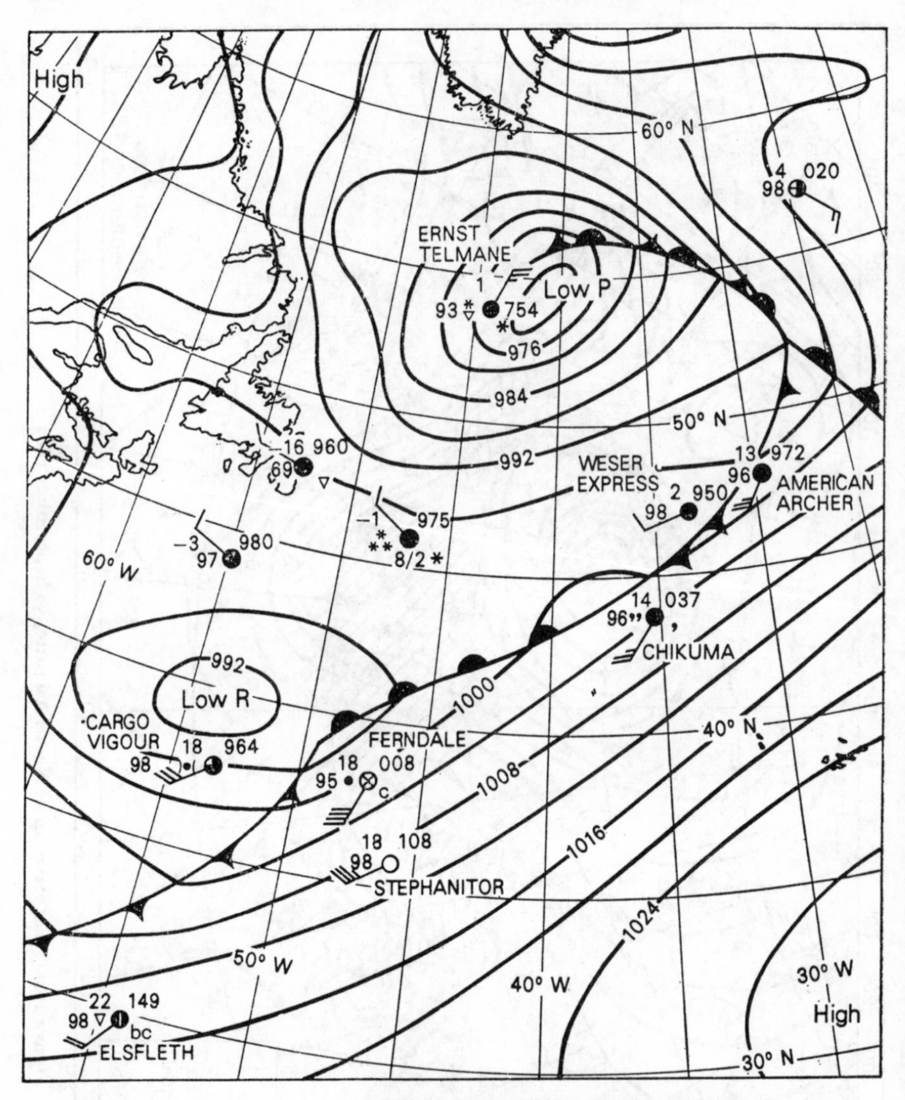

Figure 12.3. Synoptic chart of North Atlantic, 1200 GMT, 14 February 1975

report with a wind sw Force 6 and a pressure 996·4 mb, indicated that a secondary depression had developed behind the cold front. This depression developed rapidly and deepened to 972 mb by 1200 GMT on 15 February when it was a major feature of the Atlantic chart and had drawn the original front into its circulation. The valuable report from the *Cargo Vigour* enabled the subsequent development to be anticipated. Figures 12.3 and 12.4 illustrate this development from the first discovery of the secondary low.

Representative Observations

The assumption that observations made at different stations are comparable

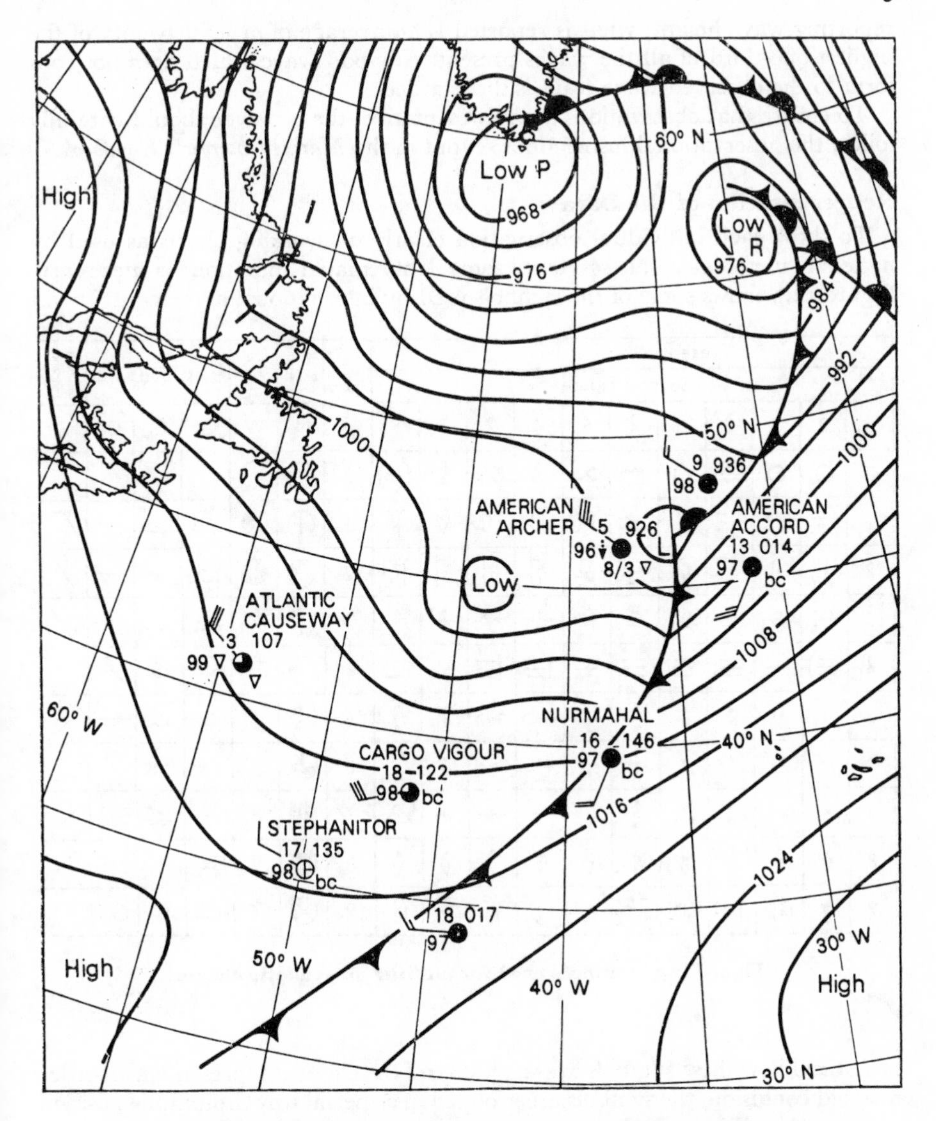

Figure 12.4. Synoptic chart of North Atlantic, 1200 GMT, 15 February 1975

and representative forms the basis of synoptic meteorology. To ensure comparable observations, both instrumental equipment and observing technique must be standardized. This is covered by the official instructions issued by the various national meteorological services, which are co-ordinated by the World Meteorological Organization. The aim is that different observers making a weather observation at the same place and time should get identical results, despite minor differences of aptitude and training. As well as being comparable, observations should, as far as possible, be representative, i.e. they should avoid unduly localized or temporary bias which would be misleading if the observation were assumed to be typical of the weather in the vicinity. For example, in reporting wave height, what is reported is an average of at least twenty of the highest one-third of all the waves present. Any odd wave which does not conform to the observed wave train is disregarded.

To ensure that observations are representative, the observer should carefully follow the observational instructions set out in the *Marine Observer's Handbook*.

Representation of the Data

To show each individual observation clearly on working charts as used by national meteorological services, some abbreviated notation is necessary. Figure 12.5 shows some of the symbols used by international agreement. It will

Present Weather (ww)											Code Figure	Cloud amount	Past Weather	Cloud			Barometer Tendency
First Figure	Second Figure																
ww	0	1	2	3	4	5	6	7	8	9		N	W	C_L	C_M	C_H	a
0											0						
1											1						
2											2						
3											3						
4											4						
5											5						
6											6						
7											7						
8											8						
9											9						

Figure 12.5. Symbols used for plotting on synoptic charts

be noticed how these symbols are suggestive of the element represented. In order to avoid confusion, the symbols are grouped in a special way around the position of the station. Figure 12.6 shows this station model together with some examples. The use of two colours (black and red) helps to avoid confusion in interpreting the plotted figures. This might otherwise occur, particularly in the case of incomplete reports. For example, in a ship report, colour helps to distinguish between dew-point temperature and sea temperature, especially if one or other value should be missing for any reason. The reader should note how concisely and conveniently this system expresses a mass of detail. The procedure is similar for the mariner wishing to plot his own synoptic chart while at sea, from the reports in the Atlantic Weather Bulletin and similar bulletins; though it may be simplified by using the Beaufort letter abbreviations instead of symbols. Details are given in the *Ships' Code and Decode Book*, Met O 509. In both these cases the method of plotting of the weather charts aims at portraying the information from a large number of coded messages in a manner in which it is as easily grasped as possible by the professional forecaster, or by the mariner for whom it is intended. When the plotting is completed the drawing up and analysis of the chart can be undertaken.

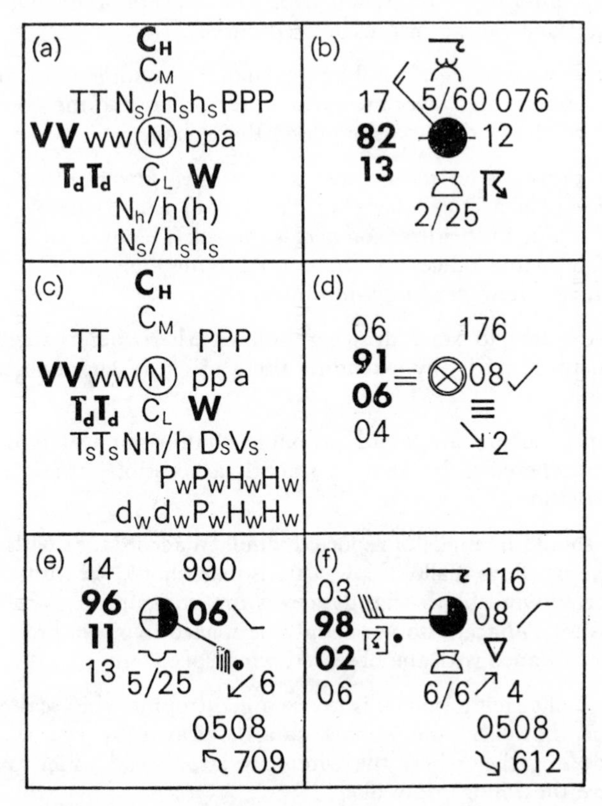

Figure 12.6. Station model with typical examples
(**bold** type represents red ink)

(*a*) Model for land station report.
(*b*) Land station report—83314 82029 07617 29563 13212 82925 85460.
(*c*) Model for ship report.

Ship reports:
(*d*) — — — 90000 91454 17606 9//// 32308 00406.
(*e*) — — — 51140 96036 99014 55500 56606 85625 01311 30508 13709.
(*f*) — — — 62735 98918 11603 69603 14108 00602 30508 31612.

When ww = 20–29, 91–94, the symbols to the left of the pecked line represent 'past hour' phenomena, and are plotted immediately to the right of the past weather symbol (W) in black.

Drawing the Isobars

In the absence of fronts, the drawing of isobars on a synoptic chart is like drawing contours on a survey map, lines of equal pressure replacing the lines of equal height. In drawing isobars the following should be borne in mind:

(*a*) Isobars are simple curved lines with loose ends at the edges of the chart, or simple closed curves; at fronts, however, they are often sharply angled.

(*b*) They must never cross, touch or join (except when two ends of the same isobar join to make a closed curve).

(*c*) Everywhere along an isobar the higher pressures must always be on one side and the lower pressures on the other, and the sides must never be interchanged on passing along the isobar.

(*d*) The pressures on consecutive isobars must always differ by the same interval on a particular chart, except at a col where they have the same value in the direction across the col. The interval in common use in forecasting offices in the United Kingdom is 4 mb. Some other countries, however, use 5-mb intervals.

(*e*) It is better to start drawing isobars where observations are more numerous, gradually extending them to areas where observations are sparse.

(*f*) Simple isobars are more probable than complicated ones. Isobars should therefore be kept as smooth as possible, consistent with the observations.

(*g*) Use should be made of reported winds in accordance with the relation given by Buys Ballot's law. An isobar should be drawn so that a reported wind blows slightly across the isobar towards the side of low pressure. Adjacent isobars should be spaced to give a pressure gradient in accordance with the observed wind speed.

One method of spacing isobars is to use a geostrophic wind scale* in reverse. With practice, however, the isobars can be drawn by eye with sufficient accuracy, close together where the winds are strong and wider apart, in proportion, where the winds are weaker.

A uniform distribution of observations is preferable to a great number irregularly grouped. In general, observations over the oceans are less numerous than over the land so that it becomes desirable for some reference to be made to charts drawn for a previous synoptic hour, when drawing isobars. Care should be taken that a continuous process of change is represented on successive charts.

Identifying the Air Masses

Before completing the analysis of a synoptic chart, any fronts which may be present must be located and drawn in. Logically, this should be done by firstly identifying the various air masses present, in the course of which the position of the air-mass boundaries (or fronts) can be determined. As was explained in Chapter 9, an air mass is a body of air which has acquired nearly uniform values of temperature and moisture content in the horizontal over an area of several thousand square miles, and through a considerable thickness. This idea is nearer reality than the reader might at first believe, because there are many source regions which are suitable for the formation of air masses, such as the subtropical oceans, snow-covered continents, ice-covered polar seas and some

---

* *See* page 31, Figure 3.5.

of the desert regions. Whenever a supply of air remains over one of these regions for a few days, a distribution of temperature and moisture content in the horizontal, sufficiently uniform to allow this air to be characterized as an air mass, is often attained.

In identifying an air mass the forecaster is really probing into the recent history of the air masses, if not actually tracing them back to their source regions. Thus he is most concerned with those physical properties of the air mass which have undergone little change in their numerical values, while the air mass has moved some distance. These physical properties which change little are known as 'conservative' properties. In practice no meteorological properties can be strictly conservative since processes such as mixing, subsidence, radiation, condensation and precipitation are continually happening within any portion of air, and the last three processes in particular tend to destroy most of the conservative properties of an air mass. Nevertheless this conception is a very useful one to the forecaster in identifying air masses, provided that he also keeps a lookout for representative properties, that is, properties which characterize an extensive region of the atmosphere adjacent to the point of observation. A good example of a representative property is an upper-air temperature (provided it is not measured at a front), though such a temperature is not necessarily conservative.

Properties which are not conservative can be used in the analysis, provided due weight is given to modifying influences. For example, surface temperature on land is of limited value as an air-mass characteristic, because it is seldom representative. At sea, however, where it is of value, the modifying influence is the heat exchange between the air and the sea surface.

Figure 12.7, (15 December 1974) illustrates the use of temperature readings combined with other considerations in making an analysis. In this case the polar front over the Atlantic is well marked. Scanning the observations one finds temperatures of 15°C and 16°C in the Azores. Three ships all with similar temperatures show that this warm air extends westwards at least to latitude 45°N, longitude 37°W. North-west of this position the temperature falls dramatically to 3°C in latitude 51°N, longitude 44°W. Obviously one would look for a front somewhere between these two positions. The ship report in latitude 50°N, longitude 30°W enables one to fix the frontal position more precisely. With a temperature of 11°C it is nearer to the temperatures in the warm air, than in the cold. Added to this the wind is backed suggesting that the front lies to the north. Overcast skies would agree with the front being close. Since the broken skies in the two first-mentioned positions indicate that the front is well clear of both these positions, there is little option but to place the front very close to the position indicated in the figure (through latitude 50°N, longitude 31°W). Further west the front is fixed by a well-marked trough between north-easterly winds in Newfoundland, and southerly winds further to the south-west. West of Ireland the temperature difference between Valentia (8°C) and the ship in latitude 48°N, longitude 18°W (14°C) coupled with a wind veer between these two positions helps to locate a warm front as indicated.

Other characteristics which serve to distinguish an air mass are those associated with its thermal stability or instability (*see* Chapter 1). In temperate regions, polar or arctic air masses are usually unstable, and tropical ones, stable. This is due to the modifying effect of the sea surface temperature as the air mass travels from its source to lower or higher latitudes. The stability or instability of an air mass can be determined by the weather phenomena and cloud types

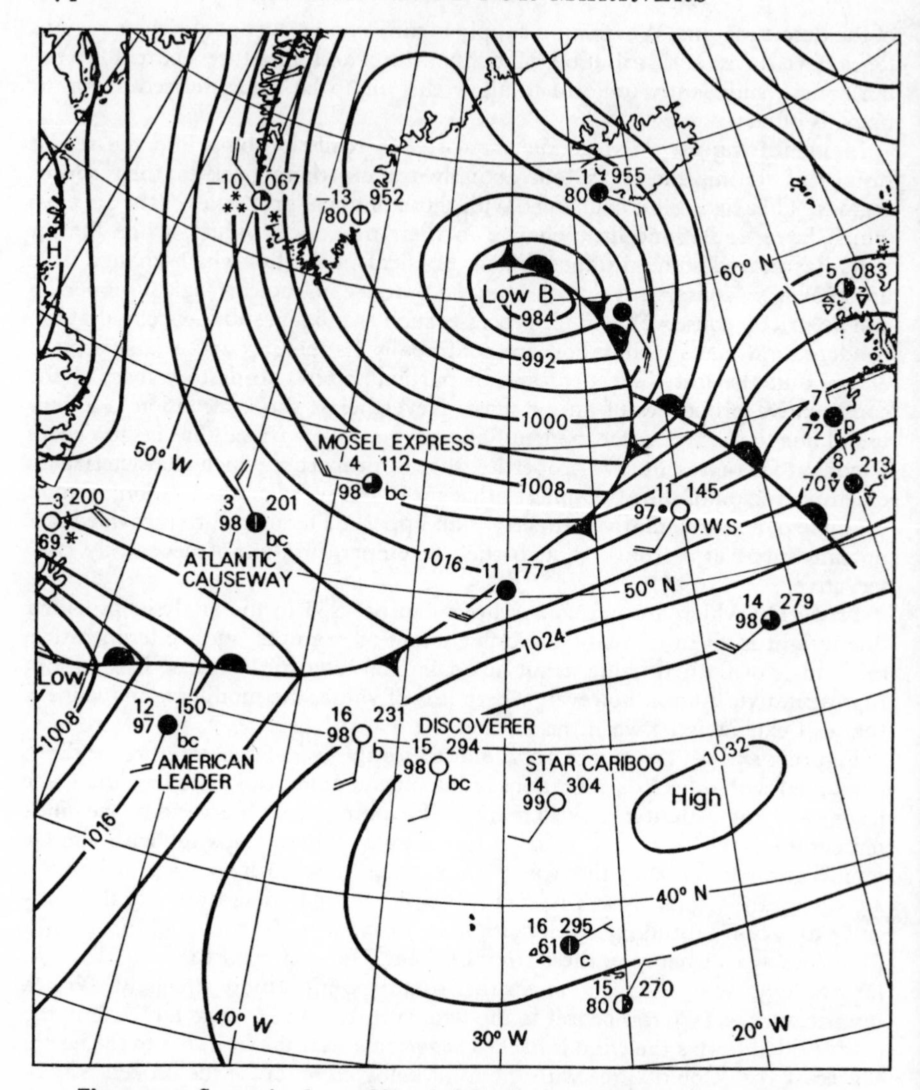

Figure 12.7. Synoptic chart of North Atlantic, 1200 GMT, 15 December 1974

associated with it. Thus, showers, thunderstorms, line-squalls and turbulence are associated with unstable air. Large cumulus and cumulonimbus clouds are also indications of instability. In contrast, stratified low cloud, fog and drizzle or slight rain are characteristic of stable air. Visibility can also help in identifying air masses. In temperate latitudes good visibility can occur in either polar or tropical air but is rather more commonly associated with polar air. If poor visibility is associated with a particular air mass on one synoptic chart, the association can help to identify the air mass on a subsequent chart.

Fixing the Positions of Fronts

These air-mass boundaries should become apparent in the course of identifying the air masses. It often happens, however, particularly when dealing with an active front, that the frontal position is more readily determined by its own associated weather phenomena than in terms of a boundary between differing air-mass characteristics. In other words, given a narrow belt of heavy rain associated with a sharp veer of wind, to quote but two of a number of possible criteria for locating the front, it is simpler to mark the front in first, before considering air-mass characteristics.

As was described in Chapter 9, fronts are generally marked by extensive weather phenomena and cloud systems that are dependent upon the type of front. The passage of a front at any place results in a characteristic sequence of changes of most of the meteorological elements. Tables 9.1 (page 120) and 9.2 (page 121) show typical changes at warm and cold fronts in the temperate zones. The changes occurring at occlusions are rather more complex but to some extent resemble those occurring at either a warm or a cold front.

In practice, the forecaster usually has available a sequence of charts with which to determine the changes occurring at each individual station. By making a detailed comparison between the latest chart which he is drawing up, and the charts for a few of the synoptic hours immediately preceding, he can mark the position of the fronts more accurately than is possible by using the latest chart alone and not referring to its predecessors.

Adjustment of Isobars to Fronts

Having drawn the isobars and located the fronts, it is usually necessary to make some adjustment to the isobars in the vicinity of the fronts. This is because the pressure gradient is discontinuous at a front and therefore each isobar changes direction more or less abruptly. A frontal analysis* is a great help in the correct drawing of isobars where observations are few. A rigid rule of drawing the fronts before the isobars should, however, not be adopted. In practice, it is far better for the two processes to proceed together, fronts and isobars being drawn tentatively and later mutually adjusted to give a satisfactory final picture.

The help given by a good analysis in drawing the isobars needs no specific example, but it may be worth while to see how drawing isobars can help to determine the position of a front. Figure 12.8 shows two ships' observations which give evidence that a cold front must be somewhere between them. Where exactly should it be placed? After drawing the isobars near each ship by deduction from the wind direction and force, an isobar of one particular value based on the observations of one ship will be found to meet the corresponding isobar based on the other ships' observations at a point. If this is done for several isobars it will be found that the points where corresponding isobars meet will lie on a line; this line must be the position of the front. This simplified construction assumes that there is no appreciable curvature of the isobars, but it can easily be extended if one knows where the isobars are likely to be curved.

In theory it is important to draw isobars as correctly as possible near a front, since this makes for a better estimate of the speed of advance of the front. But in practice such precision is not often attainable from ships' observations, unless a large number of them is available, and estimates of the speed of a front must to some extent be based upon its recent behaviour.

---

* The reader's attention is drawn to the remarks in Chapter 13 about plotting the analysis issued in the Atlantic Weather Bulletin for Shipping.

Logical Sequence of Charts

A fundamental principle in the analysis of weather charts is that every chart should follow logically from the previous one. The word 'logically' implies that the movement of the fronts should be in accordance with the wind field and that the whole process of change should be a continuous one. Where new features appear on a chart they must be accounted for logically. For example, where a new front appears it should be the result of a front-forming or frontogenetic process already shown on a previous chart. Fronts disappear only as a result of a front-dissolving or frontolytic process.

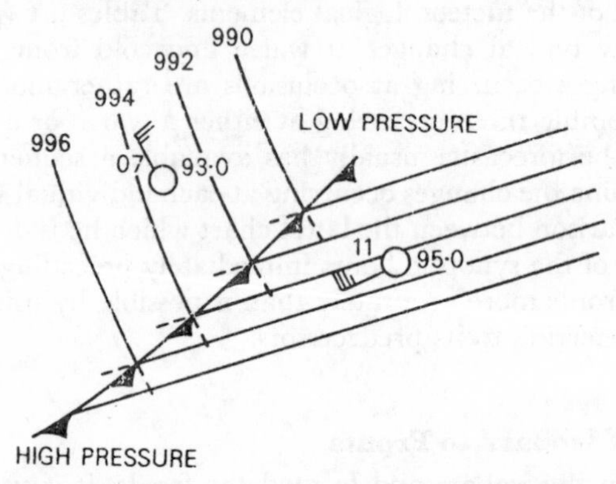

Figure 12.8. Locating a front by the intersection of isobars

Requirements of a Good Analysis

An accurate analysis is essential for the proper understanding of the weather already experienced, and such an understanding is an obvious prerequisite in any attempt to forecast future developments. The analysis should show:

(*a*) The positions of depressions, anticyclones and other features of the pressure distribution.

(*b*) The extent and identifying properties of the different air masses involved.

(*c*) The relation of the air-mass boundaries, or fronts, to the depressions.

(*d*) Accurate drawing of isobars, especially in the vicinity of fronts.

(*e*) A logical and consistent development from the previous analysis.

Over the greater part of the ocean the number of ships' observations is inadequate for the positions of barometric features to be accurately fixed. In drawing up the chart and in analysing the air mass over a certain part of the ocean, the forecaster may have to place great reliance on some particular ship report. Whereas on land a doubtful observation can usually be confirmed or rejected by comparing it with adjacent observations, this is often impossible over the sea and the observation must be accepted at face value. This is the main reason why a high standard of accuracy is required from ships that make observations for Meteorological Services.

The Use of Synoptic Charts in Forecasting

A synoptic chart provides a convenient visual summary of all the elements of the weather over a large area. A sequence of synoptic charts enables one to study the movement and development of the various weather systems involved. At meteorological forecasting offices a continuous series of charts is produced, covering at least the 'synoptic' hours of 0000, 0600, 1200 and 1800 GMT, daily.

The main features of these charts, namely low-pressure centres, high-pressure centres, troughs, ridges and fronts all tend to preserve their identity, in the short term, and can be traced from one chart to another, sometimes for a large number of charts. Accordingly their movement can be followed, taking account of acceleration, deceleration, intensification and weakening. Thus, on the basis of what has happened in the series of charts leading up to the most recent chart (say for hour HH), a forecast chart, or prognosis, can be created for HH + 6, HH + 12 hours, etc., by extrapolating the movement, acceleration and development (i.e. deepening or filling of pressure systems, intensification or weakening of fronts) of the main features of the chart. While making the necessary allowances for the trends indicated and for the important modifying effect of time of day, and of variations in terrain, one applies to the corresponding features of this forecast chart, weather similar to that characterizing those features on the most recent chart (for HH hours). Thus, for example, fine weather might be accredited to the central parts of the anticyclone on the forecast chart, and a belt of rain and low cloud might be indicated as extending for, say, 150 nautical miles ahead of the advancing warm front. While the basic principle of weather forecasting is relatively simple, the practical production of a prognostic chart is very complex and requires considerable theoretical knowledge and long experience.

So far only the surface charts have been mentioned and these represent only a part of the story. The pattern of pressure and winds changes considerably as one ascends to higher levels of the atmosphere. The air movements at all these higher levels, at least throughout the troposphere, interact with the movements at other levels and influence the pressure pattern at the surface. For this reason it is necessary to study chart sequences showing the flow patterns at a number of representative levels (typically 700 mb, 500 mb, 300 mb, 100 mb) and to produce forecast flow patterns at these levels which need to be reconciled with any forecast produced for the surface level. Pressure, wind, temperature and humidity at these higher levels are measured regularly by a large number of stations and enable charts to be drawn which are representative of these levels. The most important of these upper-level charts are those depicting the air flow. There are technical advantages in constructing these charts by plotting contours of the height of specified pressure surfaces, rather than plotting pressures at specified heights. The winds are related to the contours of these charts in the same way as to isobars on a surface chart.

In addition to these charts of the actual contours at various levels, charts are also drawn showing the difference in height from one pressure level to another. These are called 'thickness charts' and the 'thickness' values plotted are a measure of the mean temperature of the layer of atmosphere in question. The winds which can be read off from the contours on these charts are called 'thermal winds' (*see* also Chapter 9). The thermal wind for a given layer is the vector difference between the winds at the top and bottom of the layer.

To quote an example of the way the conditions at these upper levels can influence the surface prognosis, it is found that the movement of a warm-sector depression centre is often closely related to the average 1000–500-mb thermal wind above the surface centre, over the time-interval considered. Accordingly any estimate based on extrapolation of the movement of the surface centre is best modified to agree with the appropriate forecast value of this thermal wind. In this, as in many forecasting problems, the estimate finally accepted is often a compromise between differing estimates based on a number of different approaches.

In addition to the question of the movement of pressure systems, that of development is very important. Sometimes lows and highs move from one chart to the next with little change in pressure or shape. At other times they intensify or decay, as they move, and obviously such changes must be taken into account, as well as mere translation. A first approach is again by means of extrapolation. If the centre has been deepening in the intervals leading up to the most recent chart, the first assumption is that it will continue to do so. An aid in assessing development is provided by the 'barometric tendency' (symbol app) which is plotted immediately to the right of the station, or ship, circle (*see* Figure 12.6).

Before using barometric tendencies as reported by ships, it is necessary to apply a correction to take account of the pressure change which results from the ship's motion, which would otherwise falsify the reading. This can be done quite easily.

In Figure 12.9 the full lines represent isobars on the synoptic chart and **A** is the plotted position of the ship. Knowing the course and speed of the ship, which are given in the synoptic message, lay off from **A** a distance **AB** so that the vector **BA** represents the distance run by the ship in the past three hours. If there was no change in the position of the isobars during the three-hour period the true barometric tendency measured at position **A** would be zero, yet solely on account of its movement the ship registers a tendency expressed by the difference of pressure between **A** and **B**. In Figure 12.9 this amounts to a rise of 2 mb. When the isobaric pattern is changing, the true tendency is obtained by subtracting this 'spurious' component from the tendency reported by the ship.

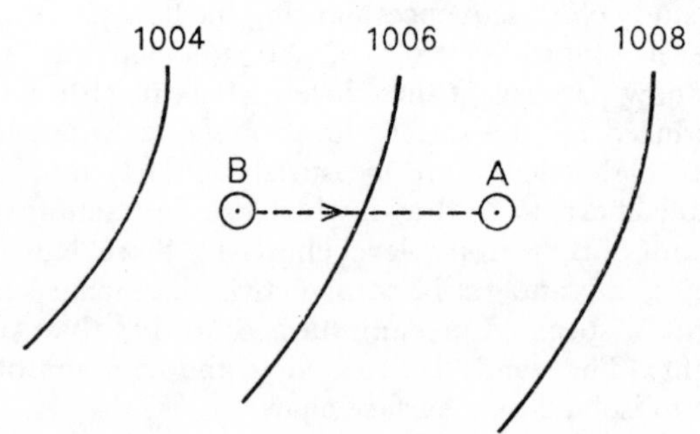

Figure 12.9. Correction of barometric tendency for course and speed of ship

The corrected tendency readings help to determine the movements of pressure systems, since the barometer usually falls in advance of a depression and rises in advance of an anticyclone. To be more precise, consider the behaviour of the barometer at a position near a depression. Suppose that the pressure at the centre of the depression (its 'depth') remains constant. Then if the centre of the depression moves towards the position, the barometer there will fall. Now consider the case when the depression remains stationary but the barometric pressure at its centre falls; in other words, the depression deepens. We may regard this deepening process as the creation of new isobars at the centre of the depression, the whole system of isobars being pushed slowly outwards. If this outward displacement of isobars reaches the position where our barometer is situated, then the pressure there will fall. Thus, in this case, the falling barometer is not evidence of the movement of the depression but of its deepening. The barometric tendency in any general case thus has two components, one due to the movement or translation of pressure systems, the other due to their change of intensity or development.

Where there are plenty of observations well distributed about a pressure centre, the movement of the latter can be inferred from the tendency field. A depression, for example, will move towards the region of greatest negative tendency.

The following rules show that barometric tendency is closely related to development in depressions and anticyclones so that it is advantageous to have as many observations of tendency as possible.

Deepening and Filling of Depressions

(*a*) Frontal depressions deepen after their birth, the rate of deepening usually increasing until some time after occlusion has begun and then decreasing.

(*b*) A depression is deepening if pressure is falling all around the centre, or if the rate of fall on one side is greater than the rate of rise on the other.

(*c*) A depression is filling up if pressure is rising all round the centre, or if the rate of fall on one side is less than the rate of rise on the other.

(*d*) When a new deepening centre moves into the circulation of an old depression, the old depression is either absorbed by the new one or the two centres rotate around each other in a cyclonic sense, i.e. anti-clockwise in the northern hemisphere.

(*e*) When a fully occluded depression deepens, it will generally be found that a new influx of either warm or cold air into the system is associated with and responsible for the deepening.

Intensifying and Weakening of Anticyclones

(*a*) An anticyclone is intensifying if pressure is rising all around the centre, or rising on one side more rapidly than it is falling on the other.

(*b*) Conversely, it is weakening if pressure is falling all around the centre, or falling on one side more rapidly than it is rising on the other.

(*c*) The intensity does not change if the pressure tendency is zero at the centre, or if the falling and rising pressures on either side of the centre have the same values.

(*d*) A warm anticyclone in which, by definition, the air at most levels in the troposphere is warmer than the air at corresponding levels outside the anticyclone, is relatively stable and depressions tend to be deflected around it. However, it is by no means rare for an apparently stable warm anticyclone to collapse quite rapidly.

(*e*) In a cold anticyclone the air at the surface and in the lower layers of the troposphere is colder than the air at corresponding levels in adjacent regions. Cold anticyclones often build up rapidly in the cold air behind a frontal depression and collapse as rapidly on the approach of another deepening depression.

From a more fundamental point of view, the deepening of a depression can be forecast insofar as one can anticipate the development of the associated patterns of pressure and temperature. If the changes presently occurring and expected to continue in the immediate future are such as to intensify the horizontal temperature gradients in the vicinity of the depression, then the latter may be expected to deepen. If the horizontal temperature gradient is expected to decrease, this will favour weakening of the depression.

Numerical Forecasting

It has long been the aim of meteorologists to make forecasts by applying the mathematical equations expressing the basic physical laws governing the motion of the atmosphere.

The formulation of the method for solving the equations numerically was proposed and first attempted by L. F. Richardson in 1922. The basic principle involved is, firstly, to define an initial distribution of values at grid points over an area to be covered. Secondly, the equations which govern the change of the variable with time are applied to each of these values so as to calculate new values for a small time-interval ahead. This process is then repeated for successive small time-intervals until the target time is reached. This produces a new set of values which defines the forecast distribution required. Because of the vast number of calculations required, the process was too laborious for the means then available and, for a long time, little progress was made.

The advent of high-speed electronic computers made it possible for the first time to perform the multitudinous calculations within a sufficiently short period.

The obvious attraction of a numerical method of forecasting is that it systematically applies the governing equations and the result does not depend upon the experience or judgement of the human forecaster. On the other hand, the process has inherent difficulties arising from the complexity of the problem. Firstly a compromise has to be struck in the choice of a grid-point network to define the field of variation. The field needs to be large—almost hemispherical, and yet the distance between the grid points needs to be small enough to represent accurately the distribution of the variable. Secondly, because the equations employed involve instantaneous rates of change they can only be used to cover small time-intervals—currently 15 minutes is used. So each of the great number of calculations has to be performed many times to obtain a forecast for time-intervals such as 48 and 72 hours ahead. A further complication concerns the variations in the vertical. Forecasting cannot be confined to a single level such as the earth's surface because the future conditions at the surface depend also upon changes occurring at higher levels. In consequence one has to represent at least the whole troposphere by selecting a number of levels to represent a 'model' of the atmosphere. Again a compromise has to be struck between using too few levels to achieve a sufficiently accurate model, and too many which would cause an unacceptable increase in the time required to perform the calculations.

At the time of writing, a ten-level model of the atmosphere is used in the British Meteorological Service. The levels used are 1000 mb, 900 mb, 800 mb, etc., to 100 mb. Instead of using pressure distribution at various height levels it is found preferable to use charts showing the contours of heights of specified pressure surfaces.

On a polar stereographic map projection covering a major portion of the northern hemisphere, the initial state of the atmosphere is specified in terms of values of contour height, horizontal wind components, and humidity mixing ratio at grid points which are about 300 km apart.

Observations received in the Meteorological Office from all parts of the northern hemisphere within a few hours of their being made are fed into the computer and, by interpolation, grid-point values are calculated which specify the contours on all the ten pressure levels. Then the computer calculates new grid-point values at all levels for a small time-interval ahead by solving the governing equations and repeats the process again and again until it reaches the forecast interval required. The latter in practice covers various intervals between 12 hours and 6 days ahead of the starting time. The computer output is obtained both as an array of grid-point values and also in the form of charts of isopleths.

While the advent of numerical forecasting has constituted a notable advance in forecasting technique, it by no means solves all the forecasting problems. For example, with the present grid spacing of about 300 kilometres it is possible for localized developments to proceed without detection. Also, the lack of observations in some sea areas means that even the initial conditions cannot be accurately specified in all areas. For these and other reasons the numerical forecast is regarded as a useful tool but is not slavishly followed by the forecaster who has to monitor its output and at times intervenes to modify the end product.

CHAPTER 13

METEOROLOGICAL ORGANIZATION AND THE PRACTICAL USE OF WEATHER BULLETINS BY SEAMEN

Collection and Distribution of Meteorological Information

This chapter discusses the meteorological organization for collecting and disseminating weather information and advises the mariner how to get the maximum benefit from the services provided, including those of 'Ship Routeing'.

Successful analysis and forecasting of the weather depends upon a vast organization functioning not only within each country but also across national boundaries. Forecasts for a given area necessitate the continuous analysis of the weather over a very much larger area. For example, forecasts for the British Isles require a running analysis of an area extending at least to the Rocky Mountains of North America. The previous chapter described how a weather map is drawn from a large number of observations taken at fixed times (synoptic hours). Each country maintains within its territory a network of reporting stations. Reports from these stations are transmitted to a central office by telephone, teleprinter or radio. Each country thus collects its own observations and then rebroadcasts them by teleprinter or radio in a collective message for the benefit of other countries. To make the observations understandable by all nationalities they are transmitted in an international code. The observations are also made available to the many subsidiary forecast offices within each country. At sea the collection of weather reports has to be dealt with internationally, for ships of many nationalities may make weather observations in a certain area and the coded results must then be transmitted to designated shore radio stations. Ships' observations from ocean areas are of value to many countries, and it is important that they be made available to all. The problem has been solved by assigning areas (*see* Figure 13.1), internationally agreed, within which ships are asked to report to individual countries through specified shore stations. Each country receiving them broadcasts the ships' reports in collective messages for the benefit of its neighbours.

World Meteorological Organization

International co-operation is essential for the collection and exchange of observations and the issue of meteorological information, not only for shipping and aviation but for all other purposes. This is fostered by the World Meteorological Organization (WMO), which is responsible for establishing international standards and procedures and for preparing the codes and specifications which are the international language of the meteorologist. The World Meteorological Organization is subdivided into Regional Associations which study the problems of particular areas, e.g. Europe, and into Technical Commissions which are concerned with particular aspects of meteorology, e.g. the Commission for Marine Meteorology.

In order to provide a network of meteorological observations in all oceans, under arrangements made by the World Meteorological Organization, selected ships of most nations voluntarily make observations at routine hours and send

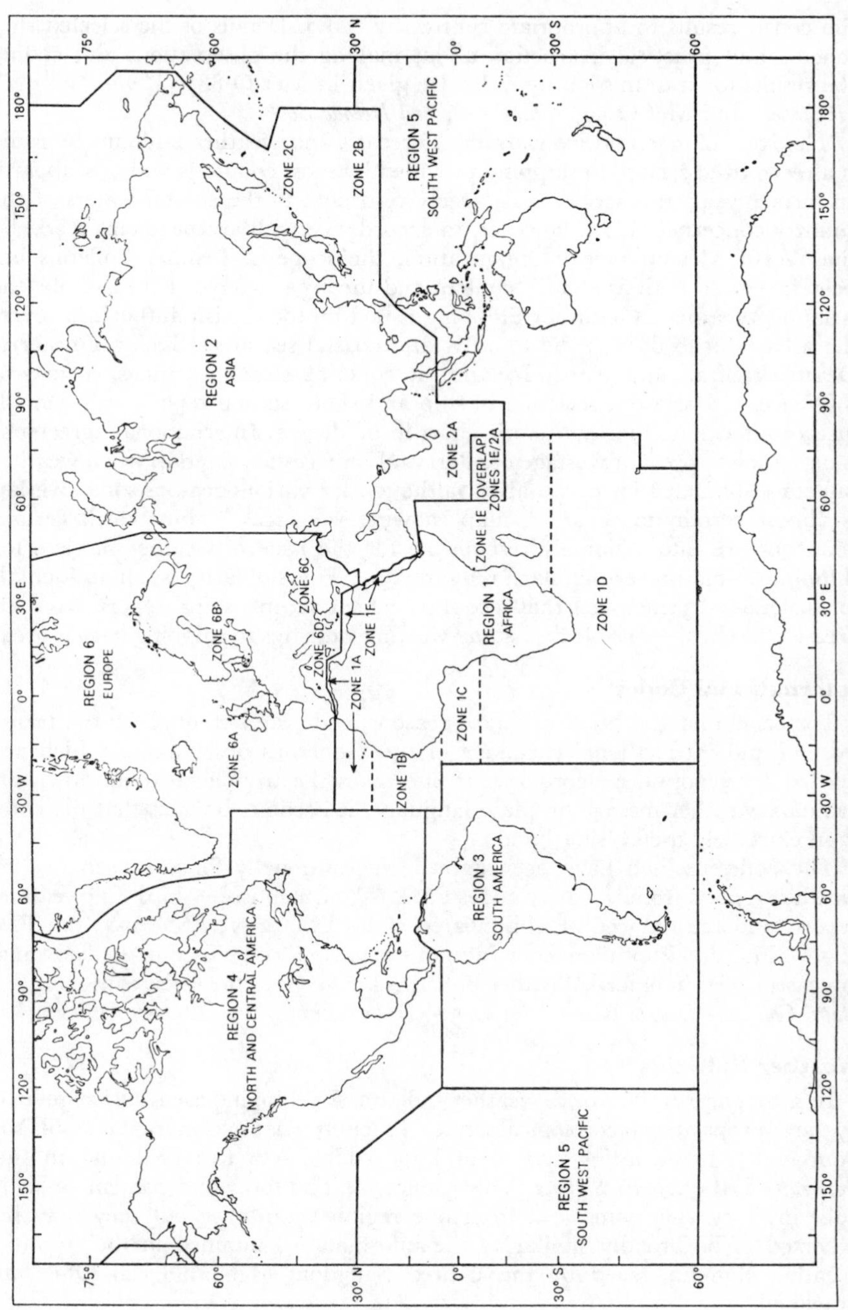

Figure 13.1. Areas in which observing ships should transmit their radio messages to the appropriate stations

the coded results to appropriate centres by radio. Details of the selected ship scheme and practical instruction about making the observations and coding the results for transmission by radio are given in Met O 887, *Marine Observer's Handbook*, and Met O 509, *Ships' Code and Decode Book*.

The issue of gale and ice warnings, forecasts and weather bulletins by radio is a recognized service to shipping provided by most countries with a seaboard. In former years this service was largely restricted to the coastal waters of the country concerned. Now, however, in accordance with arrangements made by the World Meteorological Organization, the scope of weather bulletins has been increased both as regards content and the area covered. For example, the Atlantic Weather Bulletin for Shipping issued by the British authorities covers the area 35°N to 65°N, 15°W to 40°W and coastal sea areas Biscay, Finisterre, Denmark Strait and North Iceland; it contains storm warnings, a general inference, a forecast, a selection of ship and shore station reports and enough further information for a synoptic chart to be drawn. International agreement is necessary to prevent wastage of effort and unnecessary overlap when weather bulletins are issued for ocean areas, although for various reasons some overlap is almost certain to occur. A map showing the areas within which certain countries are internationally responsible for the issue of weather services for shipping would be very similar to Figure 13.1. This similarity is not accidental, for the general principle is that a country which accepts ships' reports from an area is thereby responsible for issuing weather bulletins to shipping in that area.

International Codes

For economic and other practical reasons, codes are essential for the transmission and international exchange of the numerous observations which are needed for synoptic meteorology. If one takes the trouble to write down an ordinary weather message in plain language, it becomes obvious that the code is an extremely useful 'shorthand'.

The codes* which have been agreed internationally for the exchange of weather reports from ships are codes FM 21–V, which is the ships' full weather report, and the reduced or abbreviated codes FM 22–V, FM 23–V and FM 26–IV. Full details of these codes and of the various codes used in analyses and forecasts issued in official Weather Bulletins are given in the latest edition of the *Ships' Code and Decode Book*—Met O 509 (Her Majesty's Stationery Office).

Weather Bulletins

In most parts of the world, weather bulletins for shipping are issued regularly by the appropriate meteorological service. Precisely what information is available from which broadcasting station and for which area can be found in the *Admiralty List of Radio Signals*, Vols 3 and 3a. The form and content of such bulletins may vary somewhat from one region to another but they may be expected to be broadly similar to the information contained in the Atlantic Weather Bulletin issued by the United Kingdom Meteorological Office at Bracknell.

The Atlantic Weather Bulletin (UK)

This is issued in six parts. Parts 1, 2, and 3 are in plain language and cover,

*From 1 January 1979, with the introduction of a new ICE group, 21–V, 22–V and 23–V will become 21–VI Ext, 22–VI Ext and 23–VI Ext respectively.

respectively, Storm Warnings, a Synopsis of the Present Weather Situation, and a Forecast for the next 24 hours.

Part 4 contains the analysis, in code, of the present weather situation.
Part 5 contains a selection of ships' reports in code, and
Part 6 contains a selection of relevant land-station reports in code.

An example of an Atlantic Weather Bulletin now follows:

BRACKNELL WEATHER FOR CQ BY W/T
PRIORITY PORTISHEAD RADIO 2/4/76—0800 GMT

Part 1. Storm Warnings—ZCZC WONT54 EGRR DTG Low 998 Denmark Strait deepening and expected South-east Iceland 974 by midnight tonight. West or north-west Storm force 10 expected to develop in south-west semicircle up to 600 n. mile of centre and to affect Northern Sections. Later decreasing slowly in west and spreading eastwards.

Part 2. Synopsis of weather conditions—ZCZC FQNT74 EGRR 0000 Low 998 Denmark Strait expected 974 South-east Iceland by midnight tonight. High 1040 50 North 34 West expected 1038 45 North 28 West by midnight.

Part 3. Forecasts—Forecasts for Biscay, Finisterre, Sole and from 35 to 65 North between 15 and 40 West, Denmark Strait and north Iceland, for next 24 hours.

Biscay—Variable or south-westerly 2–4 becoming northerly 4–5 in north. Showers in north. Moderate or poor becoming good.

Finisterre—Mainly northerly 4 or 5 veering north-easterly 5 to 7. Showers. Moderate or good.

Sole—Northerly 4 to 6 backing westerly later. Showers dying out. Mainly good.

East Northern Section—South-west 7 to severe gale 9 locally Storm 10 veering west or north-west and generally gale 8 to Storm 10 later. Rain clearing to showers later. Moderate.

West Northern Section—Cyclonic variable at first in extreme north otherwise south-west gale 8 to Storm 10 veering North-westerly. Rain clearing to showers. Moderate.

East Central Section—Northerly 5 or 6 backing westerly and increasing 6 to gale 8 but lighter in extreme south-west. Showers. Moderate or good.

West Central Section—South-west 6 to gale 8 veering north-west and decreasing 5 or 6 in north but variable 3 or 4 in South. Fair. Good.

East Southern Section—North-east 5 or 6 locally 7. Showers. Good.
West Southern Section—Easterly 5 or 6. Showers. Good.

Denmark Strait—Easterly 4 or 5 increasing 7 to severe gale 9 backing north-east and perhaps Storm 10 at times. Light icing becoming severe. Temperatures minus 01 to minus 04 falling to minus 06 to minus 09 later.

North Iceland—Southerly 4 backing south-east and increasing 7 to severe gale 9. Light icing becoming moderate. Temperature minus 01 to minus 04.

Part 4.

Analysis:

10001	33300	00206=	99900	81298	31495		
00920=	81386	64267	00730=	81205	54007	00720=	
81210	41048=	81303	54073=	81510	32106=		
85242	48345	01425=	88014	36655=	88002	69375=	
88010	59051=						
99911=	66457	30757	36706	41695	45707	51655	54566
56505	59437	61408	61337	62305	64267=		
66250	64267	62215	57218	52248=	66650	70515	66427
62408=	66457	30395	32407	35445	38465	40508=	
66650	38566	40537	40508=	66250	40508	40416	36356
30327=	66450	38278	39176	42095	46047	50035	54007=
66257	54007	54073	54101=				
99922=	44008	55101	56050	55027	50035	52020	52080
52101=							
44992	62305	63246	65266	62305=			
44000	60416	60316	61217	64205	66266	64366	65426
63435	60416=						
44008	68506	65465	61467	58448	59315	59215	63175
66156	70225=						
44016	30367	34407	38465	39556	40647	45665	49685
55697	60636	61506	56498	57345	56226	62095	59065
53107	49105	39166	36167	30197=			
44024	30337	38367	42407	43506	45607	53598	56575
55547	55385	54236	55167	49175	44196	38236	30267=
44032	48516	53405	52276	48235	41315	44385	45456
48516=							
44040	49385	49336	46308	47355	49385=		
44008	30415	34456	36506	35567	30575=		
44000	30485	31478	32505	30497=			
44016	36808	34767	31746	30645=			
44008	42786	46745	43705	48697	50755=		
44012	32157	31105	34056	36106	32157=		
44012	38065	40035	42055	40077	38065=		
44008	70055	68000	63060	64100=	19191=		

Part 5. Ships' Reports

99527	70355	02061	82417	98022	34205	3//08	///// =
99470	70173	02064	73528	97028	23008	30807	///// =
99570	70206	02064	82235	97616	16605	30706	99/06 =
99359	70185	02061	80214	97022	18614	3////	///// =
99380	70710	02063	02910	98138	13412	30/02	99/0/ =
99482	70493	02061	92104	91454	34702	3////	///// =
99459	70482	02064	90918	91455	32501	30602	///// =
99418	70408	02061	50608	96021	22706	3////	///// =
99347	70436	02063	91147	94828	10019	30706	11808=
99339	70562	02061	20110	99020	06615	3////	///// =
99432	70613	02060	92204	98022	22007	30702	///// =
99388	70317	02061	40410	98012	30614	3//04	///// =
99467	70221	02061	70304	98002	26309	30503	00/00 =
99444	70286	02061	90510	97022	36009	3////	///// =
99507	70075	02064	63415	98031	12006	30402	27902=
99575	70122	02064	63310	98152	18104	30101	03004=
99488	70101	02061	43411	98010	15109	30604	///// =
99515	70235	02063	73218	97022	31606	30303	34015=

Part 6. Land Station Reports

03026	70418	68268	14803=	06011	33402	97021	15555=
03075	60213	70151	12402=	04030	91530	40697	99601=
03262	80811	61516	05703=	04390	02823	80010	02000=
03804	53611	72216	11406=	07110	83207	56616	10808=
03953	60115	80258	15505=	07510	70205	40102	13009=
03976	73622	66898	16103=	08001	52504	65030	13609=
01203	80809	15697	09901=	78016	10000	71020	14618=
01228	83216	80588	08302=				

The first three parts of the Atlantic Weather Bulletin are self-explanatory and provide a generalized summary of the present weather situation and the expected developments therefrom. Both present and future weather, however, can be more fully and more precisely understood if the mariner has before him an up-to-date synoptic weather map, the general use of which has been described in Chapter 12. Such a map can be drawn on board ship from the information supplied in parts 4, 5 and 6 of the Bulletin. The detailed technique for doing this will be described shortly, following a description of 'Facsimile' charts which, for a vessel suitably equipped to receive them, obviate the need to record and plot the coded part of the bulletin.

Facsimile charts

The facsimile recorder is a radio-telephotographic device which enables a map produced on shore to be reproduced automatically in a recorder on board ship. The recorder connected to a suitable HF radio receiver, will by the operation of a switch, reproduce, unattended, in about a quarter of an hour on electro-sensitive recording paper an exact copy of the weather map as drawn in the meteorological centre ashore. The recorder can be quite small and can easily be mounted in the average chart room. The advantages of this system are fairly obvious; corrupt groups and transmission errors are eliminated and the map, as received, is drawn by a professional meteorologist. Tests have shown that when the reception of morse or radio-teleprinter signals are difficult or even impossible, intelligible weather maps can usually be received by facsimile. Reception is independent of 'single operator' periods of watch.

In addition to the analysis, prognostic maps showing what the weather situation is expected to look like 24 hours, 48 hours, and 72 hours ahead (*see* page 188), and ice maps and wave maps can be transmitted by this method.

Facsimile maps, not necessarily prepared specially for shipping, but quite intelligible to the average ship's officer, showing surface conditions (an analysis and prognosis) are regularly transmitted by radio by various meteorological services in the northern and southern hemispheres. No doubt more will be provided as the demand increases. The North Atlantic is particularly well served. Analysis and prognostic weather maps covering the eastern North Atlantic are transmitted several times daily, on a regular schedule from the UK and from European countries. In the western part of the Ocean, similar maps are broadcast by the USA and Canada, and they also issue analyses and prognostic maps of wave conditions covering the whole Atlantic, and ice charts from May to September showing the distribution and category of ice in the Gulf of St Lawrence, Newfoundland, Labrador and Greenland areas.

Similar maps of surface weather in the North Pacific are issued by the USA and Japanese authorities.

The facsimile network in the southern hemisphere is not so extensive, but is increasing.

A specimen of a surface prognosis (FSXX) (Figure 13.2) and of a surface analysis (ASXX) (Figure 13.3) broadcast by facsimile from Bracknell show conditions in the North Atlantic at 0600 GMT, 7 April 1976; these are, of course, very much reduced in size from the original. The prognostic map is of very great value, because it shows exactly, and in many cases better than he can explain in words in a short radio bulletin, what the meteorologist had in mind when he issued his forecast. The intermediate analysis and prognostic maps conveniently

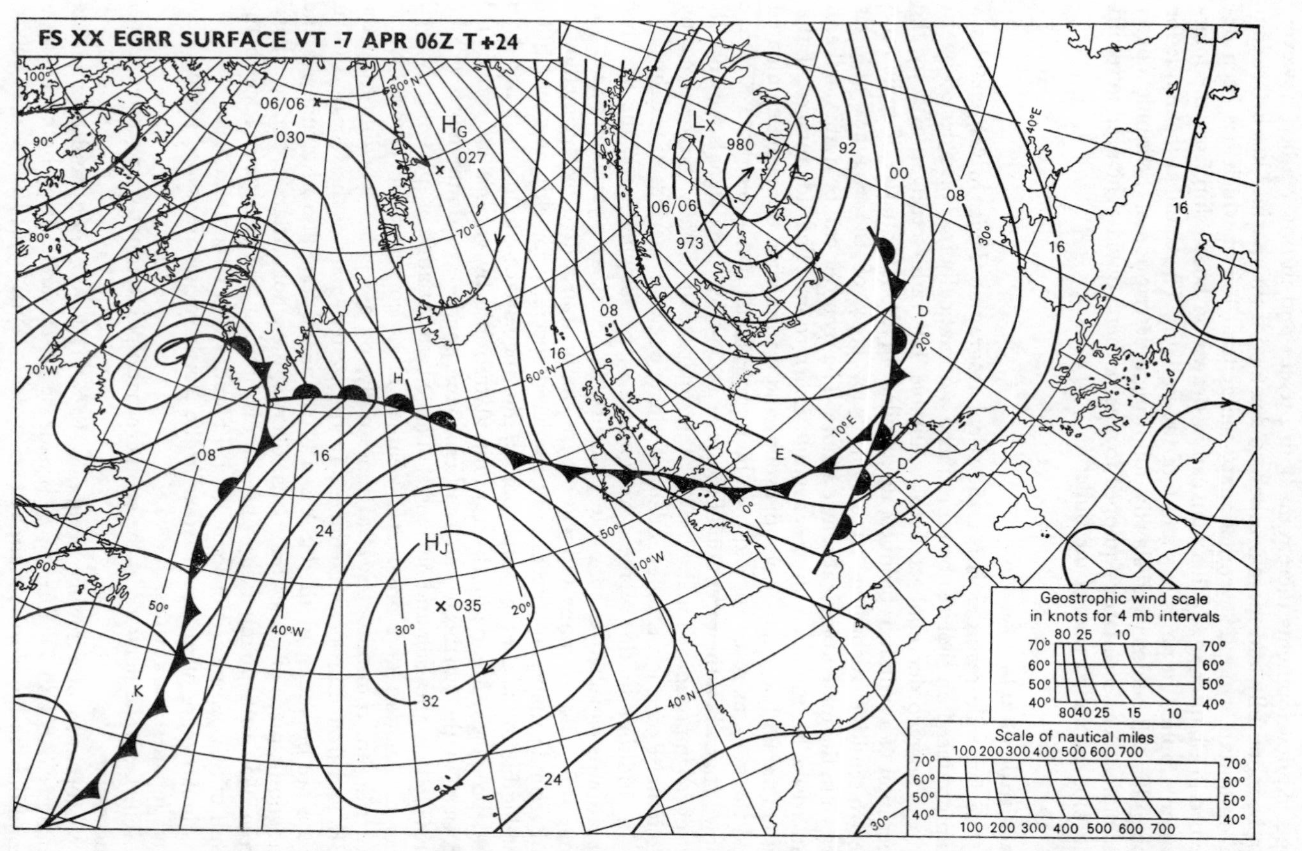

Figure 13.2. Surface prognostic chart FSXX for 0600 GMT, 7 April 1976 (as broadcast by facsimile from Bracknell)

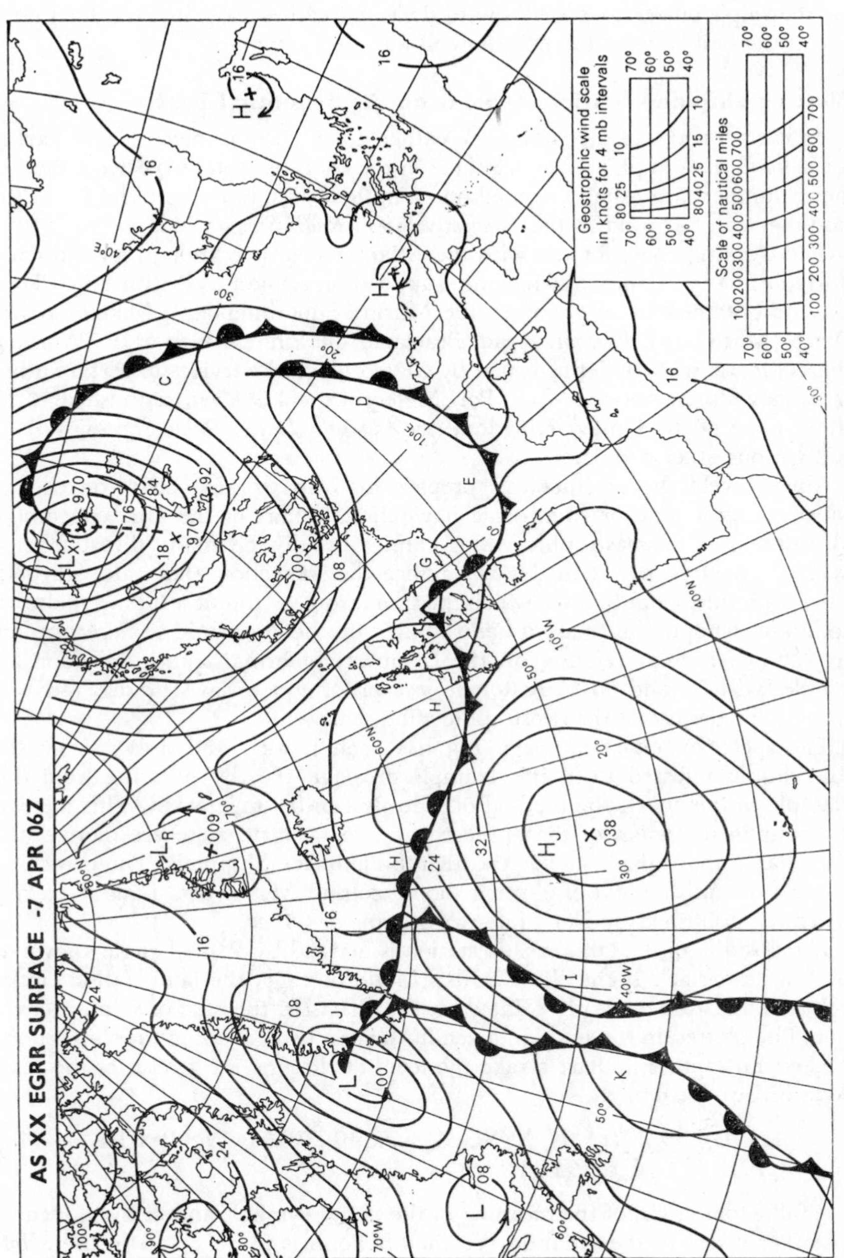

Figure 13.3. Surface analysis chart ASXX for 0600 GMT, 7 April 1976 (as broadcast by facsimile from Bracknell)

fill in the gaps between the times of radio weather bulletins and are thus a ready means of keeping 'up to date' as to the developing weather situation.

Even if facsimile comes into general use aboard ship, there will still be a need for the ship's officer to know how to draw a weather map himself, for that is one of the best ways of learning how to interpret the map.

Plotting Ship and Station Reports on the Synoptic Chart

As Parts 5 and 6 of the Atlantic Weather Bulletin are received before Part 4, it is a good plan to plot these messages on the chart as soon as received. Station index numbers are usually printed on synoptic charts, but where this is not the case they can be found in the *Admiralty List of Radio Signals*, Vol. 4.

The working chart for use with the Atlantic Weather Bulletin for Shipping is Metform 1258, copies of which are supplied on request to British ships. They may be obtained by applying to the Marine Superintendent, Meteorological Office (Met O 1a), Eastern Road, Bracknell, Berkshire RG12 2UR. Working charts for use with Bulletins issued by Meteorological Services in certain other areas may also be obtained from Port Meteorological Officers established at the larger ports of the United Kingdom (*see Admiralty List of Radio Signals*, Vol. 3) and various other countries.

Alternatively the mariner can prepare his own working chart on tracing paper, using a map or small-scale navigational chart of the area concerned. Measurements of geostrophic winds cannot usually be made direct from a working chart prepared in this way, since the projection and scale will both differ, as a rule, from that for which the standard geostrophic scales were drawn up. Blank synoptic maps are usually printed on the gnomonic or stereographic projection for polar regions, on the Lambert conformal conic projection for middle latitudes and on Mercator projection for equatorial latitudes. (*See* also page 31 for the use of the geostrophic wind scale.)

The specimen chart on page 193 has been plotted and drawn from the observations referred to in the example on page 186. By referring from this example to the chart, the method of plotting can be understood. This method is a simplified version of the one used at forecast centres, where more detail is available than given in the Weather Bulletin for Shipping. Those wishing to learn the full method of plotting should consult Met O 877, *Guide to Plotting Procedures*, published by Her Majesty's Stationery Office.

The position of the land or ship station is marked by a small circle known as the 'station circle'. If the chart used is Metform 1258, the land station circles will be found to be already printed on the chart. In the case of ships, a circle should be plotted in the position given in the first two groups of the ship report.

The ship reports in Part 5 take the form of the first six groups of the Ship Code FM 21–V, namely:

$99L_aL_aL_a$ $Q_cL_oL_oL_oL_o$ $YYGGi_w$ Nddff VVwwW PPPTT
$3P_wP_wH_wH_w$ $d_wd_wP_wH_wH_w$.

A full explanation of the meaning of the above symbols and of those used in the other codes in the Atlantic Weather Bulletin is to be found in the *Ships' Code and Decode Book*—Met O 509.

The land station reports in Part 6 of the Bulletin take the form of the first four groups of the SYNOP code FM 11–V, namely:

IIiii Nddff VVwwW PPPTT.

Having plotted the station circle, the various elements of pressure, temperature, etc. are plotted around the circle in the relative positions shown in Figure 13.4. This illustrates the schematic layout in the case of a ship report and gives an example of the plotted version.

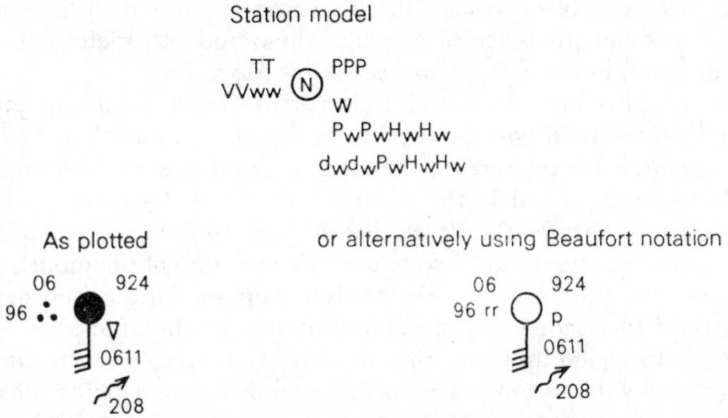

Figure 13.4. Station model for plotting reports from Atlantic Weather Bulletin

PPP = barometric pressure (mb).
TT = air temperature (°c).
VV = visibility.
N = cloud amount.
ww = present weather.
W = past weather.
P_wP_w = period of sea waves.
H_wH_w = height of sea waves.
d_wd_w = direction from which swell waves are coming.
P_w = period of swell waves.
H_wH_w = height of swell waves.

The wind is best plotted first because the arrow may otherwise conflict with figures already plotted. Once this has been done, the other figures can be placed as near as possible to their nominal positions without conflicting with the wind symbol.

The wind direction is shown by an arrow flying with the wind, and the speed by the number of feathers or pennants on the arrow, each full feather representing 10 knots and a half feather representing 5 knots. A pennant represents 50 knots. The wind indicated in Figure 13.4 is thus 180° 40 knots.

Barometric pressure, air temperature, visibility, and period and height of waves and/or swell are plotted as received. Total cloud amount, past weather and present weather can be plotted by using the appropriate symbols as in Figure 12.5 of the previous chapter, or alternatively by using the Beaufort letters which are given in Tables 28 and 29 of the *Ships' Code and Decode Book*. Land station reports are plotted similarly.

Plotting the Analysis

Having plotted the available station reports and ship reports, the next job is to plot the analysis, which will assist in completing the synoptic map. The analysis is based on much more detailed information than is available to mariners, and it gives in effect a summary of the general synoptic situation in the area concerned. Let us take, for example, Part 4 of the Atlantic Weather Bulletin given on page 186. This is coded in the International Analysis Code (I.A.C. (Fleet)) FM 46–IV, and takes the form:

10001 33300 $OYYG_cG_c$
99900 $8P_tP_cPP$ $L_aL_aL_oL_ok$ $(L_aL_aL_oL_ok)$ $md_sd_sf_sf_s$
99911 $66F_tF_iF_c$ $L_aL_aL_oL_ok$ $L_aL_aL_oL_ok$ . . . .
99922 44PPP $L_aL_aL_oL_ok$ $L_aL_aL_oL_ok$ . . . .

Although this is the form used in the Atlantic Weather Bulletin (UK), it is important to note that some meteorological services use slightly different forms of the code, for example, several authorities quote positions in whole degrees only and express them in the form $QL_aL_aL_oL_o$ where Q is the octant of the globe. To indicate this mode of expression the indicator 33388 is used for the second group of the message (in place of 33300). These and other alternative forms are set out in detail in the *Ships' Code and Decode Book*.

The group 10001 indicates 'analysis follows'. In the second group 33300, the figures 00 indicate that all positions are in the $L_aL_aL_oL_ok$ and that the latitudes are for the northern hemisphere. The figure in position k enables the latitude or longitude to be expressed to the nearest half degree by means of the code given in Table 31 of the *Ships' Code and Decode Book*. In the third group, the first figure is an indicator, the second two figures give the day of the month, and the last two figures, the time to which the analysis applies. Thus 00206 shows that the analysis is for the second day of the month and for the time 0600 GMT.

The next group 99900 indicates that pressure systems follow. In this section a group beginning with an indicator 8 is followed by a group giving the position of the pressure system specified in the 8-group and sometimes also by a group specifying the movement of the pressure system. Taking the first three groups of this section—81298 31495 00920—the first figure 8 shows that these groups give details of a pressure system; the next figure 1 indicates a 'low' (Table 32 of *Ships' Code and Decode Book*) and the next figure 2 shows 'little change' as regards deepening or filling of the pressure centre (also Table 32). The last two figures are the tens and units figures of the central pressure of the system and indicate a value of 998 mb. The next group gives the position of the centre as 31°N 49°W (using Table 31 of the same book). In the next group the last four figures show the centre to be moving towards the east (090°) at a speed of 20 knots. The first figure of the group, being 0, gives no further information.

After further groups specifying the position and movement of other pressure systems, the indicator group 99911 shows that the next groups describe fronts. The figures 66 indicate a front. The next figure 4 denotes a cold front, 5 indicates that the intensity of the front is moderate, and the last figure, 7 shows that the front is liable to 'waves'. These three figures are covered by Table 34 of the *Ships' Code and Decode Book*. The following groups show the position of the front, namely 30°N 75½°W, 36½°N 70°W, etc. After other fronts and their positions have been specified, the group 99922 indicates the commencement of the section describing the isobars. This starts with a group 44008 which indicates that the following positions define the 1008-mb isobar. These positions are 55½°N 10°E, 56°N 5°E, 55°N 2½°W, etc. The isobar is drawn as a smooth curve through the specified points. After describing a number of isobars the message ends with the indicator group 19191.

Sometimes other information may be given between the isobars section and the 'message ends' group. A group 99944 indicates that significant weather data follows; 99955 indicates 'tropical system follows'; 88800 indicates 'wave and/or sea temperature isopleths follow'. The groups 77744 and 44777 indicate the beginning and end of a plain language statement.

Notes on the Completion of the Chart

When the analysis message (Part 4) has been plotted (*see* example in Figure 13.5) the result should be studied in conjunction with the observations from

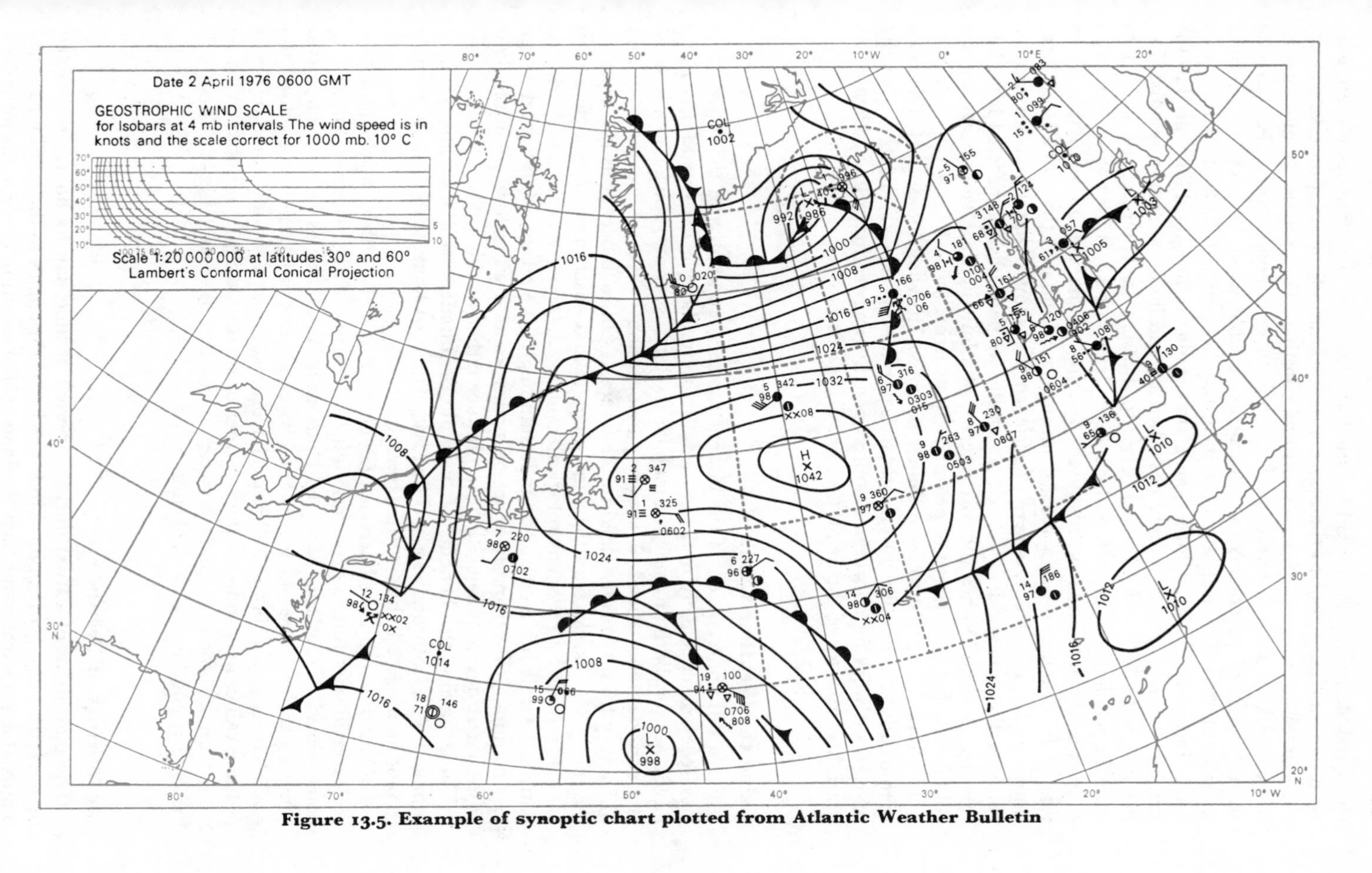

Figure 13.5. Example of synoptic chart plotted from Atlantic Weather Bulletin

ships and shore stations already plotted on the chart. In some cases there may be discrepancies, owing to the approximate nature of information given in the message or to errors of transmission and plotting. The isobars should then be drawn with as smooth curves and as uniform gradients as possible, and also should be made to conform as closely as possible with the pressure readings already plotted. The isobars as received may be specified for 8-mb rather than 4-mb intervals. This is done where the gradients are tight enough for the pressure field to be adequately described by specifying alternate isobars. The mariner must be careful in such cases to draw in the intermediate isobars so that the whole chart is covered by isobars drawn at 4-mb intervals. Otherwise estimates of wind force based on the spacing of the isobars are liable to be wrong. In drawing these isobars, it must be remembered that they often form sharp angles at a front, where experience has shown that the wind usually changes direction suddenly. The apex of this angle is always directed from low pressure to high. Within the warm sector of a low, i.e. between the warm and cold fronts, the isobars are approximately straight.

It will be found best to use ordinary pencil for drawing the isobars. They can then be adjusted readily until a satisfactory result is achieved. Fronts are usually marked on working charts in coloured pencil using ordinary lines. Details are given in Table 27 of the *Ships' Code and Decode Book*.

Using the Weather Chart

For a full appreciation of the present weather situation and the movements and developments likely in the immediate future, the mariner should preferably have available both an analysed map for the latest synoptic hour, and a forecast chart for 24 hours ahead. There may be occasions when the weather develops differently from that envisaged in the forecast. On such occasions the 'man on the spot', who experiences the wind changes and onset of weather before they can be reported to a meteorological centre, can more readily assess what is happening if he has a complete understanding of the recent analysis and future expectations. It is obviously easier to modify an analysis or forecast in the light of the latest developments, than to produce either unaided.

Those having the benefits of 'facsimile' equipment will normally be kept supplied with both analysis and forecast. Those without will normally be able to plot an analysed chart as just described.

Having drawn up the latest 'actual' chart, it may help the mariner to obtain a clearer idea of future expectations if he draws up a forecast chart for, say, 24 hours ahead of the latest analysed chart using the latter as a basis. The main centres can be moved in accordance with the forecast positions given in the plain language part of the bulletin. The forecast pressure values should be noted as these values may indicate that a larger or smaller number of isobars must be drawn on the forecast chart between particular lows and highs. These will indicate strengthening or slackening winds.

Movements of Fronts

The movement of a front during a period of time, say six hours, is obtained by measuring the geostrophic wind speed at right-angles to the front for several points along the front and then marking off the resulting movements from each point in a direction perpendicular to the front. The following rules are also useful:

(a) The speed of a warm front may usually be taken as about two-thirds of the geostrophic wind speed.

(b) The speed of a cold front is usually the same as, or slightly greater than, the geostrophic wind speed.

(c) The speed of an occlusion is best assumed to be equal to the geostrophic wind speed.

Extrapolation is also useful in estimating the movement of fronts. For example, if in Figure 13.6, **AA′**, **BB′**, **CC′**, denote successive positions of the fronts at six-hour intervals, then its position after a further six hours might with some confidence be expected to be **DD′**.

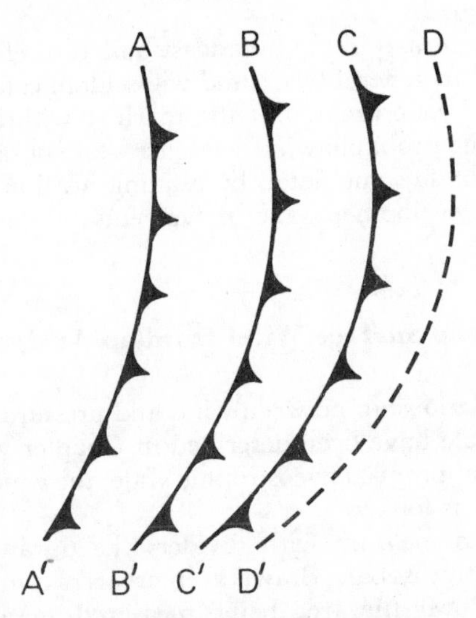

Figure 13.6. Method of extrapolation for estimating future position of front

Personal Observations and Local Forecasting

The mariner who makes regular weather observations may well be in a position to amend the analysis in his own area when it is found to be in error.

It takes a lot of time to plot a full synoptic chart and to prepare an analysis and forecast, and by the time the radio weather bulletin is received aboard the ship the data upon which the bulletin are based may be about six hours old.

For example, the analysis of the 0600 GMT chart (Part 4 of the Atlantic Weather Bulletin) is broadcast to shipping at 1130. By the time it is received aboard the ship part of the analysis may prove to have been incorrect for some reason, such as a front moving faster than was anticipated, or a ridge of high pressure building up unexpectedly over the area, though caution should always be observed before deciding that an analysis needs amendment. Whether afloat or on shore, anyone who tries to relate his own observations to each synoptic chart as it is drawn will soon find himself deriving an increasing amount of information from the chart. His own observations may enable him to correct the analysis, and in consequence, the forecast. For example, the combined analysis and forecast might imply that the present conditions in the vicinity of the vessel should be cloudy with a fresh westerly wind, and a cold front approaching from the west. If the vessel has recently experienced a period of rain, and the wind has veered north-westerly with breaking skies, the mariner can reasonably assume that the front has already passed and that he can expect post-frontal conditions to prevail for a while. It is worth noting that even though the forecast was in error in respect of timing, the mariner who can correct its errors is in a better position than one who disregards it. The forecast might indicate that, once the front has passed, no other marked features are to be expected for twelve hours or more. Alternatively it might indicate a secondary cold front, with squalls, following within about six hours of the first front. In this case, in spite of the timing of the first front having been in error, the mariner could apply his knowledge of the correct timing of the first front to make a good estimate of the timing of the second.

Since the official forecasts in the broadcast bulletins refer to quite large areas they must be framed in general terms and will seldom contain detail applicable to any small part of these areas. But the mariner with a completed synoptic chart can estimate the probable wind, weather, state of sky and other elements for the next few hours in some detail by relating his ship's position and future movements to the development and movements of the fronts and pressure systems on the chart.

Determination of the Surface Wind from an Analysis or Forecast Chart

The theoretical relationship between wind and pressure gradient and the use of the geostrophic scale have been described in Chapter 3—Wind and Waves. In the absence of an appropriate geostrophic scale, the wind may be determined from a weather map as follows.

The first step is to measure with dividers the distance in nautical miles between two consecutive isobars, drawn at intervals of 4 mb. If, and only if, the gradient is constant over the area being measured, greater accuracy can be achieved by taking a third of the distance between three consecutive isobars. The distance found is now used in conjunction with the table of geostrophic wind speeds on page 197. If the distance is, say, 100 nautical miles and the latitude is 50°, then the geostrophic wind speed is found to be 30 knots. It is important to remember that this is the speed at about 600 m—further steps have to be taken to arrive at the surface wind speed. The speeds shown in the table are correct only when the isobars are straight or very slightly curved. If there is appreciable curvature, the geostrophic wind already found requires to have a correction applied to it—the corrected speeds are shown in the table on page 34. In the present example, with a 30-knot geostrophic wind, if the radius of curvature of the isobars is 300 nautical miles and is cyclonic, then the speed of the wind at 600 m will be reduced to 25 knots. If the curvature is anticyclonic, the speed would be increased to 47 knots. The speed of the surface wind will for practical purposes be $\frac{2}{3}$ of the speed found for 600 m.

Investigations have shown, however, that when the air is very unstable, the surface wind speed may be as high as $\frac{4}{5}$ of the speed at 600 m. Unstable air conditions may be recognized by the presence of large cumulus or cumulonimbus clouds; these commonly develop in the airstream in the rear of active cold fronts.

The procedure described above may be employed to find the surface wind speed from the charts of other nations provided that the isobars are drawn for intervals of 4 mb.

Table 13.1. Geostrophic wind speeds for 4-mb isobaric intervals

Latitude	Distance (*nautical miles*)												
	20	25	30	35	40	45	50	60	80	100	150	200	300
	Geostrophic Wind Speed (*knots*)												
10°										133	89	67	44
20°							136	113	85	68	45	34	23
30°				131	115	102	92	77	57	46	31	23	15
40°			119	102	89	80	72	60	45	36	24	18	12
50°		121	101	86	75	67	60	50	38	30	21	15	10
60°	132	105	88	75	66	59	53	44	33	26	18	13	9
70°	122	98	81	70	61	54	49	41	31	24	16	12	8
80°	118	93	79	67	59	52	47	39	29	24	16	12	8
90°	115	92	77	66	57	51	46	38	29	23	15	11	8

Steep pressure gradients are not experienced in equatorial regions except in tropical storms and accordingly the strong winds shown in the table as being theoretically appropriate to low latitudes are rarely encountered in practice. Geostrophic winds in excess of 130 knots seldom if ever occur.

Forecasting of Sea Temperature

Sea surface temperature depends upon several meteorological factors which are to some extent interrelated, and at present there is no method by which it can be accurately forecast at a particular point. A knowledge of the sea surface temperature is important for the mariner wanting to forecast the development (or dispersal) of fog in his locality.

Away from regions where warm and cold currents are found closely adjacent, sea temperature is usually very conservative and seldom changes as much as 0·5°C in a day as a result of meteorological causes alone. A good estimate of sea temperature for a short distance ahead of a ship can often be made from an observation of sea temperature, combined with a knowledge of the horizontal gradient of sea temperature in the area, which can be obtained from the appropriate 'routeing charts' published by the Hydrographer of the Navy. However, in localities such as the vicinity of the Grand Banks of Newfoundland and off the coasts of Japan, where the horizontal gradient of sea temperature is on the average very steep, an estimate made in this way would have a limited value, since in such an area (as within the Gulf Stream) a ship is always liable to encounter sudden changes of sea temperature wherever detached portions of the warm Gulf Stream and cold Labrador Current happen to be brought close together. Such changes are, of course, largely unpredictable and are not shown in the atlases. The mariner will be aware that these regions also coincide with the areas of maximum fog frequency.

Forecasting of Sea Fog

The forecasting of the time of onset of sea fog is rather more difficult than forecasting radiation fog over land because, with the latter problem, more accurate estimates can usually be made of the quantities involved, such as the rate of fall of air temperature. Also the diurnal range of sea temperature is very small and the temperature of the air overlying the sea only changes markedly if it is moving from a warmer to a colder sea region or vice versa. Formation of sea fog depends on small changes in air temperature, upon the magnitude and sign of the temperature difference between sea and air, as well as upon wind speed. It may be slow in forming and in clearing.

Sea fog may be expected when the dew-point of the air is higher than the sea surface temperature. From a study of the weather chart, its formation can be anticipated in air which is moving towards a region where the sea temperature is lower than the dew-point of that air; the reason being that on coming into contact with the cold water, that air will have its temperature reduced eventually to its dew-point.

The force of the wind is important for forecasting sea fog. If the wind is strong, usually the fog will be lifted by turbulence into low stratus cloud. Wind of force 3 appears to be most favourable for the occurrence of fog in the Newfoundland region and this probably applies in other regions also. Above force 4 the chance of fog decreases markedly. There is a smaller proportion of fogs with calms and a few cases with forces 6 or 7. On the Newfoundland Banks, however, with a southerly wind, fog can occur with winds of almost any force.

The behaviour of the smoke from the funnel is often a valuable indication. If it hangs about in horizontal streaks, it shows that an inversion of temperature, caused by considerable cooling of the lower layers of air, is already present. This condition may lead to shallow fog formation.

The chance of any marked change of wind direction should be estimated from barometric readings and tendencies, and general weather signs and conditions. If the wind is likely to change so that warm and moist air is replaced by cooler and drier air, the chance of fog will be reduced or eliminated altogether if the change is rapid. In general this means that when the wind starts to blow from a more northerly direction in the northern hemisphere, the chance of fog forming is reduced, or that fog already formed will be quickly dispersed. The effect is particularly noticeable in the vicinity of the Grand Banks of Newfoundland.

The above remarks apply to the normal sea fog of the oceans. In the well-known foggiest regions of the world, fog is frequent throughout the year, in most cases because the sea temperature nearly always is below the air temperature and dew-point. A knowledge of this fact will help the seaman to anticipate fog, even if he has no special experience of the region. In principle, a change to a wind direction which brings air from a warmer oceanic region, at any time of the year, will be favourable to fog formation.

Practical Value of Weather Information to the Mariner

In the open ocean the practical use that the mariner can make of a radio weather bulletin, whether it be associated with an analysis and a plotted synoptic chart or not, depends on circumstances. There is little doubt that with the bulletin and the synoptic chart before him, he should have a reasonably good picture of the existing and impending weather for a period of about 12 hours ahead. For a relatively high-powered ship there would thus be quite a few occasions when a timely alteration of course or speed could be made to avoid bad weather, or to find favourable instead of unfavourable winds. The same may apply, in somewhat fewer cases, to a low-powered vessel. A bold alteration of course in an unfavourable situation may certainly increase the mileage steamed, but by gaining favourable weather fuel may be saved and there may be less risk of damage to the ship and cargo, and less wear and tear on crew and passengers. Figure 13.7 shows one practical instance where such meteorological navigation proved of definite value to a ship. When the arrival of dirty weather is confidently anticipated, seamanlike precautions can be taken to see that everything is secure aboard the ship in advance of its arrival. (*See* also section on 'Ship Routeing' on this page.)

The practical value of gale warnings and forecasts to all shipping when operating in coastal waters is obvious. Even if there is no bulletin and no official forecast available, the mariner can often do much in the way of anticipating bad weather by a judicious study of his meteorological instruments, and by personal observation of the existing and past wind and weather, in association with 'weather sense'. It is nearly always possible to obtain, by request, radio weather messages from other ships in the vicinity to supplement the information available to a solitary observer aboard a ship.

Ship Routeing

The routeing of ships by and from shore establishments began in the USA in the 1950s and the practice was adopted by various European Government Meteorological Services during the 1960s. Hitherto the practice had been to make use of well-known and well-established principles to select routes when crossing the oceans. Lt Maury of the US Navy painstakingly collected observations from ships' deck logbooks and used these in 1852 to recommend routes for vessels crossing the Atlantic. The main purpose of these routes was to avoid collisions but seasonal weather conditions were also taken into account. His work is perpetuated today in the well-known Pilot Charts issued by the US Hydrographic Office each month, which, as well as showing surface currents and monthly wind distribution, also recommend routes between the principal ports. The British Admiralty also publish a series of charts recommending routes based on climatological patterns, and their publication *Ocean Passages* recommends routes in great detail.

The feasibility of routeing vessels to avoid the worst weather conditions became possible when technological advances in the post-war era introduced high-speed computers into the collection and transmission of meteorological data and into the prediction of the development and movement of weather systems. Although the term 'Weather routeing' is sometimes used a more correct phrase is perhaps 'Ship routeing' since, although weather is the main factor, the principles of navigation, seamanship, naval architecture, oceanography and marine engineering must also be considered before a route can be selected.

The master of a ship is the best person to choose the route that his ship will follow if weather conditions have no influence on the situation, but for many centuries ships moved under the forces of the environment alone and today the shipmaster is still obliged to take careful note of weather likely to be experienced on passage, particularly if he has to navigate with operational economy in mind. Climatological data, radio weather bulletins and facsimile charts are indeed useful but he does not have at his immediate disposal the vast resources of a modern meteorological centre such as that at Bracknell, England. Here the task of selecting routes is entrusted to a team of master mariners with long sea-going experience, who devote their whole time to selecting the most advantageous routes for ships, which use the service, to follow. They are provided with

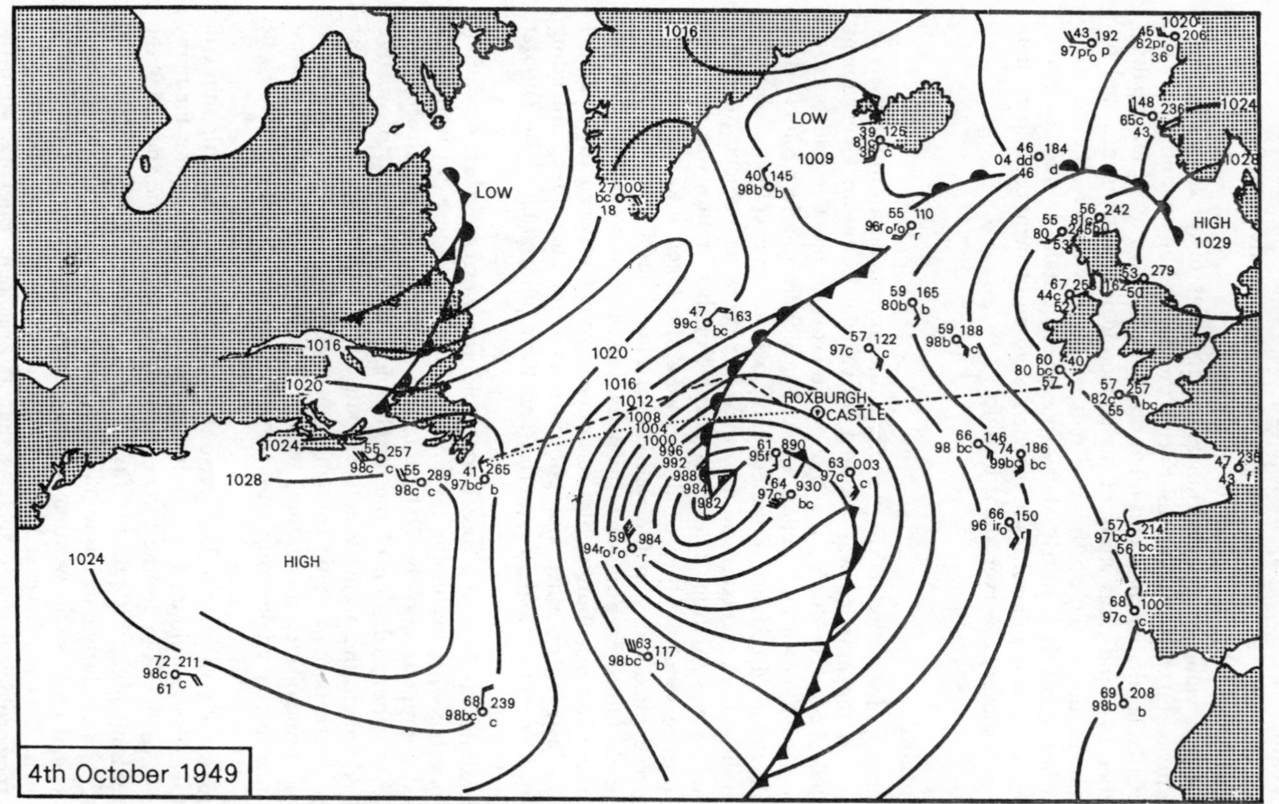

Figure 13.7. An example of meteorological navigation

The *Roxburgh Castle* (Captain J. D. B. Fisher) was *en route* from Liverpool to Newfoundland. The normal great circle track was followed as far out as 30°W. On receiving the analysis broadcast in the Atlantic Weather Bulletin, it was seen that there was a small but deep depression moving in a north-easterly direction, so that if the ship had continued on its course the centre of the depression would have passed just ahead of the ship, giving storm force NW winds. The master therefore changed course to a north-westerly direction until about 150 n. mile north of the normal route, then followed a great circle track to Cape Race; this secured the advantage of the E'ly and NE winds on the northern side of the depression.

a continuous flow of analysis and forecast charts, ice information, warnings, bulletins and satellite pictures and are briefed on the meteorological situation by experienced forecasters.

The Method

The ship's response to various wave fields is determined by extracting sufficient data from the deck logbooks to construct performance curves (Figure 13.8). The wind and wave fields must be predicted for as far ahead as possible—at Bracknell at present this is about 72 hours. This is done by feeding into the Meteorological Office main computer (one of the fastest in the world, capable of processing eight million instructions per second) a vast mass of meteorological data obtained from land stations, merchant ships and balloon-borne radiosondes.

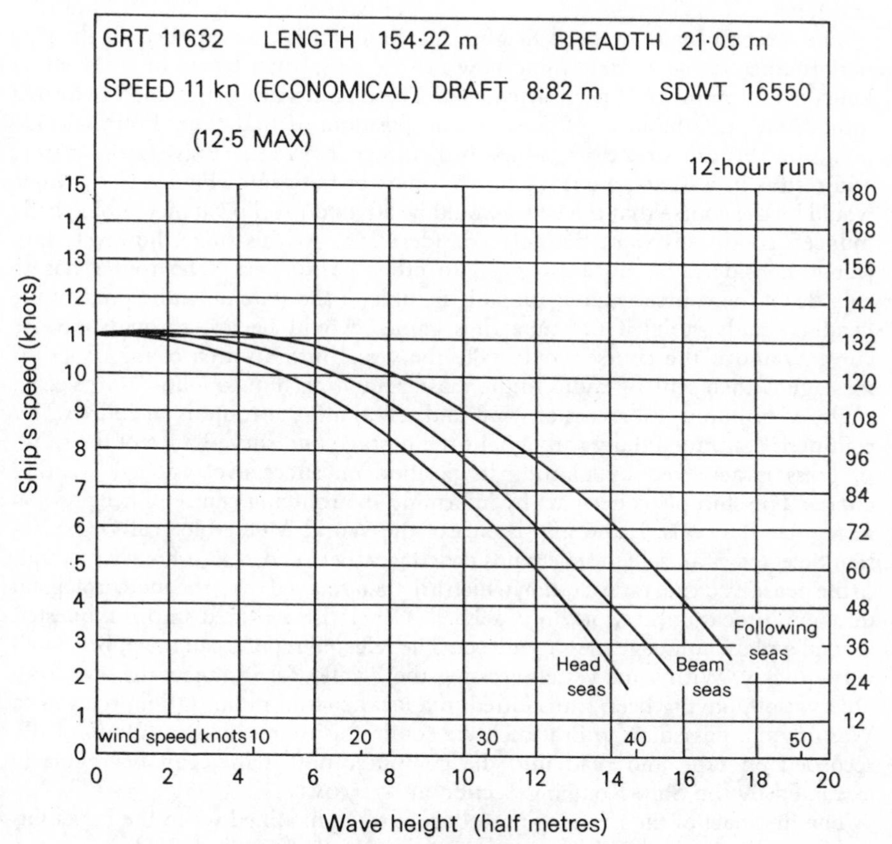

Figure 13.8. Ship's performance curves

All the data are subjected to rigorous quality control by the computer, and are then used to produce a computer analysis of fields of temperature, humidity, winds and contour heights of pressure levels over the whole of the northern hemisphere down to about 15°N, for each of 10 separate levels in the atmosphere. A forecast for the next 72 hours is then computed using the Meteorological Office 10-level forecast model.

A relationship between surface wind speed and wave height was derived by Pierson and Moskowitz to give the formula

$$H = 0{\cdot}0214\ V_o^2$$

where H = Wave height in metres
V_o = Surface wind in metres/second.

Strictly surface wind speeds are not produced by the Bracknell forecast model so the empirical relationship of Findlater and others between the surface and 900-mb wind are used. The surface wind direction is also deduced from the 900-mb wind using these relationships. The computer utilizes this formula to convert the wind field into a wave field and produces a series of charts depicting conditions up to 48 hours ahead. Subjective modifications are made for actual reports of sea and swell, taking due account of the likely variations due to time and fetch.

The forecast wave or swell height and direction is then applied to the ship performance curve to determine how far the vessel will travel in the next 12 hours over a number of possible courses. These points are then joined to form a 'time-front', i.e. the locus of possible ship positions at that time. From selected points on this time-front the process is then repeated in successive 12-hour steps and results in a least-time track for that part of the route (Figure 13.9) which would be the course that the vessel would be advised to follow if meteorologically induced conditions were the only consideration. At this stage, however, subjective consideration must be given to other parameters. The router has to consider if the course is navigationally feasible, if the state of loading makes the heading inadvisable, if the time thus gained would be lost through adverse currents and if the course would take the vessel into an area of fog or ice. A message is then sent by radio to the master advising him to follow the selected route. If required a forecast of wind and sea along the route is included. This routine is continued daily throughout the passage and surveillance of the ship's progress is achieved by plotting its position on successive 6-hourly weather charts. The ship plays her part by informing the routeing centre of her position at regular intervals. If the ship is part of the World Meteorological Organization Selected Ship scheme she sends coded weather reports at 6-hourly intervals to the nearest coastal radio station which are then relayed over the meteorological telecommunication network to Bracknell. Otherwise a routed ship is requested to send a plain language message direct. The selected reports can reach Bracknell very quickly. With some vessels crossing the Pacific, for instance, the 0600 GMT observation, having been transmitted to a local coastal radio station, relayed to Washington, passed on to Bracknell via communication satellite, automatically recorded on tape and read into the computer data bank, can be regularly extracted by the Ship Routeing Section at 0715 GMT.

The first part of the routeing procedure is often modified when the vessel has to pass through restricted waters such as the North Sea, the Caribbean or the St Lawrence. In these cases the initial advice is whether the ship proceeds via the Pentlands or the English Channel, the Mona or Sombrero passages, or via Belle Isle or Cape Race. In some cases this part of the routeing is critical. In one particular case a bulk carrier was due to leave Middlesbrough on the north-east coast of England bound for St John, New Brunswick, Canada. Being loaded with steel products consisting mostly of steel plates, she had a very low centre of gravity and was therefore prone to heavy rolling. The master was concerned that she would avoid heavy weather, particularly on the beam. The wave

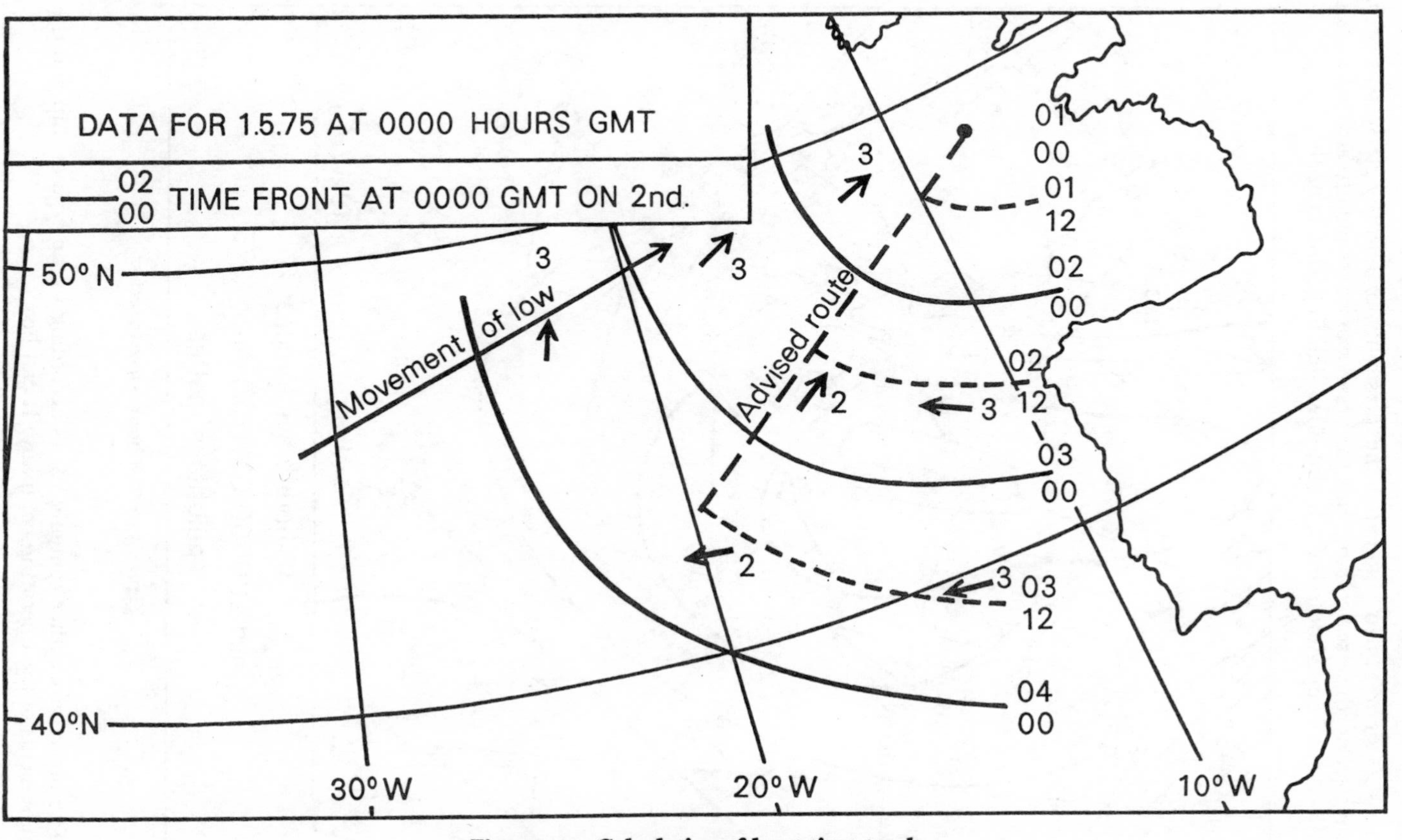

Figure 13.9. Calculation of least-time track

forecasts (Figure 13.10) indicated the possibility of north-westerly seas of 7 metres off north-west Scotland and the vessel was consequently advised to sail south-about from Middlesbrough and down the English Channel.

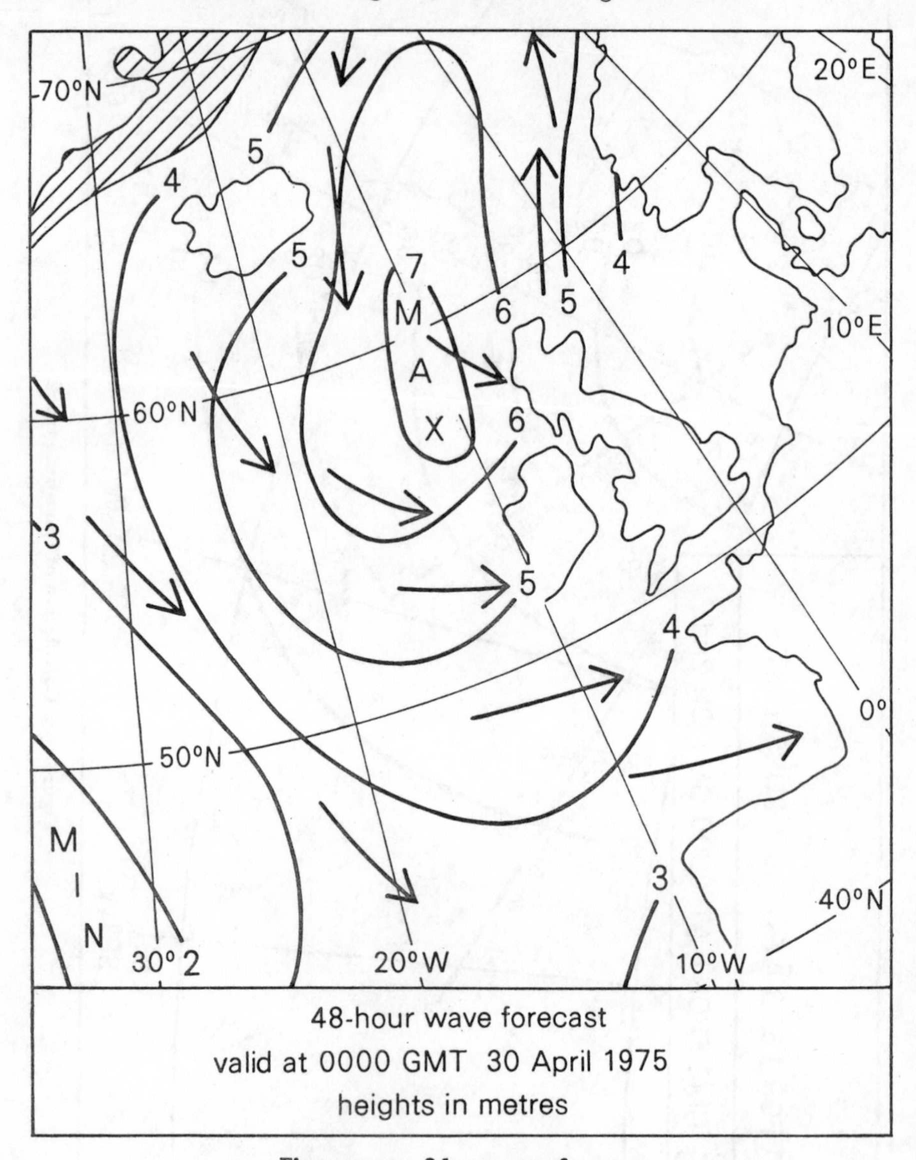

Figure 13.10. 48-hour wave forecast

The object of weather routeing varies according to the type of ship and the requirements of the operating company. It can be

(*a*) Least time. When ship routeing began the main objective was to reduce time on passage regardless of other considerations, but least time routeings are now mostly confined to oil tankers, which do not suffer cargo damage and are less susceptible to hull damage than other ships. On the basis of time saved one tanker company has calculated that during one winter of ship routeing in the North Pacific an average saving of 36 hours and 100 tonnes of bunkers per vessel per crossing was achieved.

(*b*) Least time with least damage to hull and cargo. A small fleet of ships carrying paper products from Newfoundland to the UK and liable to hull damage from pounding, particularly during the UK to Canada 'no cargo passage', used the Routeing Service and found that damage bills of £30 000 per ship per year fell to negligible amounts. The routeing charge per ship per year was approximately £300. This type of routeing is undertaken at Bracknell more frequently than any other.

(*c*) Least damage. This is requested when the vessel is carrying a particularly sensitive cargo such as livestock on deck, uncrated cars, etc.

(*d*) Constant speed. Some ship charters stipulate the maintenance of a certain speed over a certain time, with a financial penalty for failure. Routeing advice is adjusted to achieve this.

(*e*) Fuel saving. In recent years, with increased oil costs, the most significant advantage of ship routeing has been in fuel saving, although this has always been a direct spin-off from least-time/least-damage routes. The present charge for this routeing is equivalent to 2 tonnes of bunkers for an Atlantic crossing and 3 tonnes on the Pacific so that significant savings can be made on fuel bills, even in those cases where service speed has been reduced as part of the fuel economy program. In a recent pamphlet *Marine Fuels Energy Conservation Program for Steam Turbine Ships* the Chevron Oil Company published the results of an Energy Conservation Program initiated in 1973. Two of their findings significantly favoured the use of ship routeing. They found that there was a marked reduction in fuel consumption when a constant power was maintained instead of frequent throttle steam value adjustments. The figure quoted was 8% reduction in fuel consumption. Also the diagram of voyage speed economics (Figure 13.11) graphically illustrates profit reduction in bad weather. Although these principles applied to a tanker fleet, they are equally applicable to other types of vessel. An example of ship routeing is shown in Figure 13.12.

The Routeing Service at Bracknell has been adapted for advice on the movement of tows, both in and out of harbour and on passage. Sometimes the advice is limited to the right time to leave harbour only. More frequently, the progress of the tow is monitored in the same way as conventional vessels and the tow is advised when and where to shelter should the need arise. In these cases the router utilizes his knowledge of pilotage problems in restricted waters in conjunction with up-to-date meteorological information. Tows may be of the usual tug/barge type or may be oil rigs, concrete platforms, floating cranes, floating docks etc.

One of the most vital aspects of ship routeing by a meteorological service is the liaison between the particular router and the master concerned. Wherever possible the opportunity is taken to visit the ship or tow before departure and

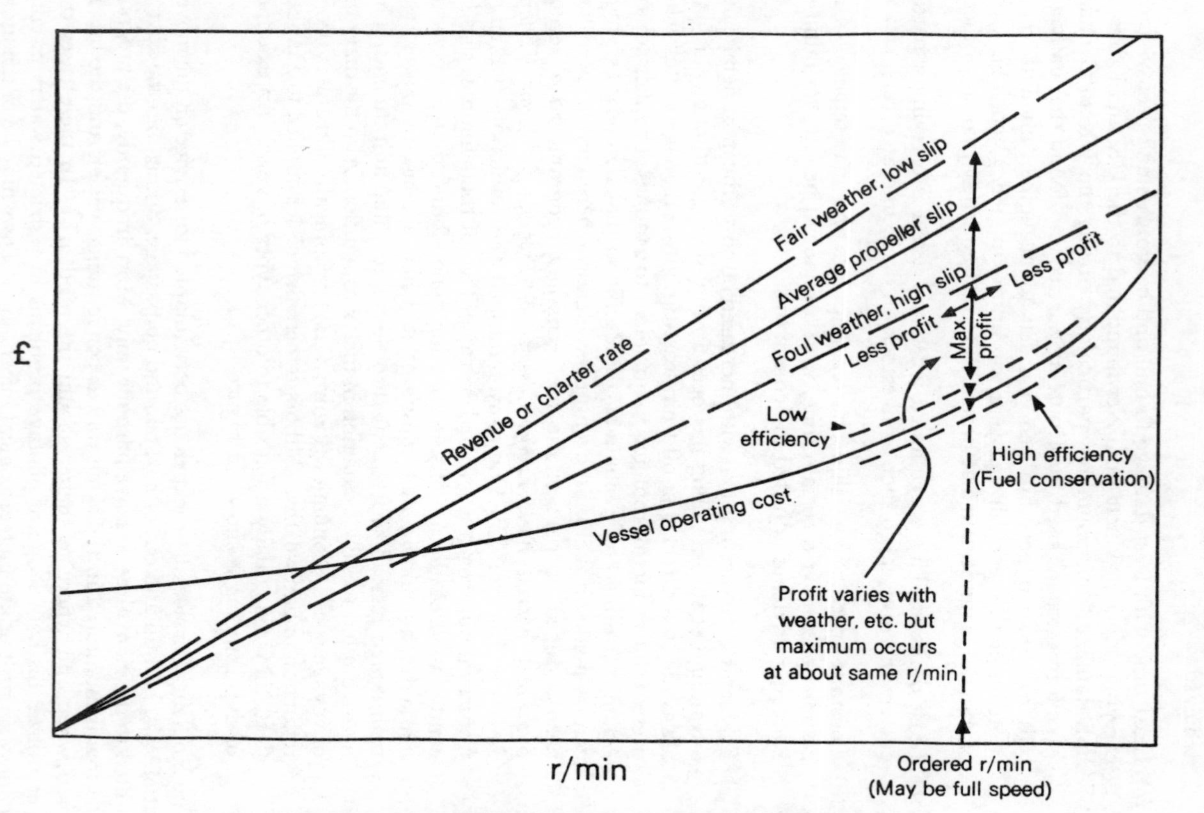

Figure 13.11. Curves illustrating voyage speed economics

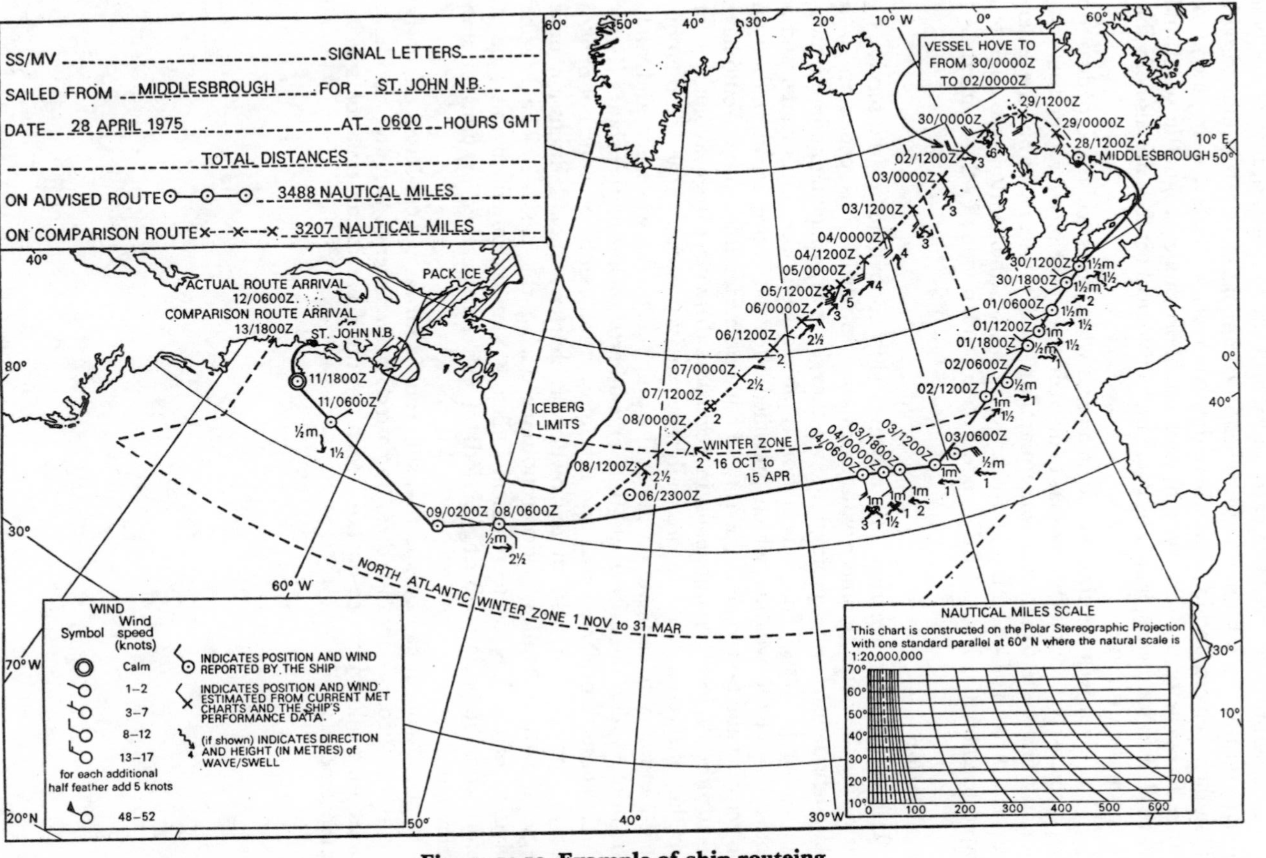

Figure 13.12. Example of ship routeing

discuss any cargo characteristics, handling problems, or preferences. Opportunity is taken to answer any question that the master may ask about the service and agree the radio stations for passing advisory messages.

Many shipmasters prefer to carry out their own routeing, and they are best able to do this, on a limited basis, when in command of vessels equipped with facsimile receivers, and in areas where they can receive charts showing medium-range forecasts of pressure systems and the short-range prognostic wind and wave charts: ice charts, issued by some meteorological centres, are also of value to the on-board router.

The 24-hour weather forecasts for sea areas which are regularly broadcast do not of themselves enable shipmasters to do their own routeing over long distances, nor is it advisable to attempt route planning on data gleaned from the International Analysis Code FLEET—changes in the weather pattern are often too rapid to allow the IAC FLEET code to be of much value in this context.

Evaluation

It has been stated earlier that a routed ship contributes to the service by providing the routeing officer with information, either in ships' code form or in plain language as to her position, the weather she is experiencing and an indication of her progress. This information is plotted along the ship's route, and a knowledge of the weather which prevailed over the entire ocean enables an objective evaluation to be made as to whether other, feasible, routes would have been more or less advantageous in regard to heavy weather avoidance, or whether the vessel could theoretically have arrived earlier at her destination. It is customary to indicate on the evaluation charts, the wave, sea and swell, conditions which the ship would have encountered on alternative routes (Great Circle, master's preferred etc.) and this information can be of value not only to the shipmaster, but also to the owner or charterer should discussion arise regarding the overall weather conditions during the passage. The Marine Division of the UK Meteorological Office is often consulted on this type of information, especially during periods of extreme weather or unusual ship behaviour.

In conclusion, it can be said that ship routeing from a shore-based establishment is a success—because the shipmaster, the weather-router and the main meteorological centre co-operate, each in their particular area of special competence.

PART V. OCEAN SURFACE CURRENTS

CHAPTER 14

THE OBSERVATION AND CHARTING OF SURFACE CURRENTS

Historical

Parts of the general surface circulation were known to the early navigators. Vasco da Gama, who called at Moçambique in 1498 after rounding the Cape of Good Hope, must have experienced the adverse force of the Agulhas and Moçambique Currents and so might be regarded as their discoverer. Columbus, in his voyage to America, encountered the North and South Equatorial Currents of the Atlantic and stated that he regarded the water movement from east to west as proved. *Dampier's Voyages*, published in 1729, includes an account of the currents experienced in his voyages in the latter part of the seventeenth century. He knew of the Guinea Current and gave relative strengths of the Equatorial Current in different parts of the Caribbean. Benjamin Franklin published the first chart of the Gulf Stream in 1770.

By the early nineteenth century seamen had a general working knowledge of the chief trends of current in areas frequented by ships. The first extensive current charts based on ships' observations were those of Rennell, published in 1832, for the Atlantic and part of the Indian Ocean. In 1845, and succeeding years, Maury published his well-known wind and current charts. These early charts and the Admiralty current charts published near the end of the century were based on plotting individual current observations from the logbooks of naval and merchant ships, the average current being shown by arrows drawn by eye estimation. It was not till after 1910 that exact statistical combination of current observations into current roses and vector mean currents was begun.

Since then many current atlases have been published and their coverage has extended to all parts of the globe. Even now, however, there are some sea areas where the observational data are insufficient for a detailed specification of the currents. With the passage of time navigational aids have improved the precision with which a vessel can determine its position. Also the vessel's speed relative to the water can now be measured more accurately than in the past. Current measurements can therefore be expected to be improving in quality and this should lead to a progressive improvement in the accuracy of current information on charts and in atlases.

Observation of Surface Current

When current charts are constructed, as many observations as possible are required during many years. The only way of obtaining enough observations is by the co-operation of merchant shipping. The method used in making the observations is to calculate the difference between the 'dead-reckoning' position of the vessel, after making due allowance for leeway, and a reliable fix (or true position). The result is the set and drift experienced by the vessel during the interval since the previous fix.

The current found by this method is that for a mean depth of about half the ship's draught. It will be correct only if the ship's true course and speed through the water are known. Knowledge of the true course involves a precise knowledge of the error of the compass, sound judgement of the leeway made and the leeway allowed for when setting the course, and good steering. A reasonably accurate estimate of leeway can only be made by experience, bearing in mind that the greatest effect will be with the wind direction abeam and the ship in a light seagoing condition. For the speed through the water a compromise between log distance and distance by engine revolutions, after making due allowance for slip, gives perhaps the best results. Much will depend on local circumstances, e.g. in rough weather with the propeller not always properly immersed it will not be possible to assess the slip with any degree of accuracy; the same is true to a limited extent when the ship is in a light condition or when there is considerable marine growth on the hull. It should be possible from time to time to test the accuracy and compute the percentage error of one or more logs by comparison with each other and with runs over a known distance in still water. In short, the dead-reckoning position to use is that which the navigator might expect the ship to be in if all considerations except currents were taken into account.

The time interval over which currents should be measured is in most cases between 12 and 24 hours. This represents a compromise between the need to have a long enough interval for the current to have a measurable effect, and the need to restrict the interval so that the measurement does not extend over varying currents whose average would have little meaning. While periods in excess of a 'ship day', which may be 25 hours, are undesirable, periods shorter than 12 hours may be acceptable in those cases in which sufficiently accurate fixes combined with a precise knowledge of the ship's performance make this justifiable. Recent advances in satellite and electronic hyperbolic position-fixing devices make possible a high degree of accuracy in position finding by ships fitted with suitable equipment.

For most vessels when in the open ocean one of the best methods of determining the true position is by means of stellar observations. A careful note should be made of the method of obtaining each fix.

Currents measured from one noon position to the next, when positions have been obtained by means of a forenoon sight and a meridian altitude, are acceptable if no better method can be used. The noon position measured in this way is a running fix where accuracy depends upon the current experienced between the forenoon sight and noon and this is the very element (albeit measured over a longer period) which we are trying to determine. Wherever possible a more accurate method, such as an electronic fix, should be used. Any period shorter than 24 hours and involving a solar noon fix is unacceptable unless the latter can be made more accurate by reference to another heavenly body such as the moon or Venus.

One of the chief difficulties in obtaining accurate current observations from ships lies in the proper assessment of leeway. Deeply laden vessels lying low in the water, such as tankers, will make comparatively little leeway, whereas high-sided ships in a light condition will make a considerable amount. Sometimes with winds of force 3 and less it is not practical to make allowance for leeway and it is recorded as negligible. With winds of force 4–6 leeway usually becomes appreciable and should be allowed for. Only the mariner, with full knowledge of a particular vessel's performance and state of lading, can be expected to be able to assess the correction to be applied. Even then it is difficult. Where the winds are changing it will be necessary to compound several assessments in respect of individual legs.

With winds of force 7 and over, leeway becomes even more difficult to estimate, but current measurements are not usually attempted in these circumstances.

Not only the correction for leeway, but also the estimation of the dead-reckoning position, needs to be as accurate as possible. This is because the current being measured in many cases only amounts to about half a knot. Unless the speed of the vessel can be judged to an appreciably higher degree of accuracy such small movements cannot be measured.

When a vessel is within range of land, currents reckoned between shore fixes are most desirable. However, currents should not be measured in regions where the tides are appreciable.

One of the most accurate means of obtaining current information whether at the surface or at depth, is by means of the moored current meter. This measures the rate at which the water passes it in much the same way as the patent log measures the rate at which it is being towed through the water. The direction is indicated by means of a compass attached to the meter. Since the instrument needs to be kept in a fixed position, its use by shipping is limited.

Use of Information on Currents

A practical knowledge of ocean currents is necessary to the shipmaster, both for the safety of his ship and to assist in its economic operation. The safety aspect is fairly obvious when one considers the vagaries to which currents are liable, and a study of shipping casualty returns shows how often an unexpected current has contributed to the casualty. Despite the modern aids at his disposal, the prudent mariner must still use lead, log and look-out, and study the information about currents in *Admiralty Pilots*, current atlases and elsewhere. By making and reporting observations of currents experienced, the mariner not only gains practical knowledge himself, but benefits shipping generally by adding to the store of observational data from which up-to-date information can be published.

Although the economic benefits which can be derived from the use of a knowledge of the relevant currents are less obvious now than in former times when vessel speeds were lower, there are still situations in which taking account of the currents can save appreciable time. These situations naturally arise when the shortest route is such as to subject the vessel to the influence of a major unfavourable current for significant periods.

With appreciable currents such as the Gulf Stream, or Kuro Shio, if the vessel's shortest route were to lie at all close to the axis of the current, time could be saved by minor diversions to utilize the current if favourable, and otherwise to avoid it.

The most obvious case in which time can be saved by navigating according to the currents is that involving long passages with little change of latitude in equatorial and tropical waters. Take, for example, the passage between Manila and Panama. This long passage traverses an ocean region where the currents are relatively strong and constant, with easterly and westerly streams lying in close proximity. By making the passage slightly further to the north or south the mariner can ensure a favourable current, rather than an adverse one, for the best part of the roughly 9000 nautical miles involved. This could save as much as two days in the overall time. It is for this reason that the publication *Ocean Passages for the World* recommends the 'Central Route' from 5°00'N, 125°30'E, south of Mindanao, to 7°00'N, 80°00'W, off the Gulf of Panama, for eastbound traffic. This gives the vessel the benefit of the Equatorial Counter-current. Westbound traffic is advised to take a Great Circle route to 13°30'N, 170°00'E, after clearing Cabo Mala, then to pass between Bikar Atoll and Taongi Atoll and thence to pass close south of Guam and then via San Bernardino Strait. This gives the vessel the benefit of the westward setting North Equatorial Current over most of the route.

In small deeply loaded ships a knowledge of currents is particularly necessary. For instance, a ship bound from Colombo to Ras Asir (C. Guardafui), in the south-west monsoon, is advised not to pass close to the southward of Socotra because the currents are such as to set the vessel towards the island. To divert the vessel further south to avoid this risk would mean that she would encounter an easterly to north-easterly set which averages about 2 knots and may at times exceed 5 knots. A diversion to the north of Socotra would avoid these very strong adverse currents and the heavy confused seas which sometimes accompany them.

Charting of Currents

Ocean currents may be depicted in a variety of ways to suit various requirements. In some parts of the world the variation of the current according to the time of year is sufficiently small to justify the production of a mean annual chart showing the general circulation. In most regions, however, it is necessary to distinguish between the different times of the year, preferably by means of separate charts for each month. Some of the most meaningful forms of representation require a large number of observations and for this reason it has often been necessary to combine monthly observations into three-monthly periods because of the lack of sufficient observations to justify a monthly representation.

Whatever the period chosen (monthly, seasonal or annual), there are three forms of display which are most commonly used. These are (*a*) the vector mean current, (*b*) the predominant current, and (*c*) the current rose.

Of these the first two represent different kinds of average current, while the third is in fact a frequency distribution which shows the relative frequency of reports in each direction and of specified speed ranges within each direction.

Vector Mean Chart (Figure 14.1)

The vector mean (or resultant) current, as the name implies, is the average of all the individual observations for the place and time intervals (e.g. month) concerned, taking both speed and direction into account. In this method of evaluating the mean, currents in opposite directions tend to cancel each other out. Accordingly, if the observations are well distributed around the compass, the resultant speed is small even though all the individual speeds may be large. Many cases occur in which the mean speed, irrespective of direction, is about one knot, but the vector mean speed is about a quarter of a knot because of the extent to which opposing currents counterbalance each other.

The vector mean value represents the overall movement of the water over a considerable period. A vector mean chart is therefore the best to consult when considering the average drift of boats, derelicts, icebergs, etc. over a prolonged period. It does not represent the most likely current on any particular occasion. On charts it is common to indicate the direction by an arrow pointing towards the direction in which the current is setting with a figure beside it to indicate the speed which is usually expressed in nautical miles per day, or in knots. Another figure may be added to give the number of observations upon which the mean is based. In some charts the length and/or thickness of the arrows is varied according to the range of speed to be indicated. Then the eye can detect at a glance regions of stronger and weaker currents.

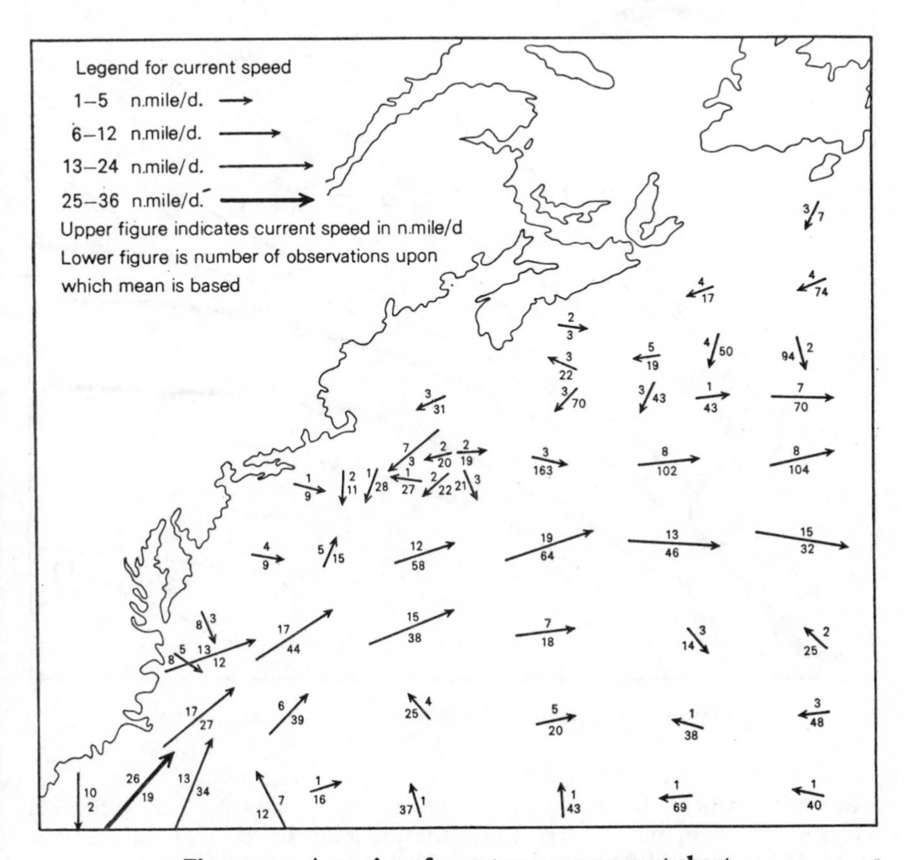

Figure 14.1. A portion of a vector mean current chart

Predominant Current Chart (Figure 14.2)

For many purposes the predominant current is more useful than the vector mean because it represents an approximation to the most frequent or most likely current. It may be defined as follows. First, the numbers of current observations whose directions fall within successive overlapping 90° sectors, each displaced 15° from its predecessor, is determined. The mid direction of the sector with the greatest number of observations is defined as the predominant direction. The predominant speed is the arithmetic mean of all the individual speeds within the 90° sector of greatest frequency. The 'constancy' of the predominant current may be defined as the percentage of the number of observations within the selected (maximum frequency) sector to the total number of observations. In Figure 14.2 constancy is indicated by the thickness of the arrow according to the legend on the chart.

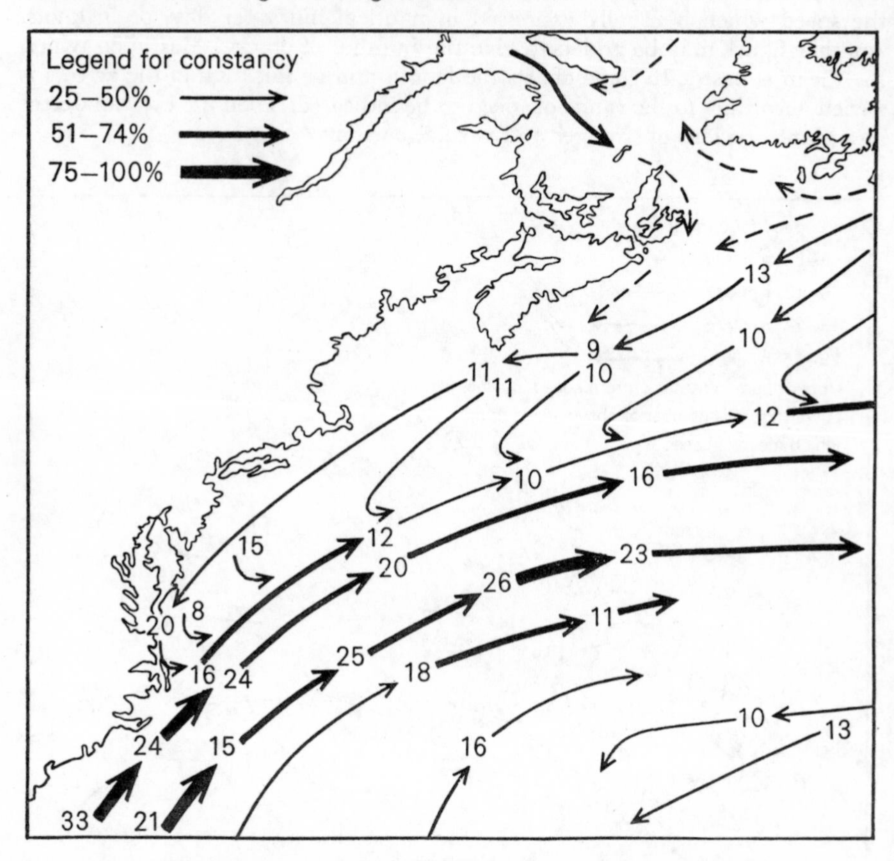

Figure 14.2. A portion of a predominant current chart

In regions where the constancy is high there is relatively little difference between the vector mean current and the predominant current. As the constancy decreases the difference between the two increases.

As in the case of the vector mean, the predominant current is normally indicated by an arrow pointing in the direction towards which the current sets while a figure beside the arrow gives the mean speed (as defined above), usually in nautical miles per day or in knots.

Current Rose Charts (Figure 14.3)

Whilst the two foregoing models of representation both have their uses, neither gives the full story regarding the possible range of variation of the current at a particular place and time of year. The current rose fills this need by showing not only the relative frequency with which the current sets in each direction, but also the relative frequency of various categories of speed within each direction.

In the current rose the total frequency of observations which fall within a particular (non-overlapping) sector of the compass determines the length of an arrow extending in this direction from the centre of the rose. Most commonly either an 8-point or 16-point compass is used so that the individual sectors cover 45° or 22½°. Each arrow is subdivided so that the relative frequency of observations within various speed categories can be indicated.

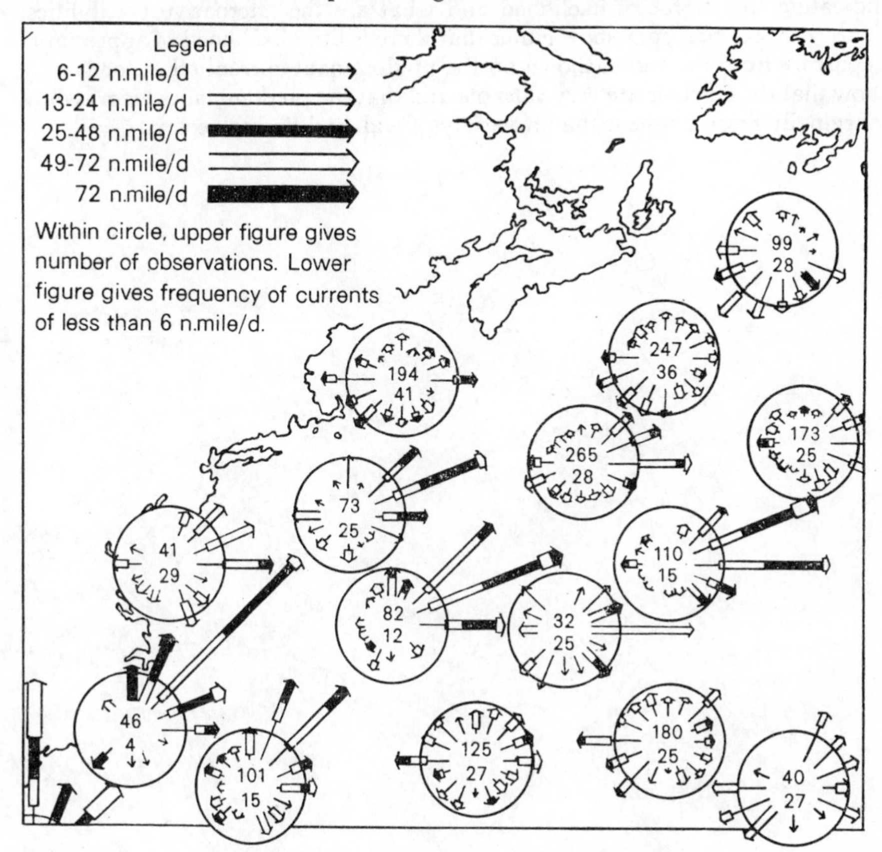

Figure 14.3. A portion of a current rose chart

Thus inspection of a current rose may indicate, for example, that most of the sets (35%) are north-easterly but that an appreciable number (25%) are in the opposite direction. Of the north-easterly sets it might show that 3% (of the overall total) were of speeds in excess of 2 knots, 12% were of speeds between 1 and 2 knots, and 20% were of speeds below 1 knot. Similar data would be available for each point of the compass.

Although the current rose provides more information than do the other common forms of representation, there are practical difficulties in its use. For a rose to show frequencies which are representative it must be based on a large number of observations, preferably several hundred. Also this large number must be concentrated within a relatively small area because there may well be variation between the different parts of the general area. Consequently, in the production of rose charts, there is often the need to compromise between the desire to make the area covered by a particular rose small enough to represent uniform conditions, and the need to make the area big enough to embrace a sufficient number of observations to produce a representative distribution.

If one wants to know what current to expect at a particular place and time both the predominant current chart and the rose chart should be consulted. The first will show what is most likely, but the second will qualify this by indicating the degree of likelihood and what are the alternative possibilities. Thus the rose chart may show either that there is little likelihood of appreciable departure from the values shown on the predominant current chart, or it may show that the currents are very variable and that the predominant vector is only marginally more frequent than a variety of widely differing vectors.

CHAPTER 15

THE CURRENT CIRCULATION OF THE OCEANS

Introduction

A summary of the general surface current circulation of all oceans is given below. The general circulation represents the trend of water movement over a long period; it emerges from the more or less variable movements of individual currents.

In each ocean the main circulation is dealt with first, followed by any extensions thereof (associated circulations). The surface circulation of the Mediterranean, for example, is not a part of the main North Atlantic circulation, but is associated with it.

The general surface current circulation of the world is shown in Figure 15.1, opposite page 218, on which the seasonal variations between the two monsoons are shown. Apart from these major changes there are some other seasonal changes in the currents which are indicated by asterisks referring to notes printed on adjacent land areas. One such is the position of origin of the Equatorial Counter-current in the North Atlantic, referred to later. It will be understood that this representation of the world circulation on a single chart is necessarily generalized. Precise detail regarding current speeds, variability, and seasonal changes can be found in the various Current Charts which deal with individual oceans.

Information about the currents in local coastal regions, so far as these are known, will be found in the latest editions of the *Admiralty Pilots*. Further information about ocean currents is to be found in the various Routeing Charts issued by the Hydrographic Department of the Navy and in their publication *Ocean Passages for the World* (N.P. 136).

NORTH ATLANTIC OCEAN

Main Circulation

The main circulation of the North Atlantic is clockwise. The southern part of this circulation includes the fairly constant west-going NORTH EQUATORIAL CURRENT south of about latitude 23°N. Further north the westerly flow gradually becomes more variable and ceases to be recognizable north of about latitude 30°N. Eastwards of the Caribbean Sea, the North Equatorial Current is joined by the SOUTH EQUATORIAL CURRENT of the South Atlantic, which flows past the north coast of Brazil. The combined Equatorial Current flows westwards through the Caribbean Sea and emerges through the Yucatan Channel. It thence flows north-eastwards along the north-west coast of Cuba and into the Florida Strait. Thence the stream flows northwards and is known as the FLORIDA CURRENT from the Florida Strait to about latitude 29°N. Between about this latitude and the southernmost part of the Grand Banks of Newfoundland the strong mainly north-easterly current forming the western flank of the main North Atlantic circulation is known as the GULF STREAM.

The northernmost part of the North Equatorial Current, together with the more variable westerly flow further north, is diverted to the north of the West Indies and flows north-westwards along the northern coasts of the Greater Antilles to converge with the Florida Current and the Gulf Stream. This north-westward flow is sometimes referred to as the ANTILLES CURRENT.

In a narrow belt north of the equator, the EQUATORIAL COUNTER-CURRENT flows eastwards between the North Equatorial Current of the North Atlantic and the South Equatorial Current of the South Atlantic. Its longitude of origin varies, being about 52°W in August to October and about 20°W in February to April. This counter-current flows past Cape Palmas and sets along the coast of the Gulf of Guinea to the Bight of Biafra and is known as the GUINEA CURRENT. The subsequent course of the Guinea Current is not known; most of it probably recurves westwards and joins the South Equatorial Current.

The warm waters of the Florida Current and Gulf Stream, between Florida Strait and Cape Hatteras, follow the course of the 100-fathom line, outside but near which lies its axis of greatest strength. Immediately north of Cape Hatteras the Gulf Stream begins to leave the 100-fathom line and gradually turns eastwards into the ocean, southwards of the Georges and Nova Scotia Banks. The inshore edge of the Gulf Stream, south of Cape Hatteras, thus gradually becomes the northern edge, northwards and eastwards of that cape. This northern edge is relatively sharply defined at all times of the year, owing to the convergence along it of the cold Labrador Current (*see* page 219).

Eastwards of about the 46th meridian the Gulf Stream ceases to be a well-defined current. It weakens by fanning out up the east side of the Grand Banks of Newfoundland. The resultant north-east and easterly flow across the ocean is towards the British Isles and the adjacent European coasts; this is known as the NORTH ATLANTIC CURRENT.

The southerly part of the North Atlantic Current turns gradually clockwise to south-easterly and, later, to south-westerly directions; it does so east of about longitude 40°W. This southerly flow forms the eastern flank of the main North Atlantic circulation. It is most marked off the western coast of the Iberian Peninsula (where it is sometimes called the PORTUGAL CURRENT) and off the north-west coast of Africa where the south-westerly flow is called the CANARY CURRENT. This finally turns westwards to pass into the North Equatorial Current in the vicinity of the Arquipelago de Cabo Verde.

Northward Extension of the Main Circulation

The more northerly part of the North Atlantic Current does not recurve southwards, but flows north-eastwards off the west coasts of the Hebrides and Shetland Isles and thence to the coast of Norway. It continues north-eastwards along this coast. In about latitude 69°N this current divides and the left branch, the WEST SVALBARD CURRENT, (formerly WEST SPITSBERGEN CURRENT) sets northwards to the west of Svalbard and thence into the Arctic Basin. The right branch is called the NORTH CAPE CURRENT and follows the coast past Nordkapp into the Barents Sea, finally setting towards the north of Novaya Zemlya; a branch of it continues along the Murmansk coast as the MURMANSK CURRENT.

The branch of North Cape Current which flows into the Barents Sea towards and around the northern extremity of Novaya Zemlya continues south-westwards in the Kara Sea along the east coast of Novaya Zemlya, where it is known as the NOVAYA ZEMLYA CURRENT. Part of this re-enters the Barents Sea along the northern shore of Proliv Karskiye Vorota, forming the LITKE CURRENT.

A small part of the warm North Atlantic Current turns northwards in the longitude of Iceland to form the IRMINGER CURRENT. Closely south-west of Iceland it divides, and the main branch turns westwards and passes into the East Greenland Current south of Denmark Strait. A smaller branch makes a clockwise circulation of Iceland.

The chief outflow of water from the Arctic Basin is the cold, ice-bearing current which sets south-westwards along the east coast of Greenland, the EAST GREENLAND CURRENT. A part of this diverges south-eastwards from the main body of the current, north of latitude 70°N. This flows south-eastwards across the region to the north-east of Iceland and gradually turns eastwards and finally north-eastwards to join the general north-easterly flow off Norway.

The East Greenland Current rounds Kap Farvel and passes northwards along the west coast, where it is called the WEST GREENLAND CURRENT. This loses volume by fanning out on its seaward side, but part of it circulates round the head of Baffin Bay and, reinforced by water flowing eastwards through the Jones and Lancaster Sounds, sets southwards along the coast of Baffin Island. This southerly flow is joined by the outflow from Hudson Strait and becomes the LABRADOR CURRENT which flows south-eastwards along the Labrador coast.

After Belle Isle Strait and the east coast of Newfoundland have been passed the Labrador Current covers the whole of the Grand Banks except, during the summer, the extreme southern part. A large branch of the current follows the eastern edge of the Banks and thus carries ice southwards to reach the trans-atlantic shipping tracks. Another branch rounds Cape Race and sets south-westerly. The bulk of the water on the Banks also sets in this direction, so that the Labrador Current fills the region between the south coast of Newfoundland, the south-east coast of Nova Scotia and the northern edge of the Gulf Stream. The Labrador Current continues southwards as a cold current along the United States coast. Its greatest southern extension is to about latitude 36°N, closely north of Cape Hatteras, in November to January; its least extension is to about latitude 40°N in August to October. The cold Labrador Current and the warm Gulf Stream converge along the northern edge of the latter.

Water also emerges from the Arctic Ocean into the northern part of the Barents Sea, forming the EAST SVALBARD CURRENT (formerly EAST SPITSBERGEN CURRENT), which flows south-westwards on the eastern side of Svalbard and, after passing the southern extremity of the latter, curves north-westwards into the WEST SVALBARD CURRENT. A similar current, further south, is directed towards Bjørnøya, and is known as the BJØRNØYA (BEAR ISLAND) CURRENT. Part of this recurves northwards into the WEST SVALBARD CURRENT and part southwards.

Associated Regions

North Sea and English Channel

This region is one of tidal streams, and variable currents depending on the present or recent wind. After subtracting the tidal streams there is, in the long run, a weak current circulation as follows. From north-eastwards of the Shetland Isles a branch of the North Atlantic Current flows southwards down the east coasts of Scotland and England to the Thames estuary, where it curves eastwards. It is there joined by a branch of the North Atlantic Current which passes up the English Channel and through the Dover Strait. The combined current flows along the Belgian and Netherlands coasts and west coast of Jutland. It continues counter-clockwise round the Skagerrak and then sets northwards along the west coast of Norway.

Only a portion of the current on the west side of the North Sea reaches as far south as the Thames estuary, for water fans out eastwards from this current all along its length.

Bay of Biscay

Off the mouth of the Bay of Biscay the current trends south-eastwards and later flows southwards along the west coast of the Iberian peninsula. A branch enters the Bay and recurves westwards along the north coast of Spain to rejoin this current near Cabo Finisterre.

Mediterranean

Part of the water from the Portugal Current enters the Strait of Gibraltar and flows along the north coast of Africa. Beyond Cap Bon it continues in a general south-easterly and later easterly direction towards Port Said. The water turns northwards at the eastern end of the Mediterranean and the counter-clockwise circulation is completed by a more variable return current along the northern coasts. In following the coast, this forms counter-clockwise loops in seas such as the Aegean and the Adriatic.

Black Sea

The general circulation is counter-clockwise. There is an almost constant surface flow of water from the Black Sea to the Aegean through the Bosporus, the Sea of Marmara and the Dardanelles. There is a sub-surface return current below this, from the Aegean to the Black Sea.

Gulf of St Lawrence

Water enters the Gulf on the northern side of Cabot Strait and flows northwards along the west coast of Newfoundland. It then turns south-westwards, along the Quebec coast and is west-going southwards of Anticosti. The GASPÉ CURRENT sets south-eastwards from the Gaspé Peninsula. Water emerges from the Strait by the CAPE BRETON CURRENT on the southern side of Cabot Strait; this is sometimes called the CABOT CURRENT.

Gulf of Mexico

Part of the water passing through Yucatan Channel turns westwards and follows the gulf coast in a clockwise direction. Another branch sets northwards across the middle of the gulf to the region of the Mississippi delta, where it turns eastwards and joins the coastal current. The combined current passes into the FLORIDA CURRENT between Cuba and Florida.

SOUTH ATLANTIC OCEAN

Main Circulation

The main surface circulation of this ocean is counter-clockwise. The SOUTH

EQUATORIAL CURRENT, flowing westwards across the ocean, extends across the equator to about latitude 4°N. South of about latitude 6°S the constancy and strength of the westerly flow gradually decreases southwards. The main concentration of the South Equatorial Current is between about latitudes 6°S and 4°N.

The eastern side of the circulation is formed by the relatively cool BENGUELA CURRENT flowing north-westerly along the south-western coasts of Africa. From about latitude 30°S, water from the Benguela Current fans out west-north-westerly and westerly on its seaward side.

While most of the South Equatorial Current flows along the north coast of Brazil and across the equator to join the North Equatorial Current, the westerly flow on its southerly flank is directed towards the Brazilian coast, southwards of Cabo de São Roque. A small part of this turns northwards along that coast and passes round Cabo de São Roque to join the South Equatorial Current; the bulk of it flows southwards along the Brazilian coast. This warm current, known as the BRAZIL CURRENT, forms the west side of the South Atlantic circulation.

The SOUTHERN OCEAN CURRENT flows all round the globe in the southermost parts of the South Atlantic, South Indian and South Pacific Oceans, and forms the completion, on the southern side, of the counter-clockwise circulation in these oceans. The Southern Ocean Current is restricted in width by passing through Drake Passage, between Cabo de Hornos and Graham Land. Eastwards of this passage it becomes very wide, its northern part fanning out north-eastwards past the southern and eastern coasts of the Falkland Islands, to reach to about the 40th parallel in the central longitudes of the South Atlantic.

Further north, the southern part of the main circulation is added to by water from the seaward side of the Brazil Current curving south-eastwards and eastwards between latitudes 28°S and 40°S. In mid ocean part of the resultant easterly flow runs north of and parallel to the colder water of the Southern Ocean Current; the remainder merges with the northerly part of the Southern Ocean Current. East of about longitude 15°W and south of the 30th parallel, the east-going water turns north-east and north to converge with the westerly flow which fans out from the seaward side of the Benguela Current.

Nearer the South African coast, between longitudes 10°E and 15°E, a branch of the Southern Ocean Current turns northwards directly into the Benguela Current. The water of the Benguela Current is, however, mainly derived by the upwelling of water off the south-west coast of Africa. A branch of the AGULHAS CURRENT of the South Indian Ocean, which rounds the south coast of Africa, also enters the Benguela Current.

The FALKLAND CURRENT does not form part of the main circulation; it branches northwards from the Southern Ocean Current near Staten Island and passes west of the Falkland Islands. Part of it continues to the Rio de la Plata estuary; the remainder branches eastwards in about latitudes 40°S to 42°S and rejoins the northern part of the Southern Ocean Current. During May to October a northerly extension of the Falkland Current continues north of Rio de la Plata. From May to July this may extend as far as Cabo Frio.

NORTH INDIAN OCEAN

Monsoonal Effects

The currents in the greater part of the North Indian Ocean, including the

Arabian Sea and the Bay of Bengal, are reversed in direction seasonally by the monsoons.

The north-east monsoon current circulation occurs during the height of that monsoon from November to January. During the latter part of this monsoon, February to April, the circulation changes. The south-west monsoon circulation prevails from May to September. October is a transitional month. The currents are therefore described below for the three periods, November to January, February to April and May to September.

North-east monsoon circulation, November to January. In the open waters of the Arabian Sea and Bay of Bengal the current sets in a westerly direction. These westerly sets extend southwards beyond the equator. Near the coasts of the Arabian Sea there is a weak circulation in a counter-clockwise sense following the coasts. In the northern part of the Bay of Bengal there is a clockwise circulation. A stronger current sets southwards down the east coast of Africa from Ras Hafun to about latitude 2°S. This, the SOMALI CURRENT, turns eastwards between about latitudes 2°S and 5°S, to form the beginning of the EQUATORIAL COUNTER-CURRENT. Between Ras Hafun and Ras Asir (C. Guardafui) the current is northerly.

Later north-east monsoon period, February to April. The flow in the open waters of the Arabian Sea and Bay of Bengal remains westerly, though the currents are more variable than from November to January. Towards the equator, westerlies which are well marked in February become less so in March, and in April the flow changes to easterly. The coastal circulation of the Arabian Sea, however, is reversed to a clockwise direction (*see* 'Gradient Currents' in Chapter 16). In February the Somali Current flows south-westwards along the African coast to about latitude 3°S. During March the flow becomes variable and by April it reverses to north-easterly.

South-west monsoon circulation, May to September. In the open waters the drift is easterly. The coastal circulation of the Arabian Sea and Bay of Bengal remains clockwise and is strengthened. The Somali Current continues to flow northwards, from Cabo Delgado to Ras Asir (C. Guardafui), and is greatly strengthened. It divides in about latitude 7°N; part continues along the coast to Ras Asir (C. Guardafui), but the bulk turns eastwards and passes south of Socotra into the general easterly current. The current south of Socotra in July to September is the strongest known in the world in the open ocean and rates up to 7 knots have been recorded.

Red Sea and Gulf of Aden

The current conforms to the monsoon. During the north-east monsoon it sets westwards in the Gulf of Aden and passes through the Straits of Bāb-al-Mandab to flow up the axis of the Red Sea. During the south-west monsoon the current in the Gulf of Aden sets eastwards; in the Red Sea the water flows down the axis of the sea and into the Gulf of Aden.

SOUTH INDIAN OCEAN

Main Circulation

The main circulation of the South Indian Ocean is counter-clockwise. The northern flank of the circulation is formed by the west-going South

Equatorial Current which occupies an analogous position to the South Equatorial Currents of the other main oceans. In this ocean, however, the South Equatorial Current lies well south of the equator and in this respect differs from the corresponding currents of the Atlantic and Pacific which extend to a few degrees north of the equator. Its northern boundary is usually between latitudes 6°s and 10°s but it varies according to longitude and season.

The South Equatorial Current, after passing the northern extremity of Madagascar, meets the African coast near Cabo Delgado. Here it divides and some of the water flows northwards along the coast. The remainder flows southwards to form a strong coastal current which, from Cabo Delgado to Lourenço Marques, in known as the MOÇAMBIQUE CURRENT. Its southward continuation is the AGULHAS CURRENT. This is reinforced by water from the South Equatorial Current setting past the southern extremity of Madagascar.

Some of the water of the Agulhas Current recurves to south-eastwards between about longitudes 25°E and 35°E and enters the northern part of the Southern Ocean Current. The remainder of the Agulhas Current continues along the coastline, and passing over the Agulhas Bank, enters the South Atlantic Ocean, where it joins the Benguela Current.

The southern side of the main circulation is formed by the cold water of the SOUTHERN OCEAN CURRENT, setting in a generally easterly direction in latitudes south of about 35°s. There is no defined northern boundary to the Southern Ocean Current; the predominance of easterly sets decreases with decreasing latitude towards the middle of the ocean. Here the centre of a rather weak counter-clockwise circulation lies somewhere between latitudes 22°s and 35°s and between longitudes 70°E and 95°E.

The east side of the circulation is not well marked. In the northern winter months the Southern Ocean Current turns northwards as it approaches Cape Leeuwin and forms a north-going current parallel to the west coast of Australia, though there is a narrow belt of south-going current close inshore. In the northern summer the Southern Ocean Current off south-western Australia sets easterly and turns southerly towards the coast in latitudes south of latitude 26°s. Between about latitudes 20°s and 26°s the current near the coast runs southwards from about March to August but is northerly in other months. The bulk of the Southern Ocean Current continues its easterly course, south of Australia and Tasmania, into the South Pacific.

Equatorial Counter-current

In the other main oceans of the world the Equatorial Counter-current consists of a relatively narrow belt of easterly currents lying between the North Equatorial and South Equatorial Currents of the ocean in question. In the Indian Ocean, also, the Equatorial Counter-current is readily distinguished during the north-east monsoon when it forms a belt of easterly currents lying between about latitudes 2°s and 8°s. In the south-west monsoon, however, the criterion for determining the northern boundary of this belt disappears, since easterlies are then more or less continuous north of about latitude 8°s. This easterly flow tends to be lighter and more variable in the south and to become more pronounced further north. In July and August the currents are light and rather variable between about latitudes 8°s and 2°N with a more-decided easterly flow further north, but in June and September the region of marked easterly flow extends southwards as far as latitudes 2°s–4°s.

Extreme Eastern Part of the Ocean

The currents here, including those of the Arafura Sea, are not well known, owing to the scarcity of observations. Eastwards of Christmas Island, between the parallels of about 10°S and 12°S, there is a predominance of westerly sets during most of the year and they form the most easterly part of the Equatorial Current.

NORTH PACIFIC OCEAN

Main Circulation

The main circulation of the North Pacific resembles that of the North Atlantic. Observations are inadequate to give detailed information about the currents over large parts of this ocean, owing to its size and the limited shipping tracks. This especially applies to the middle longitudes, both near the equator and in the variable current region further north.

The southern part of the main circulation is formed by the west-going NORTH EQUATORIAL CURRENT. Immediately south of this the EQUATORIAL COUNTER-CURRENT flows eastwards across the ocean, but its limits are not exactly known and they may be subject to some seasonal variation. During the latter half of the year the southern limit appears to be nearer the equator in the west than in central or eastern longitudes. Over most of the ocean the counter-current is usually found between latitudes 4° or 5°N and 8° or 10°N. The SOUTH EQUATORIAL CURRENT, the northern limit of which reaches to about latitude 4° or 5°N, is described under the South Pacific.

The North Equatorial Current has no defined northern limit. This west-going current lessens in strength as the predominance of trade winds decreases, until it is lost in the variable current region lying to the northwards. The latitude to which some predominance of westerly current extends appears to vary with the season. In mid ocean it is between latitude 20°N and 24°N in winter and in about latitude 30°N in the late summer or autumn.

The Equatorial Counter-current flows eastwards throughout the year across the whole ocean. During March to November, this counter-current is formed by the recurving of the South Equatorial Current northwards and part of the North Equatorial Current southwards down the east coasts of the Philippines. In December to February the North Equatorial Current is the only source of the counter-current. During these months the South Equatorial Current, north of the equator, turns south in about longitudes 140°E to 150°E and finally south-eastwards, and thus makes no contribution to the counter-current. In all seasons part of the North Equatorial Current enters the Celebes Sea and emerges in a north-easterly direction to join the counter-current. The counter-current appears to be strongest in its most westerly portion, from northwards of Halmahera (between New Guinea and Celebes) to about longitude 145°E.

To continue the main circulation, a considerable part of the water from the North Equatorial Current turns north-eastwards when east of Luzon and flows up the east coast of Taiwan to form the KURO SHIO, a warm current corresponding to the Gulf Stream of the North Atlantic. Southwards of the Japanese Islands the Kuro Shio flows north-eastwards. This current then fans out to form the NORTH PACIFIC CURRENT, which sets eastwards across the ocean toward the North American coast. It is joined by cold water from the Bering Sea, flowing down the east coast of Kamchatka and turning south-east and then east.

The whole forms a broad belt of variable current with a predominance of easterly sets, filling most of the area between latitudes 35°N and 50°N, across the ocean. The colder part of this is known as the ALEUTIAN CURRENT and in the middle longitudes of the ocean is found northwards of about latitude 45°N.

Water fans out to the south-east and south from the southern part of the North Pacific Current, and passes into the central region of variable currents. Eastwards of about longitude 150°W the remainder of the North Pacific Current and the bulk of the Aleutian Current turn southwards and south-westwards and finally merge with the North Equatorial Current. Near the coast this southward current is called the CALIFORNIA CURRENT.

The California Current does not actually meet the coast; from November to February a relatively cool counter-current, known as the DAVIDSON CURRENT, runs northwards, close inshore, to at least latitude 48°N. During the rest of the year the space between the California Current and the coast is filled by irregular current eddies.

In the extreme eastern part of the Equatorial Counter-current, seasonal variations occur off the Central American coast and numerous eddies are formed, which seem to vary from year to year. In most months the counter-current will be met between latitudes 5°N and 10°N, and it generally turns north and north-west along the Central American coast, finally to enter the North Equatorial Current. Early in the year part of the counter-current branches south and enters the South Pacific.

Associated Regions

Northern Part of the Ocean

The Bering Sea currents are not well known, but there is a counter-clockwise circulation round the coasts, northwards on the east and southwards on the west side. The cold southward current flows along the east coast of Kamchatka as the KAMCHATKA CURRENT, and then past the Kuril Islands where it is known as the OYA SHIO.* The Oya Shio continues along the east coast of the main Japanese island of Honshu, until it meets the northern edge of the Kuro Shio in about latitude 36°N. The Oya Shio thus corresponds to the Labrador Current of the North Atlantic. (*See* Chapter 8—Visibility—north-west Pacific, for the frequency of fog in this area.) It is joined by water emerging through Tsugaru Kaikyō. Water fans out south-east and east all along the course of the current, from Kamchatka southwards. The resultant easterly current flows parallel and adjacent to the North Pacific Current and forms the Aleutian Current. The more northerly part of this, as it approaches the American coast, sets north-east and then north-west past Queen Charlotte Islands and along the coast of south-east Alaska. This is the ALASKA CURRENT. It follows the Gulf of Alaska coastline to set to the westwards across the head of the gulf and then along the south coasts of the Aleutian Islands. West of longitude 155°W to 160°W, some water recurves from the Alaska Current to the south and south-east and rejoins the east-going Aleutian Current; the remainder recurves northwards and enters the Bering Sea and thence it turns north-eastwards and later northwards to form the east side of the Bering Sea circulation, referred to previously.

---

* The whole of the cold current from the Bering Sea is sometimes referred to as the Oya Shio, especially by American writers.

China Seas and other Regions Westwards of the Main Circulation

In the China Seas and Java Sea the currents are monsoonal. During the south-west monsoon the general direction of current is westerly in the Java Sea and north-easterly in the China and Eastern Seas. In the Yellow Sea the currents are complex and variable. During the north-east monsoon the currents become mainly southerly in the Yellow Sea, south-westerly in the Eastern and China Seas and easterly in the Java Sea. In the southern part of the China Sea there is a variable current area west of Borneo and Palawan in both monsoons, but a weak monsoonal current runs along the west coasts of these islands, alternating between north-easterly and south-westerly during the year. The eastern part of the Eastern Sea is occupied by the Kuro Shio.

In the China Seas, the north-easterly current is found from May to August inclusive. September is the transitional month, but the north-easterly current still persists in the southern part. In October the south-westerly current becomes established everywhere, and this continues till about the middle of March. April is the transitional month, but the south-westerly current still persists in the southern part.

In the Java Sea the westerly current runs from June to September and the easterly current from November to March. April, May and October are transitional months.

In the Japan Sea the circulation is counter-clockwise all the year, the north-going current on the east side of the sea, known as the TSUSHIMA SHIO, is a branch of the Kuro Shio which has passed through Korea Strait. Part of the Tsushima Shio branches off through Tsugaru Kaikyō and flows into the Oya Shio, and another part branches off through Sōya Kaikyō. The south-going current on the west side of the sea, past Vladivostok, is called the LIMAN CURRENT.

In the central parts of the Sea of Okhotsk the currents are variable. Nearer the coasts there is a counter-clockwise circulation, but along the west coast of Kamchatka the flow is southerly close inshore. Off the coast of Sakhalin Island the south-going current is known as the EAST SAKHALIN CURRENT.

SOUTH PACIFIC OCEAN

Main Circulation

The main circulation is counter-clockwise. Less is known about the South Pacific currents than those of the other oceans south of the equator, because of its great size and on account of the large areas, particularly in the east, which are not frequented by shipping.

The SOUTH EQUATORIAL CURRENT of the Pacific, though lying mainly south of the equator, extends to about latitude 5°N though varying according to longitude and season. Its northern limit is defined by the east-going Equatorial Counter-current of the North Pacific, immediately north of it.

South of about latitude 6°S the strength of the current gradually decreases southwards while the direction becomes more south-westerly. Some westerly component persists at least to latitude 20°S. Between about longitudes 90°W and 135°W there is a weak southerly flow between latitudes 20°S and 30°S.

On the west side of the ocean the course of the South Equatorial Current varies seasonally. In June to August the whole current follows the north coast of New Guinea to the north-westwards and then recurves north and north-eastwards, entering the east-going Equatorial Counter-current of the North Pacific. In December to February the South Equatorial Current does not pass into the counter-current; it recurves south-westwards and southwards and flows past the north coast of New Guinea in a south-easterly direction. There is thus a complete reversal of current along this coast during the year.

The west side of the main circulation is not well marked, except along the Australian coast from Great Sandy Island to Cape Howe, where the EAST AUSTRALIAN COAST CURRENT sets southwards.

In the region of the Great Barrier Reef the flow is south-easterly in the northern winter but in the northern summer, north of about latitude 18°s, the flow is north-westerly and continues westwards through the Torres Strait.

The south side of the main circulation is formed by the SOUTHERN OCEAN CURRENT which sets to the east or north-east. Observations are scanty over most of this region.

Between Australia and New Zealand easterly sets predominate. The bulk of the East Australian Coast Current mixes with the water of the Southern Ocean Current and flows eastwards towards New Zealand. The current sets north-eastwards along both the west and east coasts of South Island, New Zealand.

The bulk of the Southern Ocean Current enters the South Atlantic south of Cabo de Hornos. The northern part of this current, however, meets the coast of Chile between Isla de Chiloé and the Golfo de Peñas, where it divides; part goes northwards to form the beginning of the Peru Current and part follows the coast south-eastwards to rejoin the Southern Ocean Current south of Cabo de Hornos.

The east side of the circulation is formed by the relatively cool PERU CURRENT, formerly known as the Humbolt Current. It follows the coastline northwards to somewhere near the equator. Between the Golfo de Guayaquil and the equator the bulk of the Peru Current trends seawards and passes into the South Equatorial Current. The Peru Current has a width of perhaps 300 nautical miles or more.

A branch of the Peru Current continues northwards up the coast during most of the year and enters the Gulf of Panama.

During the northern winter in the eastern North Pacific, the east-going Equatorial Counter-current extends further south than at other seasons. A branch of this current then turns southwards along the coast of Ecuador into the South Pacific, but in most years its southern limit is only a few degrees south of the equator. This is called EL NIÑO, or the HOLY CHILD CURRENT, as in some years it begins to flow about Christmas-time, although it is more regularly observed in February and March. In exceptional years it extends down the coast of Peru, occasionally to beyond Callao, when the intrusion of this warm water into a region normally occupied by the cool Peru Current kills fish and other marine life.

Central Oceanic Region

Between about latitudes 20°s and 40°s, and longitudes 90°w and 180° there is a vast area where observations are scanty and where, as far as is known, the currents are mainly light and variable. The centre of the general counter-clockwise circulation of the South Pacific lies somewhere in this region and there may well be several eddies with centres located not far from latitude 30°s.

ARCTIC OCEAN

The main inflow of water into the Arctic Basin is from the West Svalbard Current. A much smaller quantity enters through Bering Strait. Fresh water is added to the Arctic Basin from rivers, notably those of Siberia, and by an excess of precipitation over evaporation.

The EAST GREENLAND CURRENT forms the main outflow of water from the Arctic Ocean. Small outflows occur due to the EAST SVALBARD and BJØRNØYA (BEAR ISLAND) CURRENTS which flow south-westwards in the northern part of the Barents Sea and the current which flows eastwards between the islands of the Arctic Archipelago towards Baffin Bay.

Within the eastern longitudes of the Arctic Basin there is a weak westerly current, as shown by the drift of the Fram and other ships in the ice. This current emerges to the south-west between Svalbard (Spitsbergen) and Greenland to form the East Greenland Current.

Along the Siberian coast the current generally flows eastwards from the Kara Sea to Bering Strait.

SOUTHERN OCEAN

The east-going SOUTHERN OCEAN CURRENT completes, on the south side, the counter-clockwise circulation in each of the South Atlantic, South Indian and South Pacific Oceans. The southern limit of this current is not clearly defined. In general the easterlies decrease and become more north-easterly by about latitude 60°S. Some easterly component continues southwards to latitudes, varying with longitude and season, between about latitudes 62°S and 67°S. Further south, westerly components increase and there is finally a westerly current setting around the coasts of the Antarctic continent. These coastal westerlies are interrupted in the north of Graham Land where the restricted nature of Drake Passage produces a north-easterly flow throughout its width.

CHAPTER 16

THE CAUSES AND CHARACTERISTICS OF OCEAN CURRENTS

Introduction

The water of the oceans is in a state of continual movement, not only at the surface but at all depths. Ocean-current circulation, in its widest sense, takes place in three dimensions but the strongest currents occur in an upper layer which is shallow compared with the ocean depth. Apart from the relatively small-scale vertical movements associated with waves, the motion near the sea surface is largely horizontal, but, at depth, it may have a vertical component.

The navigator is only concerned with the currents in that depth of water in which his vessel is floating, i.e. in a layer extending to a depth equal to the vessel's draught. As, in general, the current varies with depth, a vessel's response to the varying currents will represent some compromise between the response to the current literally at the surface and that at the depth of the ship's draught. Consequently currents measured by determination of a vessel's set and drift are normally regarded as being applicable to a depth equal to half the vessel's draught. Most of our information about so-called surface currents comes from a variety of ships of various draughts. Although such observations may not all be strictly comparable it has been customary to treat them all as representing 'surface' currents although in the main they apply to depths varying between about 3 metres and 10 metres.

The Main Causes of Ocean Currents

The processes which cause ocean currents are complex and are not yet fully understood. In most cases there is more than one factor contributing to the existence of a particular current. Frequently there is a long chain of recurrent cause and effect wherein it is difficult to identify a beginning or an end.

Two main causes of ocean currents may be distinguished. These are, firstly, wind stress acting on the water surface and, secondly, pressure gradients within the water. The currents resulting from these two causes are described, respectively, as wind-drift currents and gradient currents.

Wind-drift Currents

Wind blowing over a water surface tends to drag the uppermost layer of water in the direction towards which the wind is blowing. As soon as any motion is imparted, however, the effect of the earth's rotation (the Coriolis force) is to deflect the movement towards the right in the northern hemisphere and towards the left in the southern hemisphere. Although theory suggests that the resulting effect should be to produce a surface flow (or wind-drift current) in a direction inclined at 45° to the right (left) of the wind direction in the northern (southern) hemisphere, observations show the angle to be less in practice. Various values between 20° and 45° have been reported. An effect of the movement of the surface water layer is to impart a lesser movement to the layer immediately below, in a direction to the right (left in the southern hemisphere) of that of the surface layer. Thus, with increasing depth, the speed of the wind-induced current becomes progressively less while the angle between the direction of wind and current gradually increases.

Many investigators have endeavoured to determine the ratio between the speed of the surface current and the speed of the wind responsible. This is a complex problem and many different answers have been put forward. An average empirical value for this ratio is about 1:40 (or 0·025). Some investigators claim a variation of the factor with latitude but the degree of any such variation is in dispute. In the main the variation with latitude is comparatively small and, in view of the other uncertainties in determining the ratio, can probably be disregarded for most purposes. The production of a current in response to the wind is not an instantaneous reaction. Initially the response is slow and it takes time for a steady state to become established. This time varies according to latitude but is something like 24 hours.

It would seem reasonable to expect that hurricane-force winds might give rise to currents in excess of 2 knots, but it is rare for such winds to persist for more than a few hours without a change in direction. With such high winds observations of wind-drift currents are not usually available. Strong currents have indeed been reported in connection with hurricanes but these have been complicated by secondary effects, such as the piling-up of water against the coast, which leads to the production of a gradient current.

Gradient Currents

These are caused by pressure gradients in the water. They occur whenever the water surface develops a slope, whether under the action of the wind, or through a juxtaposition of waters of differing temperature and/or salinity. The initial water movement is downslope but the effect of the earth's rotation is to deflect the movement through 90° to the right (left in the southern hemisphere) of this direction.

An interesting example of a gradient current occurs in the Bay of Bengal in February. In this month the current circulation is clockwise around the coasts of the Bay, the flow being north-easterly along much of the east coast of India. With the north-east monsoon still blowing, the current here sets against the wind. The explanation of this phenomenon is that the cold wind blowing off the land cools the water at the head of the Bay. A temperature gradient thus arises between the cold water in the north and warm water in the south. Because of the density difference thus created a slope develops, downhill towards the north. The resulting northward flow is deflected to the right, i.e. eastward, and so sets up the general clockwise circulation.

Complex Currents

While it is convenient to distinguish between the two foregoing types of current in terms of their basic causes, the currents encountered in practice are frequently complex in character. We have seen how the effect of the wind (in the northern hemisphere) is to produce a surface flow in a direction up to 45° to the right of the wind direction, also that successive lower layers of water are deflected progressively further to the right. This angle of deflection increases while the speed of motion decreases until the latter becomes negligible by the time the angle of motion has become diametrically opposed to the original wind direction. This implies a bulk transport of the water within the affected layer roughly at right angles to the direction of the wind. This transport leads to an accumulation of water to the right of the direction towards which the wind is blowing. This in turn tends to produce a gradient current flowing downslope from the region of accumulation. The effect of the earth's rotation diverts this through 90° to the right (northern hemisphere) and thus produces a flow in the direction of the original wind. In this way the original wind-drift current is strengthened by a gradient current caused indirectly by the wind.

In many cases a wind-drift current is superimposed upon a gradient current which may be due to temperature and/or salinity differences. The resulting current is then a combination of the two effects.

Near the ocean boundaries further complications arise. For example, over a wide belt in the equatorial regions of the Atlantic, the trade winds produce a west-going current setting towards the Caribbean Sea and Gulf of Mexico. Because of the configuration of the latter there is an accumulation of water in the Gulf so that water levels are higher on the western coast of Florida than on the eastern coast. The resulting eastward flow through the Florida Strait constitutes the Florida Current which, after turning northwards past the Bahamas becomes the Gulf Stream.

Effect of Evaporation

Evaporation may in some cases contribute to current formation. For example, in a relatively shallow sea like the Mediterranean, the rate of evaporation is high, and the inflow of water from rivers is not sufficient to maintain the level of the sea. Water therefore flows in from the Atlantic, through the Strait of Gibraltar, to make good the deficiency. The effect of the earth's rotation is to tend to divert the east-going flow through the Strait to the south. There can be no actual deflection within the Strait because of its narrowness, but further east the inflow is thus deflected against the African coast and so produces a counter-clockwise circulation. By evaporation, the Mediterranean surface water becomes saline and dense. It therefore sinks and the excess of this denser bottom water emerges over the sill forming the shallow Strait of Gibraltar and below the incoming water.

Effect of Wind Blowing over a Coastline

This may be illustrated by reference to the Benguela Current off SW Africa. There the coast is orientated NNW–SSE and the prevailing wind is SE'ly. The direct effect of the wind is to produce a surface current setting roughly towards the west (say 30° to the left of the direction towards which the wind is blowing). However, the indirect effect, taking account of the sub-surface layer, is to cause a transport of water towards the SW and this by establishing a gradient from SW to NE leads to a gradient current setting NW'ly (90° to the left of NE). The resultant of this and the direct current is a roughly WNW'ly current which removes water from the coastal regions. This is made good by the upwelling of water from lower levels. The water thus brought to the surface is cold in comparison with the sea surface temperatures appropriate to these latitudes. The area of upwelling is revealed on maps of sea surface temperature as an area of low temperature surrounded by higher temperatures.

Relation of General Current Circulation to that of Wind

Although there are many local exceptions, there is in general a close relation between the large-scale patterns of prevailing currents in the oceans, and the corresponding patterns of the prevailing winds. Figure 16.1 shows the mean surface current circulation for the Atlantic Ocean. For comparison Figure 16.2 shows the pattern of resultant winds in August. It will be seen that, in both these figures, there is a clockwise circulation or gyre in the North Atlantic and a counter-clockwise gyre in the South Atlantic. In many places the direction of the current is related to that of the wind being most commonly diverted to the right (in the northern hemisphere) of the wind by an angle not exceeding 45°.

In a general way it can be seen that the main features of the wind circulation correspond with those of the current circulation. For example, in the North Atlantic, the NE'ly trade winds correspond with the westward-setting North Equatorial Current, while the variable westerly winds of the more northern latitudes (40°N–50°N) correspond with the E'ly sets of the North Atlantic Current.

Another parallel between the water circulation and the atmospheric circulation may be seen in the existence of lines or zones of demarcation between water masses of differing characteristics, such as those existing between air masses. Temperature, for example, may change little over hundreds of miles and then suddenly, within the space of a few miles, may change rapidly and thereafter continue with little further change. A good example of such a discontinuity is provided by the boundary between the warm Gulf Stream and the cold Labrador Current. This is so well marked that it has been called the 'Cold Wall'. Not only does the water temperature fall abruptly as one crosses this boundary when approaching the American continent, but the two water masses can often be distinguished visually, the dark blue water of the Gulf Stream contrasting with the olive or bottle-green water of the Labrador Current. In the region of the Tail of the Grand Bank of Newfoundland a change in water temperature of 12°C has been recorded in less than a ship's length.

Current Characteristics

Currents vary considerably from one occasion to another. In so far as a large proportion of surface currents are wind-induced, and the winds are known to be highly variable, this might seem to follow naturally. However, recent detailed measurements seem to indicate a higher variability than can be ascribed to the wind. Thus significant current variations occur even when the winds are comparatively constant. Even in the open ocean the water flow may show rapid variations in both speed and direction such as might be expected in the random eddies of a shallow stream.

Some of the stronger currents which are associated with thermal gradients, such as the Gulf Stream, are liable to considerable variation both in position and strength from time to time. These streams are subject to meanders which sometimes develop into eddies.

Whereas, in the open ocean in regions of variable winds, the current may set with varying speeds in any direction, in some other regions the direction of set is comparatively stable. Near extensive and relatively straight coastlines, particularly in the trade-wind belts, the water is constrained to flow almost parallel with the coast. In equatorial regions, also, largely as a result of the converging trade winds, current directions are comparatively stable, with westerly sets predominating over extensive latitudinal belts. In some regions such as the northern Indian Ocean the regular seasonal (monsoon) reversal of

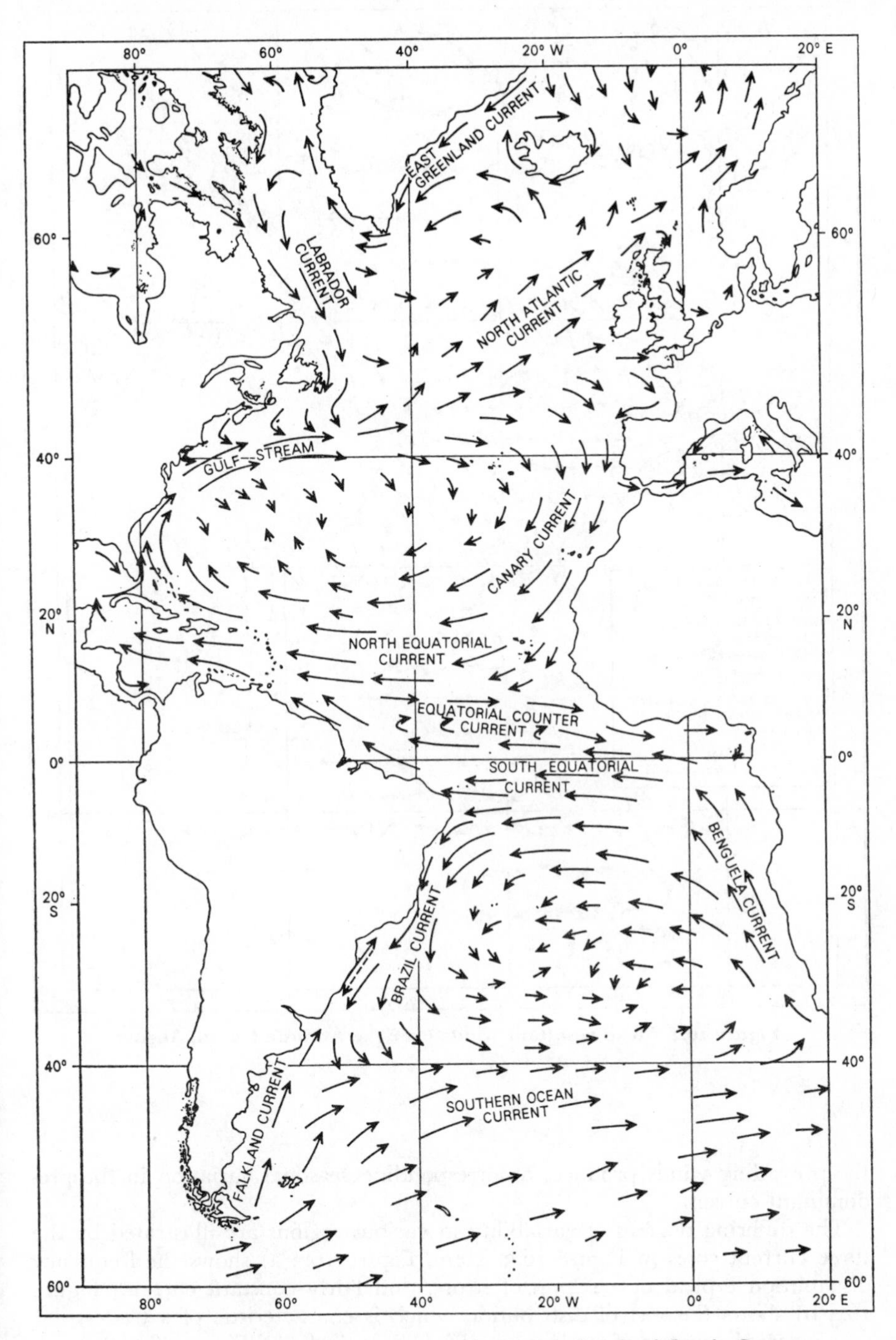

Figure 16.1. Mean surface current circulation for the Atlantic Ocean

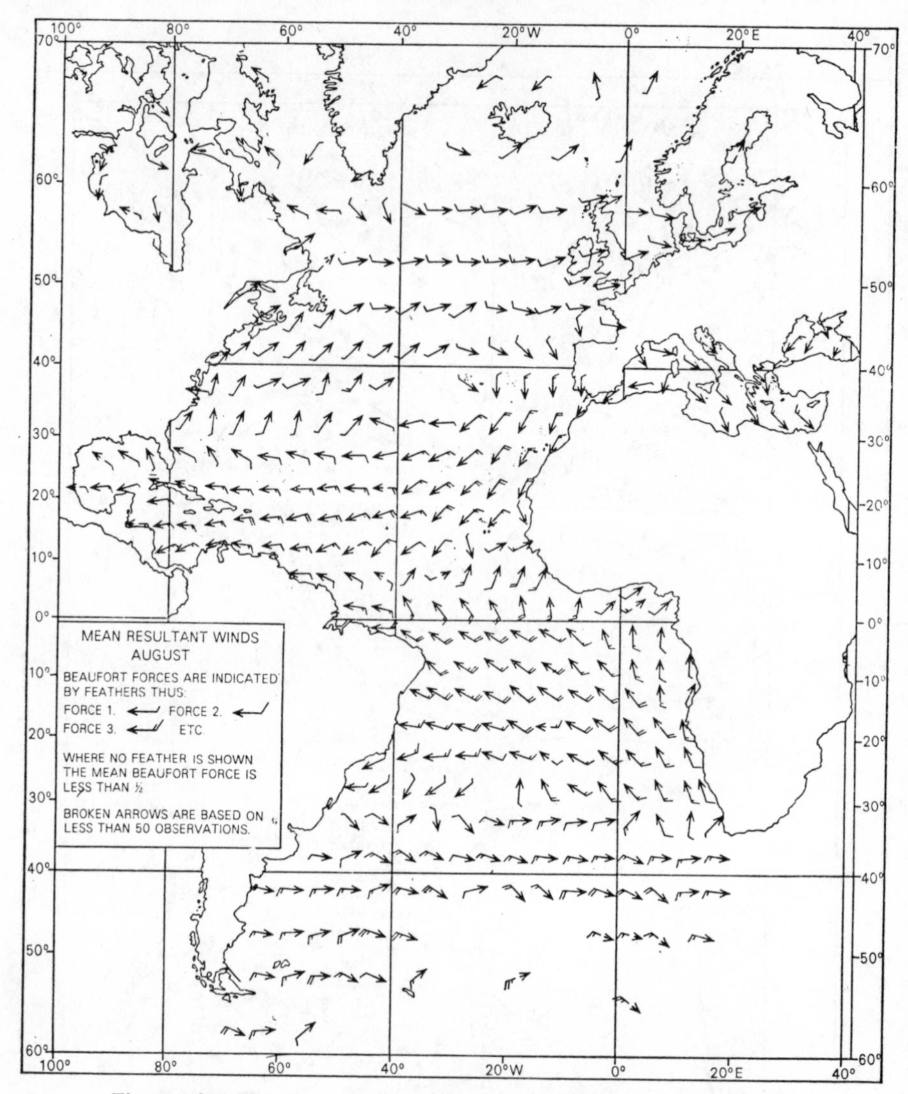

Figure 16.2. Mean resultant winds over the Atlantic Ocean, August

the prevailing winds produces a corresponding seasonal variation in the predominant current.

The differing degrees of variability in various regions are illustrated by the three current roses in Figure 16.3. Here, Figure 16:3(a) shows the frequency distribution typical of a region of strong and fairly constant current, Figure 16.3(b) shows the sort of distribution which is characteristic of a trade-wind area where the degree of predominance is less marked, and Figure 16.3(c) shows that characteristic of a region of variable currents where there is no decided predominance.

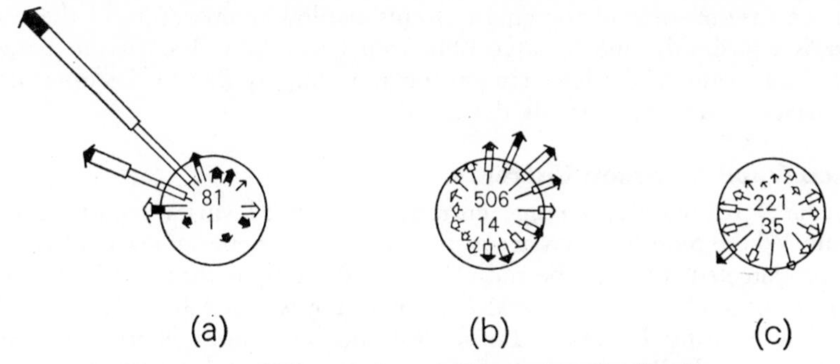

Figure 16.3. Three characteristic current distributions

Strength of Current

Over large areas of the open ocean in the middle and higher latitudes the most frequently reported current speed is around $\frac{1}{2}$ kn, and most of the individual reports vary between nil and $1\frac{1}{2}$ kn. In certain other regions the general level of speed is higher and the extremes correspondingly so. For example, in some equatorial regions (notably in the mid Pacific and western Atlantic) the mean speed is in the range 1–$1\frac{1}{2}$ kn and speeds in excess of 3 kn have been reported. Rather similar values occur in the Gulf Stream but speeds up to $5\frac{1}{2}$ kn have been recorded in Florida Strait. Still higher speeds are reported in the Somali Current during the south-west monsoon. Here from about June through September average speeds are above 2 kn and on one occasion a speed of 7 kn was reported.

Warm and Cold Currents

In general, currents which set continuously eastwards or westwards acquire temperatures appropriate to the latitude concerned. Currents which set northwards or southwards over long distances, however, transport water from higher to lower latitudes, or vice versa, and so advect lower or higher temperatures from the region of origin. The Gulf Stream, for example, transports water from the Gulf of Mexico to the central parts of the North Atlantic where it gives rise to temperatures well above the latitudinal average. Between the Gulf Stream and the American coast the water is much colder since it derives from Arctic regions by way of the Labrador Current. The transition from this cold water to the much warmer water of the Gulf Stream is marked by a very strong gradient of sea surface temperatures. Both here and elsewhere strong temperature gradients indicated by sea temperatures isotherms can be used to detect the boundaries between currents.

Among the principal warm currents may be listed:

Gulf Stream
Kuro Shio
Brazil Current
Moçambique Current
Agulhas Current
East Australian Coast Current

The principal cold currents are:

Labrador Current
East Greenland Current
California Current
Kamchatka Current
Oya Shio
Falkland Current
Peru Current
Benguela Current

In the case of some of the cold currents the low temperature of the surface water is not simply due to advection from lower latitudes. In the Benguela current for example the low temperatures are largely due to the upwelling of sub-surface water as previously described.

Seasonal and Monsoon Currents

In many regions the average current experienced varies according to the seasons. Where monthly data are available many varieties of seasonal variation may be detected. One of the most familiar patterns is that in which there is relatively little change for several months in the winter followed, after a brief transition in spring, by a period of several summer months wherein the current flow differs markedly from that of the winter period. A brief transition period in autumn leads back to the winter conditions.

In some regions three main seasons may be detected, while in others there would seem to be a fairly steady cycle of development throughout the months of the year.

The most striking seasonal variation is that induced by the Asian monsoons. The great annual pressure oscillation which replaces the summer low pressure over southern Asia by the winter high pressure over central Asia produces an almost complete seasonal reversal of the prevailing winds over the northern parts of the Indian Ocean and off the Pacific coasts of Asia. In accord with these wind changes the principal currents in these areas undergo a seasonal reversal. Thus, in the Indian Ocean, in the northern winter, the currents between the equator and latitude 15°N set in a westerly direction in accord with the north-east monsoon. In the northern summer, however, during the south-west monsoon the currents of this region are reversed and set in easterly directions. This reversal is the more notable because the mean speed is comparatively high (about 1–1½ kn) in both directions. Off much of the Pacific coast of Asia, also, there is a complete reversal of current from winter to summer. Thus, off the south-east of Vietnam, the average current is south-westerly (1–1¼ kn) in winter and north-easterly (about 1 kn) in summer.

Annual and Secular Variation

It is highly probable that the average current in a given month and position varies from one year to another. In general, however, the number of observations available each year in a given location, even considering a one-degree square, is insufficient to determine the variation satisfactorily. There may well be variation also on a longer-term scale, for example when comparing one thirty-year period with another. Here again any real variation is likely to be obscured by the paucity of observations, especially in earlier periods. Because of the high variability of currents and the inherent difficulty of measuring them accurately, there are barely sufficient observations in many areas to determine the current regime even using all the observations extending over a period of more than 100 years. In these circumstances it is doubtful if comparisons over shorter periods can be regarded as meaningful.

PART VI. ICE AND EXCHANGE OF ENERGY BETWEEN SEA AND ATMOSPHERE

CHAPTER 17

THE FORMATION AND MOVEMENT OF SEA ICE AND ICEBERGS

Introduction

Ice is found at high latitudes in both hemispheres but because of their physical dissimilarities the climatic and ice regimes of the Arctic and Antarctic regions differ greatly. The Arctic Basin is an area of ocean about 3000 metres deep which is covered by a thin shell of ice about 3·5 metres thick. Antarctica, similar in area, is a continent covered by an ice cap which is up to 3000 metres thick. The annual mean temperature at the South Pole is −49°C (the lowest temperature yet recorded in Antarctica is −88·3°C), whereas at the North Pole the annual mean temperature is estimated to be −20°C (the lowest temperature yet recorded in the Arctic Basin is only a little below −50°C).

The ice cap covering the Antarctic continent accounts for more than 90% of the earth's permanent ice. The ice constituting the ice cap is constantly moving outwards towards the coasts where many thousands of icebergs are calved each year from the glaciers and ice shelves which reach out over the sea. As a consequence, large numbers of icebergs are to be found in a wide belt which completely surrounds the continent. In contrast, the icebergs of the Arctic region are almost entirely confined to the sea areas off the east and west coasts of Greenland and off the eastern seaboard of Canada. The Arctic Basin remains almost completely covered by pack ice throughout the year whereas the greater part of the pack ice surrounding Antarctica melts each summer. This is because the Antarctic pack ice is located in lower latitudes than its Arctic counterpart.

Sea ice is a complex substance varying in shape, size, age, thickness and many other characteristics so that any description must use terms which are readily understood and accepted by the reader. The terms used in the following description conform to the *Sea Ice Nomenclature* published by the World Meteorological Organization in 1970 (incorporating amendments up to 1975). This nomenclature is also published in the *Marine Observer's Handbook* and in *The Mariner's Handbook*, N.P. 100.

Classification of Ice

The many forms of ice which may be encountered at sea are conveniently classified under three main headings. Of these the most important is called SEA ICE, which is formed by the freezing of sea water. The other major categories are ICEBERGS and RIVER ICE, which are derived from fresh water. Icebergs are a serious hazard to navigation around Antarctica and also in some parts of the northern hemisphere, notably the eastern seaboard of Canada. They will be described later. River ice is sometimes encountered in harbours and off shore close to estuaries during the spring breakup. As it is then in a state of decay it generally presents only a temporary hindrance to shipping. The description which now follows is confined to sea ice.

Formation of Ice

Fresh water and salt water do not freeze in the same manner. This is due to the presence of dissolved salts in the sea water. The salt content, or the salinity as it is known, is usually expressed in parts per thousand: sea water, typically, has a salinity of 35 parts per thousand though in some areas, especially where there is a considerable discharge of river water, the salinity is much less. In the Baltic, for example, the salinity of the surface water is less than 10 parts per thousand throughout the year.

When considering the freezing process, the importance of salinity lies not only in its direct effect in lowering the freezing temperature, but also in its effect on the density of the water. The loss of heat from a body of water takes place principally from its surface to the air. As the surface water cools it becomes more dense and sinks, being replaced by warmer, less dense water from below which in turn is cooled, continuing the process of overturning (or convection as it is more commonly known). The maximum density of fresh water occurs at a temperature of 4°C; thus, when a body of fresh water is cooled to this temperature throughout its depth convection ceases, since further cooling results in a slight decrease in density. Once this stable condition has been reached, cooling of the surface water leads to a rapid drop in temperature and ice begins to form when the temperature falls to 0°C.

With salt water the delay, due to convection, in the lowering of the temperature of the water to its freezing point is much more prolonged. In some areas where there is an abundant supply of relatively warm water at depth, for example off south-west Svalbard, convection may normally prevent the formation of ice throughout the entire winter despite very low air temperatures. This delay is, in part, due to the considerable depths of water found in the oceans, but it is mainly due to the fact that the density of salt water continues to increase, with cooling, until the surface water freezes. In fact the theoretical maximum density of sea water of average salinity (which can be achieved by supercooling in controlled laboratory conditions) occurs at a temperature which is well below its freezing point. Figure 17.1 shows the relationship between freezing point, salinity and temperature at maximum density.

It can be seen that in water with salinity of less than 24.7 parts per thousand the maximum density is reached before the freezing temperature and where the salinity is greater than 24·7 parts per thousand the freezing point is reached before the density attains its theoretical maximum value.

The greatest delay in reaching the freezing temperature occurs when the sea water, throughout its depth, is initially at an almost uniform density. In some areas, however, the density profile is not uniform. In these cases, discontinuities occur where a layer of lower salinity overlies a layer of higher salinity. (At temperatures between 3°C and freezing, variations in density are more dependent on variations of salinity than on changes in temperature). The increased density at the surface of the upper layer, achieved by cooling, may still be less than the density of the lower layer. The salinity discontinuity between the two layers then forms a lower limit to convection and the delay in reaching the freezing temperature is then dependent upon the depth of the upper layer. This is particularly so in the Arctic Basin where there is a salinity discontinuity between the surface layer (the Arctic water) and the underlying more saline Atlantic water. Cooling of the surface water around the periphery of the Basin, and within regions of open water, leads to convection in a shallow layer which may extend to only 50 metres in depth.

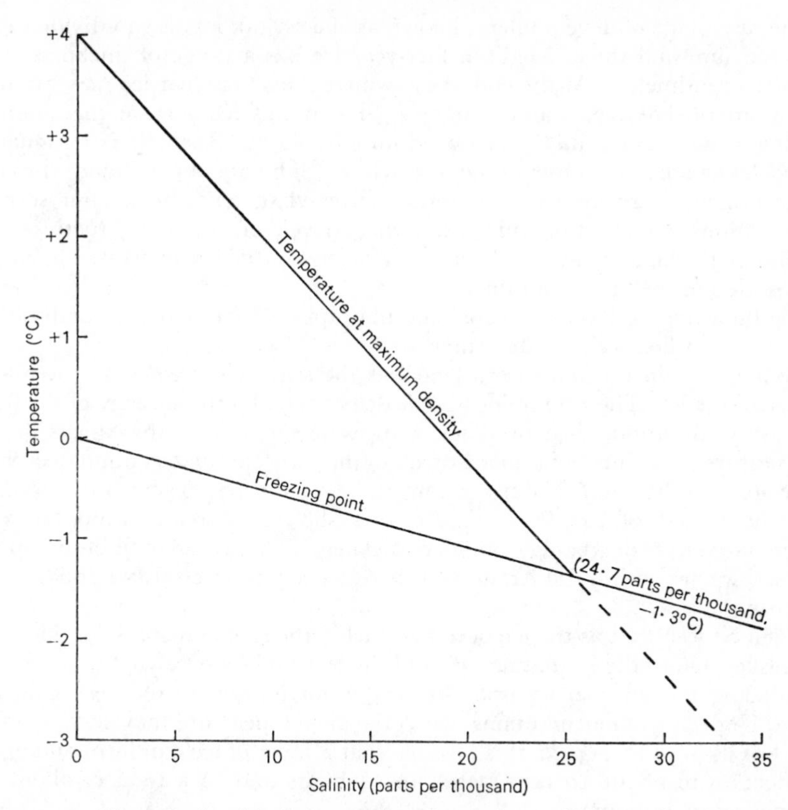

Figure 17.1. The relationship between freezing point, salinity and temperature at maximum density

The first indication of ice is the appearance of ice spicules or plates, with maximum dimensions up to 2·5 centimetres, in the top few centimetres of the water. These spicules, known as FRAZIL ICE, form in large quantities and give the sea an oily appearance. As cooling continues the frazil crystals coalesce to form GREASE ICE which has a matt appearance. Under near-freezing, but as yet ice-free conditions, snow falling on the surface and forming SLUSH may induce the sea surface to form a layer of ice. These forms may break up under the action of wind and waves to form SHUGA. Frazil ice, grease ice, slush and shuga are classified as NEW ICE. With further cooling, sheets of ICE RIND or NILAS are formed, depending on the rate of cooling and on the salinity of the water. Ice rind is formed when water of low salinity freezes slowly, resulting in a thin layer of ice which is almost free of salt, whereas when water of high salinity freezes, expecially if the process is rapid, the ice contains pockets of salt water giving it an elastic property which is characteristic of nilas. This latter form of ice is sub-divided, according to thickness, into dark nilas (less than 5 cm thick) and light nilas (5–10 cm thick).

Once again the action of wind and waves may break up ice rind and nilas into PANCAKE ICE which later freezes together and thickens into GREY ICE and GREY-WHITE ICE, the latter attaining thicknesses up to 30 centimetres. These forms of ice are referred to as YOUNG ICE. Rough weather may break this ice up into ICE CAKES or FLOES.

The next stage of development, known as FIRST-YEAR ICE, is sub-divided into thin, medium and thick. Medium first-year ice has a range of thickness from 70 to 120 centimetres. At the end of the winter thick first-year ice may attain a maximum thickness of about 2 metres. Should this ice survive the summer melting season, as it may well do within the Arctic Basin, it is designated SECOND-YEAR ICE at the onset of the next winter. Subsequent persistence through summer melts warrants the description MULTI-YEAR ICE which, after several years, attains a maximum thickness, where level, of about 3·5 metres; this maximum thickness is attained when the accretion of ice in winter balances the loss due to melting in summer.

The buoyancy of level sea ice is such that approximately one seventh of the total thickness floats above the water.

Ice increases in thickness from below as the sea water freezes on the under surface of the ice. The rate of increase is determined by the severity of the frost and by its duration. The need for a measure combining the effects of air temperature and time has given rise to the concept of 'accumulated frost degree-days'—the total of daily mean air temperatures, below 0°C, summed over the period of the frost. Figure 17.2 shows the relationship between accumulated frost degree-days and ice thickness. (This is based on observations from a few sites within the Arctic and therefore is not necessarily applicable to all areas.)

It will be seen that as the ice becomes thicker the rate of increase in thickness diminishes due to the insulating effect of the ice (and its overlying snow cover) in reducing the upward transport of heat from the sea to the very cold air above. Under extreme conditions, when the air temperature may suddenly fall to as low as −30 to −40°C, it is possible that a layer of ice can form and grow in thickness to about 10 centimetres in 24 hours and to a total of about 18 centimetres in 48 hours.

Two other factors which contribute to the growth of sea ice are more applicable to the Antarctic region, than to the Arctic. The first concerns the effect of snow cover. Where this is relatively deep, say upwards of 50 centimetres, the sheer weight of the snow may depress the original ice layer below sea level so that the snow becomes water-logged. In winter the wet snow gradually freezes, thus increasing the depth of the ice layer.

The other effect is due to the supercooling of water as it flows under the deep ice shelves which are a typical feature of the Antarctic coastline. The supercooled water is prevented from freezing by the pressure at this depth. Observations have shown that the flow of water under the ice shelves is often vigorous, the consequent turbulence resulting in some of the supercooled water rising towards the surface as it leaves the vicinity of the ice shelf. The consequent reduction of pressure may lead to the rapid formation of frazil ice in the near-surface water. The same mechanism can also result in the accumulation of a relatively deep layer of porous ice beneath an original ice layer. In this way recently broken fast ice (*see* page 242 for a definition) over 4 metres thick, encountered on the approaches to Enderby Land in the southern autumn (March) was observed to consist of only 30 centimetres of solid ice and 4 metres of porous ice, the whole layer offering little resistance to forward progress. This effect is almost entirely confined to the fast-ice zone. (For distribution of fast ice *see* Chapter 18.)

At the first stage of its development sea ice is formed of pure water and contains no salt. The downward growth of ice crystals from the under surface

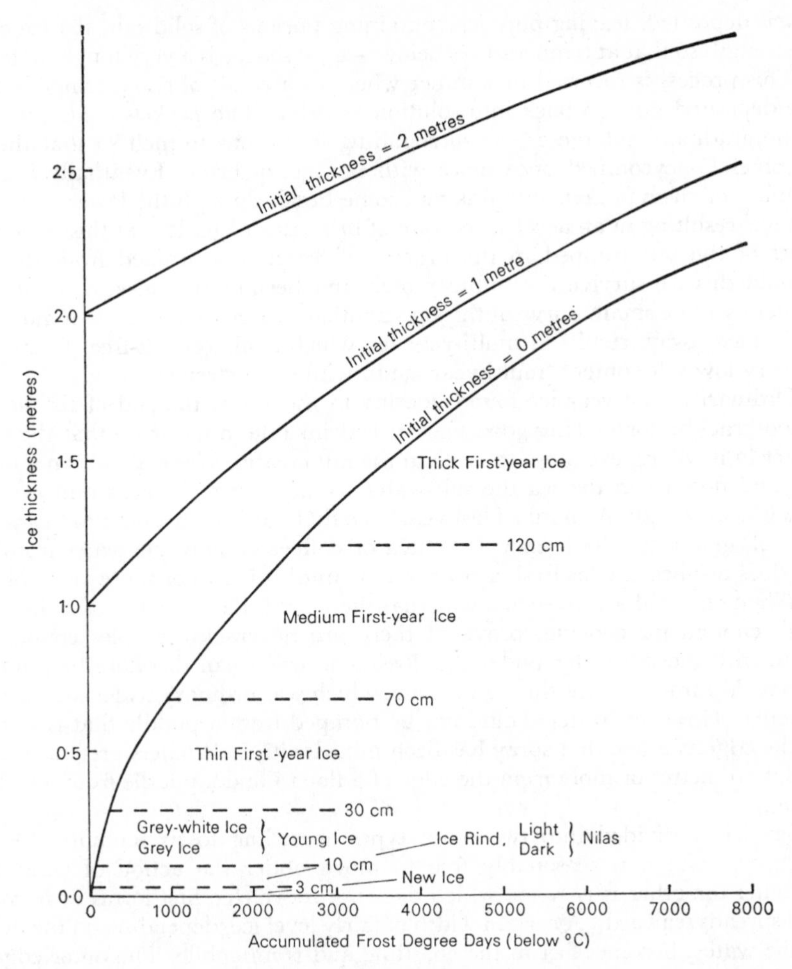

Figure 17.2. The relationship between ice growth and accumulated frost degree-days (below 0°C) for various initial ice thicknesses

of the ice results in a network of crystals and small pockets of sea water. Eventually these pockets become cut off from the underlying water and, with further cooling, they shrink in size as some of the water in these pockets freezes out. The residual solution (brine) now has a higher salt content, its salinity being highly dependent on temperature. Since there exists, at least in winter, a substantial positive temperature gradient downwards through the ice, it follows that the temperature at the top of a pocket of brine is lower than at its base. This leads to freezing at the top of the brine pocket and melting at the base, resulting in a slow downward migration of the brine through the ice. Thus brine is drained from the ice at a very slow rate.

As cooling continues the salt content is gradually crystallized out of solution. There are certain critical temperatures in this process, namely −8°c, the temperature at which the first crystalline deposit (sodium sulphate) occurs, and −23°c, the temperature at which common salt begins to be deposited. As the salt is deposited, leaving pure ice containing pockets of solid salt, the ice gains in strength so that at temperatures below −23°C sea ice is a very tough material.

This process is reversed in summer when, as a result of rising temperatures, the deposited salts go back into solution as brine. The pockets containing the brine gradually enlarge as the surrounding ice begins to melt so that the ice becomes honeycombed once more with pockets of brine. Eventually a great number of these pockets interlink and some break through the lower surface of the ice, resulting in an accelerated rate of brine drainage. It is at this stage that most of the salt trapped in the process of freezing is drained from the ice. Should this ice survive the summer melt and become second-year ice its salt content will be small. Survival through another summer season, when more salt is drained away, results in multi-year ice which is almost salt-free. Because of its very low salt content, multi-year ice in winter is extremely tough.

Ordinarily, first-year ice found floating in the sea at the end of six months is too brackish for making good tea but is drinkable in the sense that the fresh water in it will relieve more thirst than the salt creates. When about ten months old and floating in the sea the salt-water ice has lost most of its milky colour and is nearly fresh. A chunk of last year's ice that has been frozen into this year's ice will give water fresh enough for tea or coffee. Usually the water from sea ice does not become 'as fresh as rain water' until it is two or more years old.

When sea ice thaws in such a way that there are puddles on top of it these are fresh enough for cooking, provided there are no cracks or holes connecting them with the salt water under the floes. The water can therefore be pumped into a ship from the ice through a hose, which was ordinary sealer and whaler practice. However, water should not be pumped from a puddle that is so near to the edge of a floe that spray has been mixed with it. Whalers preferred to go about 10 metres or more from the edge of a floe to find a puddle from which to pump.

Sea ice is divided into two main types according to its mobility. One is PACK ICE, which is reasonably free to move under the action of wind and current; the other is FAST ICE which does not move. Ice first forms near coasts and spreads seawards. A certain width of fairly level ice, depending on the depth of the water, becomes fast to the coastline and is immobile. The outer edge of the fast ice is often located in the vicinity of the 25-metre contour. A reason for this is that well-hummocked and ridged ice may ground in these depths and so form off-shore anchor-points for the new season's ice to become fast. Beyond this ice lies the pack ice which has formed, to a small but important extent, from pieces of ice which have broken off from the fast ice. As these spread seawards they, together with any remaining old ice floes, facilitate the formation of NEW, and later YOUNG ICE in the open sea. This ice, as it thickens, is constantly broken up by wind and waves so that the pack ice consists of ice of all sizes and ages from giant floes of several years' growth to the several forms of new ice whose life may be measured in hours.

Deformation of Ice

Under the action of wind, current and internal stress the pack ice is constantly in motion. Where the ice is subjected to pressure its surface becomes DEFORMED. In new and young ice this may result in RAFTING as an ice sheet overrides its neighbour; in thicker ice it leads to the formation of RIDGES and HUMMOCKS according to the pattern and strength of the convergent forces.

During the process of ridging and hummocking, when large pieces of ice are piled up above the general ice level, vast quantities of ice are forced downward to support the weight of ice in the ridge or hummock. The downward extension of ice below a ridge is known as an ICE KEEL, and that below a hummock is called a BUMMOCK. The total vertical dimensions of these features may reach 55 metres, approximately 10 metres showing above sea level. In shallow water ice floes piled up against the coastline may reach 15 metres above mean sea level.

CRACKS, LEADS and POLYNYAS (open areas within an ice field) may form after pressure within the ice has been relaxed. When these openings occur in winter they rapidly become covered by new and young ice which, given sufficient time, will thicken into first-year ice and cement the older floes together. Normally, however, before the first-year stage is reached, the younger ice is subjected to pressure as the older floes move together, resulting in the deformation features already described.

Off-shore winds drive the pack ice away from the coastline and open up a SHORE LEAD, which is a navigable passage between the main body of the pack ice, and the shore. In some regions where off-shore winds persist through the ice season, localized movement of shipping may be possible for much of the winter. Where there is fast ice against the shore, off-shore winds develop a lead at the boundary, or FLAW as it is known, between the fast ice and the pack ice: this opening is called a FLAW LEAD. In both types of lead, new-ice formation will be considerably impeded or even prevented if the off-shore winds are strong. On most occasions, however, new or young ice forms in the leads and when winds become on-shore the re-frozen lead closes up and the younger ice is completely deformed. For this reason the flaw and the coast, especially when on-shore winds prevail, are usually marked by tortuous ice conditions.

Clearance of Ice

The clearance of ice from a given area in summer may occur in two different ways. The first, applicable to pack ice only, is the direct removal of the ice by wind or current. The second method is by melting *in situ* which may be achieved in several ways. Wind again plays a part in that where the ice is well broken (open ice or lesser concentrations) wave action will cause a considerable amount of melting even if the sea temperature is only a little above the freezing point. Where pack ice is not well broken or where there is fast ice the melting process is dependent on incoming radiation.

In the Arctic in winter the ice becomes covered with snow to a depth of about 30 to 60 centimetres. While this snow cover persists, almost 90% of the incoming radiation is reflected back to space. Eventually, however, the snow begins to melt as air temperatures rise above 0°C in early summer and the resulting fresh water forms puddles on the surface. These puddles now absorb about 60% of the incoming radiation and rapidly warm up, steadily enlarging as they melt the surrounding snow and, later, the ice. Eventually the fresh water runs off or through the ice floe and, where the concentration of the pack ice is high, it will settle between the ice floes and the underlying sea water. At this stage the temperature of the sea water will still be below 0°C so that the fresh water freezes on to the under-surface of the ice, thus temporarily reducing the melting rate. Meanwhile, as the temperature within the ice rises, the ice becomes riddled with brine pockets, as described earlier. It is considerably weakened and offers little resistance to the destructive action of wind and waves. At this stage the fast ice breaks into pack ice and eventually the ice floes, when they reach an advanced state of decay, break into small pieces called BRASH ICE, the last stage before melting is complete. Wind, waves, and rising temperatures combine to clear the ice from areas which are affected by first-year ice. In other areas, mainly within the Arctic Basin, the summer melting probably accounts for a reduction in ice-floe thickness of about 1 metre.

The breakup of fast ice by puddling seems to be limited to the Arctic region; it has not been observed in the Antarctic where snow depths are usually greater (50 cm–1 m). In the latter area, due to the presence of the surrounding turbulent ocean, the fast ice is more often broken up by the action of ocean swell, particularly after the pack ice has been removed by the off-shore winds which prevail in these regions. It is said that another factor of no little importance is the presence of diatoms in the lower layers of the fast ice which, because of their dark colour, absorb solar radiation passing through any snow-free ice, leading to weakening and melting from within.

Movement of Ice

Pack ice moves under the influence of wind and current; fast ice stays immobile. The wind stress on the pack ice causes the floes to move in an approximately downwind direction. The deflecting force due to the earth's rotation (the Coriolis force) causes the floes to deviate to the right of the surface wind direction in the northern hemisphere and to the left in the southern hemisphere. Since the surface wind direction is normally backed relative to the geostrophic wind direction by an amount roughly equal to the deflection (in the opposite sense) due to the Coriolis force, the ice movement due to wind drift can be considered approximately parallel to the isobars.

The speed of movement, due to wind drift, varies not only with the wind speed but also with the concentration of the pack ice and the degree of ridging. In very open pack ice (1/10–3/10 cover) there is much more freedom to respond to the wind than in close pack ice (7/10–8/10) where free space is very limited. Also, if there is considerable ridging, the wind will obviously have more effect in inducing motion than if the ice surface were smooth. The degree of ridging may be expressed in terms of a scale of ten points. The ratio of ice drift to geostrophic wind speed producing the drift is known as the 'wind-drift factor'. Table 17.1 gives approximate values of wind-drift factor for certain concentrations and degrees of ridging.

Table 17.1. The ratio of ice drift to geostrophic wind

Degree of Ridging	Concentration of ice		
	2/10 (Very Open Pack)	5/10 (Open Pack)	8/10 (Close Pack)
0/10	1/240	1/350	1/480
3/10	1/55	1/80	1/140
6/10	1/30	1/41	1/70
>6/10	1/27	1/39	1/63

The total movement of pack ice is the resultant of wind-drift component and current component. As regards the latter, since most of the ice is immersed in the sea it will move at the full current rate except in narrow channels where it may form an ice jam. When the wind blows in the same direction as the current, the latter will run at an increased rate and therefore the ice movement, under these conditions, due to wind and current, may be considerable. This is particularly so in the Greenland Sea and to a lesser extent in the Barents Sea and off Labrador.

Another effect of the wind is that when it blows from the open sea on to the pack ice, it compacts the floes into higher concentrations along the ice edge which now becomes well defined. Conversely, a wind blowing off the ice moves the floes out into the open sea at varying rates, dependent on their size, roughness and age, resulting in a diffuse ice edge.

Movement of Pack Ice within the Arctic Basin

Figure 17.3 displays a schematic representation of surface currents within the Arctic Basin and adjacent Atlantic Ocean in so far as they are known. The main flow of ice occurs across the pole from the region of the East Siberian Sea towards the Greenland Sea. On the Eurasian side of this transpolar stream the pack ice moves under the influence of the counter-clockwise current circulations within the peripheral seas and on the North American side the pack ice drifts in a clockwise direction within the Beaufort Sea current gyre.

The bulk of the ice is carried out of the polar basin by the East Greenland Current although some passes into Baffin Bay through Smith Sound. Ice formed in the East Siberian Sea takes from 3 to 5 years to drift across the polar basin and down to the coast of Greenland. Ice of this age, therefore, becomes pressed and hummocked to a degree unknown in ice formed in lower latitudes.

Movement of Pack Ice in the Antarctic

The action of wind and current around the coast of Antarctica imparts a northerly component to the movement of pack ice so that it is eventually carried into the warmer waters of the Southern Ocean Current where it melts. The pack ice of this hemisphere is therefore mainly first-year ice. It is only in a few areas, such as the Weddell and Bellingshausen Seas, where the action of wind and current carries the ice on to the coasts, that second- or multi-year ice is commonly found.

ICEBERGS

Icebergs are large masses of floating ice derived from floating glacier tongues or from ice shelves. The specific gravity of iceberg ice varies with the amount of imprisoned air and the mean value has not been exactly determined, but it is assumed to be about 0·900 as compared with 0·916 for pure fresh-water ice, i.e. about 9/10 of the volume of an iceberg is submerged. The depth of a berg under water, compared with its height above the water, varies with different types of bergs. Table 17.2 has been derived from actual measurements of bergs south of Newfoundland by the International Ice Patrol.

The colour of bergs is an opaque flat white, with soft hues of green or blue. Many show veins of soil or debris; others have yellowish or brown stains in places, due probably to diatoms. Much air is imprisoned in ice in the form of bubbles permeating its whole structure. The white appearance is caused by surface weathering to a depth of 5 to 50 centimetres or more and also to the effect of the sun's rays, which release innumerable air bubbles.

Icebergs diminish in size in three different ways: by calving, melting, or

Figure 17.3. Surface currents in the Arctic Basin and the adjacent North Atlantic Ocean

erosion. A berg calves when a piece breaks off; this disturbs its equilibrium, so that it may float at a different angle or it may capsize. In cold water melting takes place mainly on the water-line. In warm water a berg melts mainly from below and calves frequently. Erosion is caused by wind and rain.

Table 17.2. Ratio of heights of icebergs above water to their depths under water

Type	Ratio Exposed height/Submerged depth
Blocky, precipitous sides	1:5
Rounded	1:4
'Picturesque' Greenland berg	1:3
Pinnacled and ridged	1:2
Last stages, horned and winged	1:1

(All of these types, except blocky bergs, are classified as 'glacier bergs'.)

Arctic Bergs

In the Arctic the irregular GLACIER BERG of varying shape constitutes the largest class. The height of this berg varies greatly and frequently reaches 70 metres; occasionally this is exceeded and one of about 170 metres above sea level has been measured. These figures refer to the height soon after calving, but the height quickly decreases. The highest berg so far measured south of Newfoundland was about 80 metres high, and the longest about 520 metres. Glacier bergs exceeding 1000 metres in length have been seen further north.

An entirely different form of iceberg is the blocky, flat-topped and precipitous-sided berg which is the nearest counterpart in the Arctic to the great tabular bergs of the Antarctic (*see* below). These bergs may originate either from a large glacier tongue or from an ice shelf. If of the latter origin they are true tabular bergs, but in either case they are tabular in form.

A third class of Arctic bergs is the ICE ISLAND, a name popularly used to describe a rare form of tabular berg found in the Arctic. Ice islands originate by breaking off from ice shelves which are found principally in north Ellesmere Island and north Greenland. They stand about 5 metres out of the water and have a total thickness of about 30 to 50 metres; in contrast, the tabular bergs of the Antarctic commonly stand about 30 metres out of the water and have a total thickness of about 200 metres.

The larger ice islands have hitherto been found only within the Arctic Ocean where they drift with the sea ice at an average rate of from 1 to 3 nautical miles per day. The best known, named T3 or Fletcher's Ice Island, was sighted in 1947 and was occupied intermittently by United States scientific parties for a number of years between 1952 and 1971. In 1962 its horizontal dimensions were estimated to be 3 nautical miles by 6 nautical miles. For most of its known life, T3 was drifting in a clockwise direction in the Beaufort Sea current gyre.

Small ice islands have been sighted in the waters of the islands of the Canadian Arctic and off Greenland where they have been carried out of the Arctic Ocean by wind and current. In addition, tabular bergs, some of which may well be very small ice islands, have been reported in the vicinity of Svalbard and in waters north of the USSR.

Antarctic Bergs

The breaking away of ice from the Antarctic continent takes place on a scale quite unknown in the Arctic, so that vast numbers of bergs are found in the adjacent waters. Bergs are formed by the calving of masses of ice from ice shelves or tongues, or from a glacier face.

Antarctic bergs are of several distinctive forms. The following descriptions should be regarded as covering only those terms which are likely to be of interest to mariners.

TABULAR BERGS. This is the most common form and is the typical berg of the Antarctic, to which there is no exact parallel in the Arctic. These bergs are largely, but not all, derived from ice shelves and show a characteristic horizontal banding. Tabular bergs are flat-topped and rectangular in vertical cross-section, with a peculiar white colour and lustre, as if formed of plaster of Paris, due to their relatively large air content. They may be of great size, larger than any other type of berg found in either of the polar regions. Such bergs, exceeding 1 nautical mile in length, occur in hundreds. Many have been measured up to 20 or 30 nautical miles in length, while bergs of more than twice this length have been reported. The largest berg authentically reported is one about 90 nautical miles long, observed from the whaler *Odd I* on 7 January 1927, about 50 nautical miles north-east of Clarence Island in the South Shetlands. This great tabular berg was about 35 metres high. According to recent data the average height of tabular bergs above the water is about 50 metres.

The number of tabular bergs set free varies in different years or periods of years. There appears to have been an unusual breakup of ice-shelf in the Weddell Sea region during the years 1927–33, when the number and size of the tabular bergs in that region was exceptional. The giant berg described above was one of these. It is probable that the calving of the larger tabular bergs is achieved by sea surface disturbances caused by underwater movements of the earth's crust (i.e. tsunamis) or, indeed, directly by earth tremors originating within Antarctica. Should these be particularly vigorous then the configuration of the ice shelves on long stretches of the coastline may be drastically altered in a matter of hours.

The glacier bergs of the Antarctic are similar to those of the Arctic region.

BLACK AND WHITE BERGS. A unique form of iceberg, called black and white bergs, has been observed north and east of the Weddell Sea. They are of two kinds, which are difficult to tell apart at a distance: (a) morainic, in which the dark portion is black and opaque, containing mud and stones, (b) bottle-green, in which the dark part is of a deep green colour and translucent, mud and stones appearing to be absent. In both kinds the demarcation of the white and dark parts is a clear-cut plane, and the dark portion is invariably smoothly rounded by water action. Such bergs have frequently been mistaken for rocks.

WEATHERED BERGS. This name is given to any berg in an advanced state of disintegration in either hemisphere. Large variations occur. The length of life of a floating berg is determined partly by the time spent in the pack before it emerges into the open sea. Thereafter its period of survival is determined largely by the rapidity of its transport to lower latitudes.

Melting of the underwater surface is a continuous process and this, aided by the mechanical action of the sea, produces caves with intervening spurs near the water-line. This finally leads to the scaling off of a portion of the berg or to a change in its equilibrium, whereby tilting or even complete capsizing may occur, thus presenting new surfaces to the sea and the weather. The presence of crevasses, earth particles or rock debris greatly enhance the processes of melting or evaporation and produce planes of weakness, along which further calving occurs. In grounding, a much crevassed berg may be wrecked. Other bergs, in passing over a shoal, may develop strain cracks, which later accelerate their weathering.

Movement of Icebergs

Since about 9/10 of the volume of an iceberg is submerged, it follows that its movement is chiefly controlled by the water movement or current. However, strong winds may exert a considerable influence on the movement of icebergs, partly by direct action on the berg, and partly through their effect upon the current. Off Newfoundland the chief factor governing the severity of any iceberg season (March to July) is the frequency of north-north-westerly winds along the Baffin Island and Labrador coasts during the months immediately preceding and early in the season. For example, in 1972 north-westerly winds prevailed from late December to May, resulting in 1587 bergs drifting south of the 48th parallel on the Grand Banks of Newfoundland. In 1966, however, strong north-easterly (on-shore) winds occurred in January, February and March, resulting in the grounding and eventual decay on the Labrador Coast of the 1966 iceberg 'crop'.

CHAPTER 18

THE DISTRIBUTION OF SEA ICE AND ICEBERGS

Introduction

Though observations of ice in the polar regions have been collected over several centuries, the data have been fragmentary and consequently insufficient to define with any accuracy the ice limits in either hemisphere. Reports from vessels might determine the position of the ice edge in a particular place and time but such reports were widely scattered and interpolation to fill the gaps between them was, of necessity, largely a matter of guess-work. Even with the introduction of routine aerial reconnaissance in the post-war years in the Arctic, and the dramatic increase in the reports from Antarctica during and since the International Geophysical Year (1957–58), there still remained large areas wherein the data were inadequate to specify the ice conditions.

The introduction of meteorological satellites in near-polar orbits has completely altered the availability of data, at least on the hemispherical scale, so that a much clearer understanding of ice conditions around both polar regions is rapidly unfolding. The sea-ice discussion which follows is based to a very large extent on satellite data. For more detailed information on sea-ice conditions in any limited region the reader is referred to the appropriate *Admiralty Pilot.*

The distribution and seasonal variation of sea ice differ greatly from the one polar region to the other. In the Arctic, sea ice covers the greater part of the Arctic Ocean throughout the year and spreads southwards, sometimes a considerable distance, into the adjacent sea areas in the winter season. The average reduction from winter to summer in the area affected by sea ice is about 25%. In the Antarctic, sea ice occurs in a belt around the continent. Here, in spite of the fact that the area covered by sea ice (in the southern winter) is half as much again as the area covered by the Arctic sea ice in the northern winter, the average reduction from winter to summer in the area affected by sea ice is about 85%. This is partly explained by the fact that the Antarctic sea ice occurs in lower latitudes than its Arctic counterpart.

The long-term variations in mean sea-ice conditions on a hemispherical, or global, scale are extremely difficult to assess at this stage. Severe ice conditions in one region of either hemisphere are usually compensated by light conditions in another. For example, a new extreme maximum pack-ice limit was established in the Barents Sea in late winter 1966. In the same season very light sea-ice conditions prevailed over the Gulf of St Lawrence and off the Labrador coast. Only when many years of satellite data have been accumulated will it be possible to determine the long-term variations in mean ice conditions.

SEA ICE IN THE NORTHERN HEMISPHERE

Figure 18.1 shows the position of the extreme maximum and minimum ice edges during the period 1962–74. The concentrations used to define these edges are, respectively, 1/10 concentration (i.e. the ratio of the area covered by ice to the total area of ice plus water, reckoned over a small area, is 1/10)

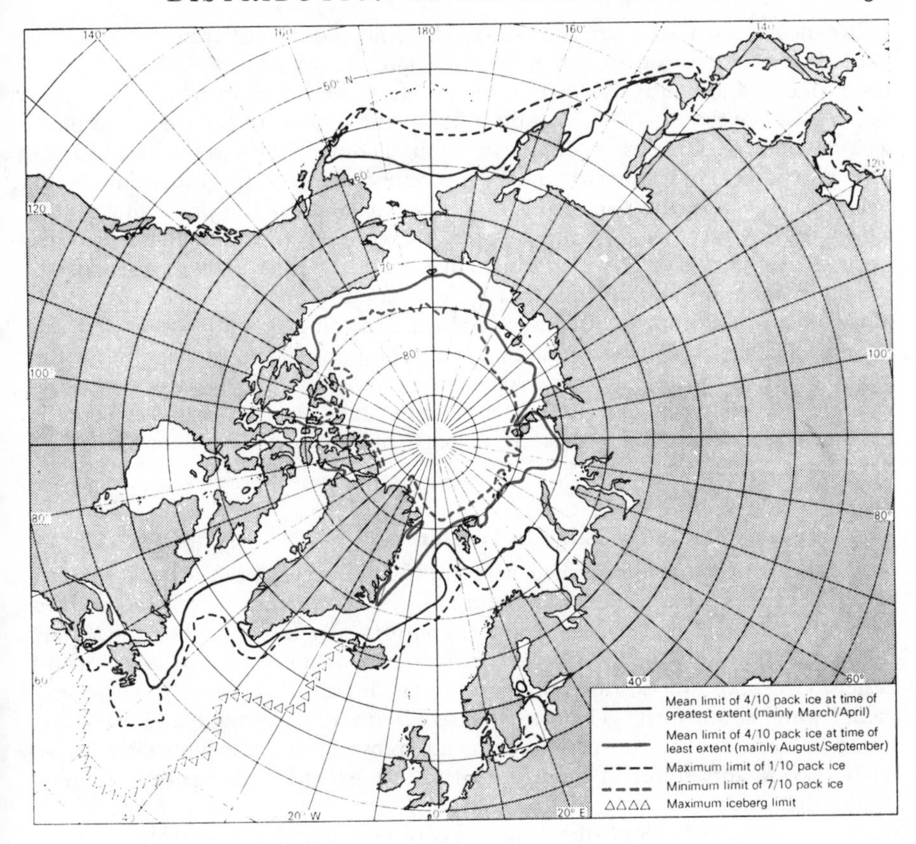

Figure 18.1. Northern hemisphere pack-ice and iceberg limits

and 7/10. The outer of these limits gives an idea of the extent of the region within which there is some risk of encountering sea ice at the worst time of the year, and the inner limit indicates the region within which sea ice can be expected to be a serious hindrance to conventional shipping, even at the most favourable time of the year. Also shown in the figure are the average positions of the ice edge (4/10 concentration) at the times of its greatest (usually about February/March) and least (usually about August/September) extent. Except for the area around Japan and North China, these average positions are based upon the monthly sea-ice charts published by the Meteorological Office and cover the period 1966–74. It will be appreciated that these limits, particularly the extremes, may need to be modified when a longer period of observations becomes available.

Distribution of Pack Ice in the Season when it is Least Extensive

Sea ice within the Arctic region recedes to its least extent in about August or September in an average year. Though pack ice covers the greater part of the Arctic Basin at this time, the peripheral coastlines are mostly ice-free except for the north and east coasts of Greenland, the northern parts of the Canadian Arctic Archipelago and parts of Zemlya Frantsa Iosifa and Severnaya Zemlya.

It can be seen that there are considerable latitudinal differences along this limit. It lies furthest north in the region around Svalbard and Zemlya Frantsa Iosifa due to the relatively warm currents of southerly origin which prevail in this region (*see* Figure 17.3). The limit lies furthest south in the east Greenland sector where pack ice is continuously brought out of the Arctic Basin in the cold East Greenland Current, and in the central parts of the Canadian Arctic Archipelago where the pack ice is driven southwards by the prevailing north-west winds of summer. Despite this latter exception an almost ice-free route exists through the North-West Passage in early September in the average year.

The extreme minimum limit (7/10) lies a considerable distance northwards of the mean minimum (4/10) just described. It should be understood that this extreme is a composite limit based on the minimum ice conditions observed in any area in any year; the displayed extreme limit is unlikely to occur everywhere in any one year. During the period of the data, the only regions where the ice has never cleared the coasts are in parts of the far north of the Canadian Arctic and Greenland.

The Advance of the Ice Edge in Winter

As winter becomes established over the Arctic regions the ice edge moves southwards in all longitudes. The nature of the ice edge advance is determined to a large extent by the ocean current pattern. By late October the Arctic Basin is usually completely covered with sea ice while in Baffin Bay and off eastern Greenland the south-going currents (*see* Figure 17.3) bring progressively colder water southwards. The atmosphere over these latter regions is becoming much colder because the prevailing winds blow from some northerly point. These two factors result in the lowering of the sea surface temperature which eventually leads to the formation of ice. This effect spreads a considerable distance southwards along the east coasts of Greenland and eastern Canada. By contrast, due to the prevailing south-west winds and north-east-setting currents, the ice edge makes much slower southward progress in the Barents Sea.

A somewhat similar situation obtains in the Pacific sector where the ice edge advances further south to the west of longitude 160°E than it does to the east of that meridian (*see* Figure 18.1).

Distribution of Sea Ice in the Season when it is Most Extensive

At the time of its greatest extent which, in most regions and years, occurs in February or March, the Barents Sea ice edge has spread south and east to engulf all the coasts eastwards of about longitude 40°E. Svalbard is completely encircled and the Greenland Sea ice edge has advanced southwards and south-westwards and encloses Jan Mayen. It almost closes the Denmark Strait and ultimately rounds Kap Farvel (*see* Figure 18.1). Further west, the Baffin Bay ice edge has spread southwards along the coasts of Baffin Island and Labrador to reach Newfoundland, though the advance in the east of Baffin Bay normally terminates at about the Arctic Circle. The Canadian Arctic ice edge has spread south to cover the whole of Hudson Bay. Meanwhile, ice has formed over most of the Gulf of St Lawrence, the sea-way and the Great Lakes. In the European sector the northern parts of the Baltic are affected by sea ice at this time as are the northern parts of the Black Sea.

In the Pacific sector, by this time of greatest extent, the ice edge has advanced southwards through the Bering Strait to cover the northern portion of the Bering Sea (*see* Figure 18.1). All but a small south-eastern part of the Okhotsk Sea is ice-covered and ice also affects parts of the Asian mainland coasts southwards to Po Hai. North of about latitude 46°30′N, under average conditions, all the Asian mainland coasts are affected. South of this latitude, ice is largely restricted to the inner parts of the more sheltered bays and inlets, including Vladivostok, and parts of Po Hai as indicated in Figure 18.1. The Gulf of Alaska remains ice-free except within the many inlets where ice is abundant in the inner parts.

The extreme maximum (1/10) lies well to the seaward of the mean maximum (4/10) in most places. In some places, notably off Newfoundland, in parts of the Davis Strait, in the Barents Sea and near the Aleutian Islands the extreme limit (1/10) extends to some 250 to 300 n. mile beyond the average (4/10) limit. In a severe winter some sheltered parts of the eastern coast of USA are affected as far south as Cape Hatteras. The Baltic is another region where the extreme maximum limit lies a considerable distance from the mean maximum (late winter) limit. In a severe season the whole Baltic region, including its approaches, is ice-covered apart from an off-shore area in the southern Baltic Sea (*see* Figure 18.1). In such a season ice affects the southern coasts of Norway, the western coasts of Denmark and the coast of Holland. In fact, in a very severe season ice may form in sheltered locations along the eastern and southern coasts of Britain, along the north and west coasts of France and even in some northern parts of the Mediterranean Sea. The extreme limit shown in Figure 18.1 is a composite derived from the worst conditions observed in any area in any year during the period of the data. As in the case of the extreme minimum limit (7/10), extreme maximum conditions are highly improbable in all areas in the same season.

Distribution of Fast Ice

During the winter some ice becomes firmly attached to the coast. The width of this fast-ice belt is greatly dependent upon the water depths. The outer limit often coincides with the 25-metre contour so that, where the sea bottom shelves steeply, the fast-ice belt is very narrow or sometimes non-existent. Where the water is shallow for a considerable distance off shore the fast-ice belt may be extensive as in the Laptev Sea where it reaches from the continental coast to a position just northwards of the Novosibirskiye Ostrova.

The fast-ice limits are important as the recurring polynyas (*see* Chapter 17), which form there due to off-shore winds, have a considerable effect on the summer breakup.

The Retreat of the Ice Edge in Summer

The erosion of the ice edge by mild southerly winds and the enlargement of recurring polynyas at the edge of the fast ice, together with the process of puddling, result in the retreat of the ice in the summer months. Though this retreat occurs relatively slowly at first the rate of melting is greatly accelerated in the warmer months of July and August. By about September the ice edge in the average year has retreated to the mean minimum position shown in Figure 18.1.

SEA ICE IN THE SOUTHERN HEMISPHERE

Though sea ice extends a considerable distance from the Antarctic coastline in winter to cover a wide belt of the Southern Ocean, most of this ice melts during the following summer, so that many parts of the coastline are accessible to shipping from December to March in an average year.

Figure 18.2 displays the mean limits of sea ice (about 5/10 concentration) at the time of least extent (usually February/March) and greatest extent (usually September/October). It also includes the ice edge for 12 December 1969, as being representative of the 'mid-season' conditions, in order to help the reader to understand the complex nature of the breakup of sea ice which is later described. Of these limits, the mean at the time of least extent, and the actual

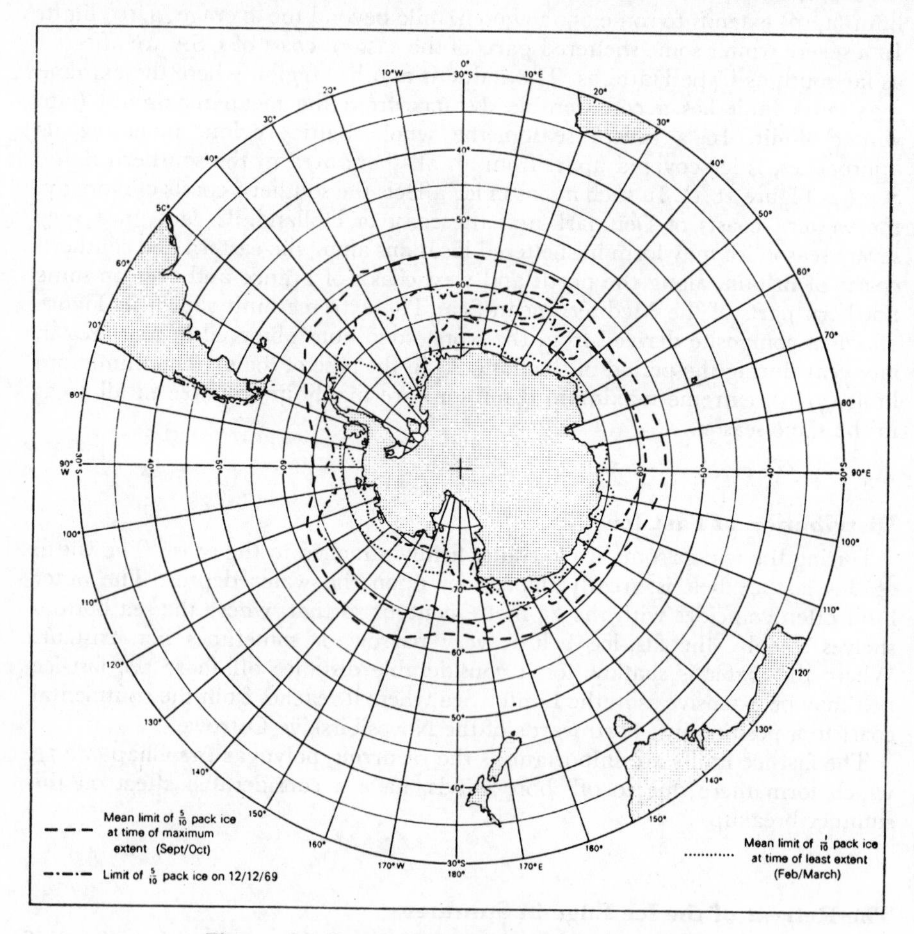

Figure 18.2. Southern hemisphere pack-ice limits

limit for 12 December 1969, are based largely on satellite pictures for four years between 1969 and 1973. The mean at the time of greatest extent is based on a longer period of earlier data. It should be noted that the mean limit at the time of least extent is appreciably different from that shown in previous publications.

The data on which the earlier position was based is known to have been inadequate and the newer limit can be regarded with more confidence as representing the average for the years 1969–73. What must remain in some doubt for the time being is the extent to which this mean can be taken as representing the long-term average. As more satellite data become available a more representative picture of the mean minimum conditions will be revealed; meanwhile the minimum limit shown in Figure 18.2 should be interpreted with caution.

The available data are insufficient to define extreme maximum and minimum limits.

Distribution of Sea Ice at the Time of Least Extent

The mean limit of pack ice (about 5/10 concentration) at the time of least extent (usually February or March) is shown in Figure 18.2. It can be seen that there are two main regions where considerable quantities of sea ice survive the summer melt. One is located in the Weddell Sea where a vast area of pack ice extends eastwards from the Antarctic Peninsula as far east as about longitude 35°W at latitude 74°S, though normally a wide lead occurs on the southern side of this pack-ice zone reaching westwards to about longitude 45°W. The persistence of pack ice in this region is thought to be chiefly due to south-easterly winds causing the Weddell Sea pack ice to pile up against the Antarctic Peninsula. The other main ice region lies to the west of the Antarctic Peninsula between the meridians of 80°W and 175°W. The action of currents and the absence of off-shore winds combine to prevent the clearance of pack ice from this region. Most parts of the coast within the Indian Ocean sector remain affected by sea ice throughout the summer (*see* Figure 18.2). Since the outer limits of the pack ice in this sector approximate to the position of the 300-metre depth contour it may be that the presence of numerous grounded bergs obstructs the clearance of sea ice in this region. In all coastal regions sea ice may pile up on the eastern sides of land-ice tongues owing to the west-going currents which predominate along most parts of the coast.

It is to be expected that ice conditions at this time of least extent will show considerable variability but the limits of this variability are not yet known. In a very light ice year the lead on the southern side of the Weddell Sea may exceed 100 n. mile in width and reach westwards to longitude 60°W, and the Bellingshausen Sea may be almost ice-free westwards to about longitude 90°W. In a bad ice year, at this time of least extent, the Weddell Sea may remain largely ice-covered and parts of its eastward-reaching ice tongue will survive through the summer, perhaps as far east as the Greenwich meridian, while the Bellingshausen Sea remains ice-covered.

The Advance of the Ice Edge in Winter

By mid March new ice has usually formed around the coast, indicating the onset of the new freezing season. The advance of the ice edge seems to occur at a somewhat slower rate than its retreat in summer owing to the longer winter season (March to September). It may well be that the Weddell Sea pack-ice tongue advances across the Greenwich meridian, between latitudes 60°S and 65°S before the region immediately on its southern flank freezes over later in the winter. The advance of the ice edge continues until late September or early October when it reaches its maximum extent.

Distribution of Sea Ice at the Time of Greatest Extent

The Antarctic sea ice usually reaches its maximum extent in late September or early October when its northern limit extends as far north as 54°S in the Atlantic sector, to 56°S in the Indian Ocean sector and to 60°S in the Pacific sector of the Southern Ocean (*see* Figure 18.2). The width of the ice belt surrounding the continent at this time, excluding the Drake Passage ice belt, varies from about 1400 nautical miles in longitude 30°W, to about 500 nautical miles in longitude 90°W. Almost all of this ice is pack ice drifting under the influence of wind and current but a small proportion is fast to the coast.

Distribution of Fast Ice

The fast-ice belt around Antarctica is discontinuous and relatively narrow. Its exact width is difficult to determine since its inner edge is often fast to ice shelves which sometimes extend a considerable distance off shore. The outer limit of the fast ice, however, approximates to the position of the 300-metre depth contour, within which vast numbers of icebergs ground, thus forming anchoring points for the sea ice to become fast. The shape of the fast-ice boundaries is therefore liable to change in detail from time to time as the configuration of the ice shelves is altered by calving and as the distribution of grounded bergs is altered as some bergs eventually drift away.

Polynyas occur in places at the outer limit of the fast ice (or at the ice front, or coast where there is no fast ice). They occur throughout the winter owing to the action of the off-shore winds which prevail around the coasts of Antarctica. These off-shore winds are sometimes interrupted in places by northerly winds associated with the depressions which frequently affect the pack-ice zone. Where they occur, these northerly winds temporarily prevent the formation of polynyas and close those which have already formed. Thus at any one time there is no continuous ring of open water close in to the coast; there are instead a number of unconnected polynyas surrounding the continent. The polynyas which have formed due to the off-shore winds become covered over with new and young ice whenever wind speeds become light. Subsequent strengthening of the off-shore wind reopens the polynyas, resulting in the northward displacement of the pack ice, thus directly contributing to the northward advance of the outer ice edge during the winter.

The Pattern of Breakup of the Sea Ice

The refreezing of the polynyas ceases when warming begins in the spring. Through the action of the off-shore winds and owing to the warming of the open water, the polynyas extend northwards and along the coasts, sometimes joining together to form vast, almost ice-free regions between the fast ice or coast and the off-lying pack ice. Meanwhile melting is occurring at the outer edge of the pack ice as air and sea temperatures rise above freezing point.

The most spectacular example of the Antarctic polynyas is that which forms off the Ross Sea ice barrier. During the 1969/70 season, which may not have been exceptional, satellite pictures revealed that by late November a vast, almost ice-free region existed between the meridians 160°W and 170°E stretching from the ice barrier to 72°S on the 180° meridian and being narrower in the east than in the west. During December the polynya extended northwards between longitudes 175°W and 175°E at the rate of about 8 nautical miles per day and by early January it had broken through to the open waters of the Southern Ocean.

A peculiarity of the breakup of the pack ice in the region east of the Weddell Sea is the melting ice, leading to low concentrations of ice cover, which occurs over a large area centred around 65°S, 10°E in early summer. This melting zone later extends east-north-eastwards and west-south-westwards so that by late December, in the 'average' year, it has reached the open sea in about longitude 30°E and the Weddell Sea coast in about longitude 20°W (*see* Figure 18.2). The line of this melting zone is more or less coincident with the axis of a relatively warm sub-surface current which runs from the south Indian Ocean into the Weddell Sea. It is thought that this melting is due to the upwelling of the warmer sub-surface water which is possibly caused by the divergence of the surface water associated with the depressions which range over the area. The tongue of ice which remains on the northern side of this melting zone (*see* Figure 18.2) is gradually eroded, often becoming detached into ice-fields before eventually disappearing.

A similar effect takes place in the pack-ice region to the east of the Ross Sea. Lanes of broken ice conditions occur in spring and summer within a zone centred roughly along 65°S, 125°W to 75°S, 160°W. At times these lanes may extend over the whole length of this zone while at other times they are discontinuous or even non-existent. In some months the zone may be displaced up to 100 nautical miles north-west or south-east of the line described. In December 1971 the American Coast Guard Cutter *Staten Island* entered the main pack-ice belt in the longitudes of the Amundsen Sea and steamed south-west through a little over 1000 nautical miles of pack ice (mostly less than 5/10 concentration) to arrive 6 days later in the open water of the Ross Sea polynya. The *Staten Island* averaged 7 knots through the pack-ice belt.

Thus the greater part of the pack-ice zone surrounding Antarctica is eroded at its northern and southern boundaries and, in the area east of about 20°W, from within. The greater part of the melting occurs in November and December and then continues at a reduced rate until melting normally ceases in early March, by which time a very large proportion of the area affected by sea ice during the previous winter has become almost ice-free.

ICEBERGS

Distribution of Arctic Icebergs

The icebergs of the Arctic region originate almost entirely from the Greenland ice-cap which contains about 90% of the land ice of the northern hemisphere. Large numbers of icebergs are produced from the east coast glaciers, particularly in the region of Scoresby Sund and are carried south in the East Greenland Current. Most of those surviving this journey drift round Kap Farvel and melt in the Davis Strait, but some follow south or south-east tracks from Kap Farvel, particularly in the winter half of the year, so that the maximum limit of icebergs (occurring in April in this region) lies over 400 nautical miles south-east of Kap Farvel (*see* Figure 18.1). However, a much larger crop of icebergs is derived from the glaciers which terminate in Baffin Bay. It has been estimated that more than 40 000 bergs may be present in Baffin Bay at any one time. By far the greater number of bergs are located close in to the Greenland coast between Disko Bugt and Melville Bugt where most of the major parent glaciers are situated. This is clearly shown in Figure 18.3 which displays the results of an iceberg census taken in Baffin Bay during the autumn of 1970. Some of this vast number of bergs become grounded in the vicinity of their birthplace and never leave their source region; others drift out into the open waters (in summer) of Baffin Bay and steadily decay, but a significant proportion each year is carried by the predominant current pattern in a counter-clockwise direction around the head of Baffin Bay. Of these, some ground in Melville Bugt and along the eastern shores of Baffin Island and there slowly decay. The remainder slowly

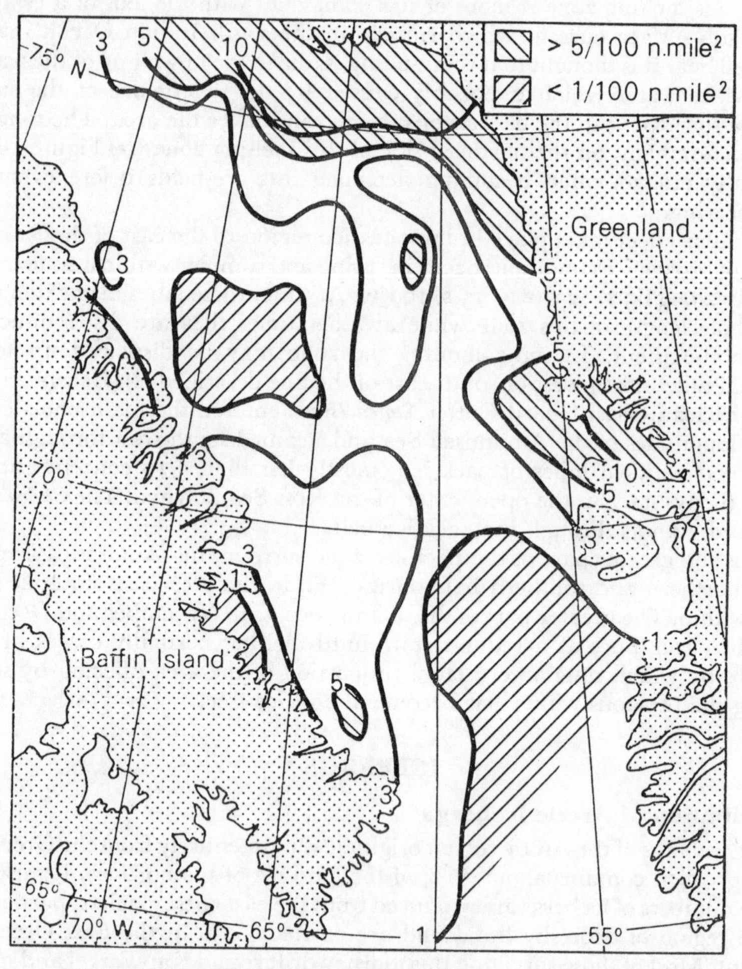

Figure 18.3. Baffin Bay—Iceberg population per 100 n. mile²
(27 September to 5 October 1970)

drift south with the Baffin Land Current and later the Labrador Current, their numbers constantly decreasing by grounding or, in summer, melting in the open sea so that, on average, only a little over 200 icebergs pass south of the 48th parallel on the Grand Banks of Newfoundland each year. Table 18.1 shows the average number of sightings south of 48°N by months, based on data for the period 1946–70.

Table 18.1. Average number of icebergs sighted south of 48°N in vicinity of the Grand Banks

	Sept.	Oct.	Nov.	Dec.	Jan.	Feb.	Mar.	Apr.	May	June	July	Aug.	Year
Average 1946–70	0	0	0	0	0	3	22	74	62	42	9	1	213

To give some indication of the year-to-year variations which can occur, in the iceberg season of 1957 a total of 931 bergs passed south of the 48th parallel, while in 1958 only 1 berg was sighted south of that limit. A reference to a record 1587 sightings in the 1972 season is given on page 249.

The iceberg season on the Grand Banks of Newfoundland begins, on average, in February and ends in July. The mean iceberg limit moves southwards from latitude 48°N in February to its mean furthest south limit at latitude 42°N in May. Most of the Grand Banks icebergs perish on meeting warm Gulf Stream water in the vicinity of the Tail of the Bank; the survivors fan out in directions between west through south to north-east, chiefly under the influence of the meandering Gulf Stream. The maximum iceberg limit, which is defined as the limit beyond which icebergs have been sighted at only very rare intervals, is displayed in Figure 18.1. These very rare sightings sometimes occur at a considerable distance from the Grand Banks area. For example, icebergs or their remnants have been seen in the vicinity of Bermuda, the Azores and the UK.

Little is known about the production of bergs in European and Asiatic longitudes. With the exception of small glaciers in Ostrova De-Longa, it is probable that not a single berg is produced along the north Siberian coast east of Mys Chelyuskina. Severnaya Zemlya probably produces more bergs than Svaldbard or Zemlya Frantsa Iosifa; bergs from its eastern coasts are carried by the current south to Proliv Borisa Vil'kitskogo and down the eastern side of Poluostrov Taymyrskiy. The small bergs typical of Zemlya Frantsa Iosifa and Svalbard which do not reach a height of more than about 15 metres are probably not carried far by the weak currents of this region, though some may enter the East Greenland Current. Svalbard bergs, probably those from the east coast of Nordaustlandet, also drift south-west in the East Svalbard and Bjørnøya Currents and are usually found in small numbers in the Bjørnøya neighbourhood from May to October. The northern half of Novaya Zemlya produces some small bergs.

The fjords and their immediate vicinity on the coasts of the Gulf of Alaska are sometimes affected by locally calved icebergs; this is the only region in the Pacific sector where icebergs sometimes occur.

Distribution of Antarctic Icebergs

It has been estimated that the ice-cap covering the continent of Antarctica contains about seven times as much ice as that which covers Greenland. Since these ice-caps are the source of icebergs it is not surprising, then, that many more icebergs occur in the Antarctic region than in the Arctic. On approaching the continent, isolated icebergs at a considerable distance from its coastline give the first indication of the generally hazardous ice conditions which lie ahead. It is not possible to quote total numbers of icebergs, but it is likely that a ship making a direct approach to the continent of Antarctica may expect to sight upwards of 300 bergs.

Figure 18.4 displays the extreme maximum iceberg limit, based on all data since Cook's circumnavigation in the late 18th century. It also displays the average limit of icebergs of a specified concentration based on data for the summer seasons 1954–60, as indicative of the mean maximum conditions. The specified concentration is such that the bergs have an average separation of 45 nautical miles.

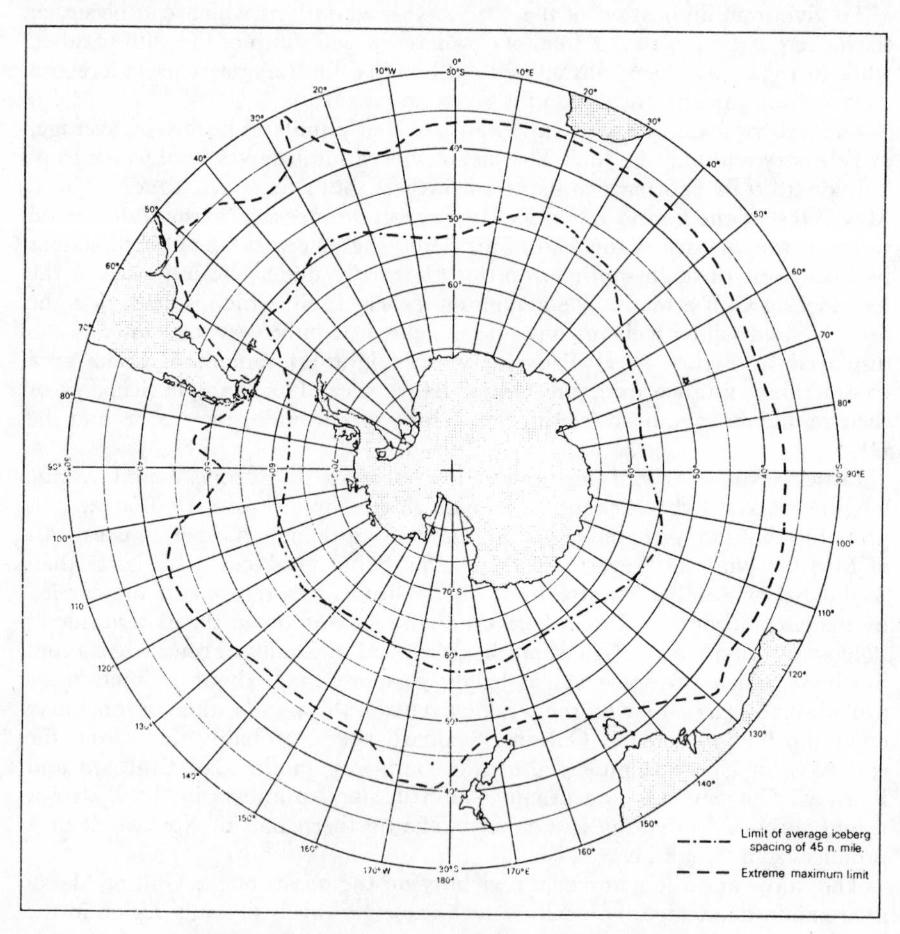

Figure 18.4. Southern hemisphere—Iceberg limits

The largest concentrations of icebergs are to be found close in to their sources—the ice-shelves, ice-tongues and glaciers which are a frequent feature of the Antarctic coastline where, owing to their great draught, many bergs will be grounded in depths less than about 300 metres. Beyond these depths there is a sharp decrease in iceberg concentration though still many thousands of icebergs will be found floating in deeper water. For some considerable distance off shore there is little melting of icebergs since the sea temperature remains at, or just below, 0°c. Their concentration, however, decreases as the distance between the meridians increases. The main melting occurs nearer to the 'mean maximum' limit shown in Figure 18.4 than to the coastline.

The prevailing current pattern of the region carries the bergs in a west to west-north-west direction away from their source region until the latitude of about the Antarctic Circle where their tracks become more and more northerly. North of about latitude 63°s the icebergs move in a more east to east-north-easterly direction under the influence of the east-setting Southern Ocean Current.

The extreme maximum limit lies furthest north in the central South Atlantic where an iceberg has been sighted in latitude 26°s. It encloses the extreme southern coast of South America, almost grazes the Cape of Good Hope and lies close to the south coasts of Australia and New Zealand. The 'mean maximum limit' lies between latitudes 46° to 53°s in the Atlantic sector (although it extends to latitude 60°s in the Drake Passage), latitude 46° to 59°s in the Indian Ocean sector and between latitudes 57° and 63°s in the Pacific sector. In fact both iceberg limits and pack-ice limits lie farther south in the Pacific sector than elsewhere. Though oceanographical and meteorological factors undoubtedly influence the positions of these limits, this spatial variation is in part due to the fact that the cold continental land mass is not evenly distributed about the South Pole; the coastline, in general, lies much further south in the Pacific than in the Atlantic and Indian Ocean sectors.

Little is known of the seasonal and annual variations in the positions of the 'mean maximum' limit. Though the calving of the larger tabular bergs is probably related to earthquakes and tsunamis, (*see* page 248) other bergs are probably calved at a more or less uniform rate throughout the year. Owing to the vast numbers of bergs which are to be found around the Antarctic coastline and to the fact that they probably survive for several years before disintegration in lower latitudes, it is thought that there is little seasonal variation in the position of the 'mean maximum' iceberg limit. It is more likely that there are significant variations in the position of this limit from one year to another as is suggested by the relatively great distances between the 'mean maximum' and extreme maximum limits (*see* Figure 18.4).

CHAPTER 19

SOME INTERRELATIONSHIPS BETWEEN OCEANOGRAPHY AND METEOROLOGY

General Remarks

Oceanography is the study of the oceans and meteorology is the study of the atmosphere. It is obvious that the two studies are intimately linked—one has only to think of such a phenomenon as sea waves to realize this, but the degree and complexity of the linkage is not always appreciated. Mariners are advised to study textbooks on oceanography not only because of its direct importance to navigation but also because an understanding of the physical processes occurring within the oceans is necessary for a proper appreciation of the processes occurring in the atmosphere. Some aspects of oceanography, for example sea surface temperature, are so vital to the study of meteorology that they are commonly discussed at length in meteorological writings. A fuller understanding of such matters, however, is obtained by studying them in the context of oceanography. In the case of sea temperature, not only must one consider heat exchange between ocean and atmosphere, but also that between the surface layer and the lower layers within the ocean. This involves a study of the thermal structure of the ocean in the vertical as well as the horizontal, and the movements occurring within the water mass.

In meteorological processes the prime mover is solar radiation. As the atmosphere is largely transparent to the incoming short-wave radiation from the sun, the greater part of this radiation passes through the atmosphere and is absorbed, in the form of heat (both latent and sensible) at the earth's surface. The sensible and latent heat passes to the atmosphere and so sets in motion most of the meteorological processes. These, therefore, although deriving their energy from the sun, do so largely by way of the earth's surface. Since about seven-tenths of this surface is water, the importance of the oceans becomes apparent.

The part played by the oceans in the global heat budget is, however, far in excess of what might be expected from the simple ratio of sea surface to land surface. Because of a number of properties which are peculiar to water and differ markedly from the corresponding properties of the materials which constitute the land surface, the sea plays a disproportionately large part in controlling the heat exchange between the earth's surface and the atmosphere. Firstly, water has a large heat capacity. It requires about five times as much heat to produce a given rise of temperature in a specified mass of water as it does to produce the same rise in the same mass of the land surface. Incidentally, the amount of heat required to raise the temperature of a given volume of water is about 5000 times as much as that required to produce the same temperature change in the same volume of air. Secondly, water allows radiation to pass through it and so to warm the water at depths within the penetration region. Thirdly, the mobility of water in the vertical plane means that heat can be exchanged between the surface and lower levels. Fourthly, the mobility of water in the horizontal plane means that vast quantities of heat received and stored in one region, for example the tropics, can be transported to higher latitudes by the agency of ocean currents and can there serve to heat the atmosphere long after the original water heating took place.

In contrast, the land has a relatively low heat capacity. Its thermal conductivity is also low and it is neither mobile nor permeable to radiation. This means that solar radiation affects only a thin surface layer of the land in comparison with the much greater depth of the affected layer of the sea. The land surface accordingly rises to higher temperatures by day than does the sea surface, and falls to lower temperatures by night. Since the quantity of radiation from a surface varies as the fourth power of the absolute temperature of the surface, considerably more energy is re-radiated from the land surface, than from the sea surface, so that the sea retains a greater amount of heat than does the land.

Because of the special properties of water, the oceans serve a dual purpose, namely to accumulate solar energy and to redistribute it, in the form of heat, in time and space.

Owing to the greater heat capacity of water, the extreme range of sea surface temperature is only about one-third of that of the air at a few metres above the surface. This range is from about −2°C in polar seas to about 35°C in regions such as the Persian Gulf, a difference of about 37°. The extreme range of air temperature is from about −68°C to about 58°C, a difference of about 126°.

H. Petterson has compared the oceans to a kind of 'savings bank' for solar energy, in that they receive deposits during excessive insolation and pay them back in seasons of want.

The Energy Exchange between Sea and Atmosphere

In Chapter 1 it was stated that the mean temperature of the earth's surface varies little from year to year, implying that a balance must exist between the magnitude of the incoming and outgoing streams of radiation over the whole earth. An understanding of the factors involved in reaching this balance is needed in order to appreciate the problems of the energy exchange between the atmosphere and the sea.

All the incoming solar radiation arrives as short-wave radiation of which about 35% is reflected back to space and lost. (This figure expresses the loss as an average over the whole surface of the earth and is known as the earth's 'albedo'.) Of the remaining 65%, about a quarter is absorbed by the earth's atmosphere and the remainder is absorbed by the earth's surface. Of the latter, some is returned to space in the form of long-wave radiation and some, in the form of heat, establishes a meridional circulation in the atmosphere which helps to reduce the excess of heat which would otherwise accumulate in equatorial regions, by transferring some of it to higher latitudes where there is a heat deficit.

Although heat passes from the sea to the atmosphere by way of conduction and convection, a large proportion of the atmospheric warming occurs as a result of the liberation of latent heat when moisture derived from evaporation of the sea surface condenses in the atmosphere. This proportion has been claimed to be around 50%.

The fact that the amount of evaporation from the sea surface depends upon the wind speed is further evidence of the intimate interdependence which exists between atmospheric and oceanographic processes.

The rate of evaporation from the sea surface depends not only upon the wind speed at the locality in question but also upon the difference between the humidity of the air at the sea surface and the humidity of the air measured, for example, at the level of the ship's bridge. This difference becomes largest when cold and dry air lies over a relatively warm sea surface.

Accordingly, the ocean regions in the northern hemisphere which are subjected to the most intense evaporation lie a short distance to the east of the American and Asian continents between latitudes 30°N and 40°N and not, as might have been expected, in the tropics. The reason is that powerful outbreaks of very cold continental air frequently overrun the ocean off Japan and eastern North America in the regions of the Kuro Shio and the Gulf Stream. In winter these are also the regions where the greatest total energy exchange (i.e. the sum of the energy provided by evaporation and convection) is occurring, while in summer the regions where the total energy exchange is at a maximum are found in the tropical Atlantic and Pacific. The winter regions of greatest total energy exchange between sea and atmosphere coincide with the area where the largest proportion of temperate latitude depressions are initiated, while the summer regions broadly coincide with those where most tropical revolving storms begin their existence.

Effect of the Sea upon the Atmosphere

The effects may be summarized as follows:

(*a*) Direct heating or cooling of the lowest layers of the atmosphere according to whether the sea is warmer or cooler than the overlying air.

(*b*) The addition or removal of water vapour to or from the atmosphere by evaporation or condensation at the sea surface. Heat is extracted from the sea surface during evaporation and later supplied to the atmosphere at a higher level when condensation occurs.

The processes referred to in (*a*) and (*b*) above form the basis of most meteorological activity. Chapter 9 described how the dramatic manifestations of the common depression are powered by the contrasting properties of the participating air masses. These air masses are created by heat and moisture exchange with the oceans over protracted periods during which the air acquires temperature and humidity appropriate to a particular region. Later, the contrasting air masses are brought together so that their potential energy can be converted into the dynamic energy of a cyclonic depression. This bringing together depends upon the action of winds whose existence can be traced back to differential heating and cooling of the atmosphere in different regions.

A further effect of the sea upon the atmosphere becomes apparent when we consider the movement of an air mass over an ocean wherein the surface water temperature is progressively increasing. In the North Atlantic, for example, it is common for cold air, originating over the polar seas or the Canadian Arctic Archipelago, to be brought by a north-westerly gradient over Newfoundland and then towards the Azores. After passing over the Gulf Stream this cold air becomes rapidly heated in its lower layers, and progressively more so as it moves to lower latitudes. Ultimately it will become convectively unstable (*see* Chapter 1) and as a result, heat and humidity originally supplied to the lower layers only, become distributed throughout the troposphere. Thus the sea can determine the degree of stability or instability of an air mass.

Turning to item (*b*) above, apart from the heating effect on the atmosphere due to the liberation of latent heat released when water vapour derived from the sea condenses, another important effect of the oceans consists of providing the water supply (via evaporation) for the formation of the clouds which furnish the world's rainfall.

Largely because of the ocean currents, the geographical distribution of mean sea surface temperature departs considerably from the simple latitudinal distribution wherein temperature decreases with increasing latitude. Because large areas of ocean have temperatures appreciably above or below the latitudinal mean, the climate of places lying downwind from such areas is correspondingly modified. An obvious example is the climate of north-west Europe which is so much warmer and more equable than that of eastern Canada in the same latitude. This is largely due to the effect of the warm water supplied to the Atlantic in these latitudes by the Gulf Stream.

Apart from such climatic variations from one region to another, temporary changes in sea temperature distribution over a period of a few days may cause corresponding variations in the weather in a given place.

The Effect of the Atmosphere upon the Sea

One of the most obvious of these effects is the waves on the sea surface produced by wind. (Waves are discussed in Chapter 3.) When one considers that the winds can be traced back to temperature differences in the atmosphere caused by corresponding differences in the underlying surface temperatures, the intricacy of the interdependence between the atmosphere and the sea can be appreciated. Apart from inducing this small-scale wave motion, the winds are also responsible for some of the large-scale drifts of water from one region to another which constitute ocean currents. These are described in Chapter 16. Consideration of currents also indicates the close interrelation between meteorological and oceanographic processes. This is because not only do temperature differences in the water lead directly to water movement, but also these temperature differences are communicated to the air and there give rise to winds which affect the sea surface and in turn give rise to current. For this reason most ocean currents must be regarded as being complex in origin.

Besides the above mechanical effects of the atmosphere upon the sea, the sea surface is heated in some regions and cooled in others by the effect of warm or cold winds. In the winter months cold winds blowing from the continents over the relatively warm waters of the oceans in middle latitudes cause considerable cooling of the water surface, partly by direct heat transfer, and partly as a result of the loss of heat by evaporation. In summer, in regions where the winds blow over the sea from the heated land, the surface water may be heated by the air in cases where the direct heat transfer is not outweighed by the cooling produced by evaporation, but this effect is of much less significance in terms of heat transfer because, with the air being cooled from below, the conditions in the air are thermally stable (*see* Chapter 1) so that only a thin layer of air is affected and this is soon cooled to a temperature near to that of the sea surface.

Another, less direct, effect of the winds is to give rise locally to the phenomenon of 'upwelling' (described in Chapter 16). By this means water is caused to rise to the surface from lower levels and by bringing with it temperatures appropriate to those levels, effectively reduces the sea surface temperature in the area.

An effect which is sometimes overlooked is that of atmospheric pressure on the level of the sea surface. Very intense atmospheric depressions produce a lifting of the water level in the central region relative to that in the outer regions where the air pressure is higher. The increased water level moves with the depression and produces what is known as a 'storm surge' which on occasions can cause serious flooding. Flooding due to local increase in sea level is also produced when storm winds blow on to a coastline, particularly if this is so shaped as to present a concave profile relative to the wind direction. Then the water displaced by the action of the wind has no ready escape and a low-lying coastline can be inundated (*see* Chapter 11—Bay of Bengal).

The Difference between Air and Sea Temperature

The average temperature of the surface sea water is slightly greater than that of the overlying air. In the tropics the mean value of this excess is about 0·8°C. In middle latitudes there are large seasonal and regional differences. In the western North Atlantic and North Pacific the surface water in winter is, as a rule, warmer than the air by about 3 to 4°C. In the eastern parts of these oceans the difference is small. In spring and summer the sea surface is colder than the air in many regions, and on the Grand Banks of Newfoundland the difference may amount to 1 or 2°C.

The air at the sea surface quickly adopts a temperature near to that of the sea; thus the air temperature over the ocean will show on the average much the same distribution as the sea surface temperature, but comparatively slight differences between water and air are of great importance to the atmospheric processes, and are particularly significant for the purposes of weather forecasting. If the sea is warmer than the air instability occurs, whereby heat and moisture are readily propagated upwards into the atmosphere. The general slight excess of sea surface over air temperature therefore means that in most regions and during most of the year the sea is energizing atmospheric processes by giving up heat and water vapour to the air. The bulk of the heat transfer, however, takes place in winter over rather well-defined areas as described above.

We have seen that the opposite effect of cooling of the air by the sea has little quantitative effect upon the atmosphere. Accordingly the main influence of the sea upon climate is to produce a levelling upwards, i.e. the effect is to produce a higher average temperature (as well as a smaller range of variation) and a greater level of precipitation.

Atmospheric conditions tend to be uniform where there are only slight differences between the air and sea temperatures. These differences are liable to become greatest in regions of rapid oceanographical changes. Such a region occurs along the northern border of the Gulf Stream. Here because of the large gradient of sea surface temperature a comparatively small horizontal displacement of air can rapidly change the air/sea temperature difference and this, no doubt, is linked with the high incidence of cyclogenesis in this region. A similar area of high cyclonic activity occurs in the western North Pacific where cold continental air overruns the warm Kuro Shio.

The moderating effect of the Oceans

Having discussed the close interdependence which exists between the oceans and the atmosphere it remains to consider the overall effect of this linkage upon the climate. The main effect is a moderating one in the sense that it tends to suppress the development of extreme conditions. Long-term (i.e. multi-year) averages of the main climatic elements show a considerable degree of constancy. This is because the supply of energy from the sun is comparatively constant and the ocean/atmosphere linkage operates in such a way as to minimize departures from average conditions.

As an example, let us assume that an anomaly develops in the form of abnormally low cloud amounts in low latitudes. This will lead to increased sea temperatures in these latitudes. This will result in increased air temperatures and so to an increased latitudinal air temperature gradient which will cause increased winds. These will lead to increased evaporation from the sea surface and so to a reduction in sea surface temperature through the extraction of latent heat. In addition the increased evaporation leads to increased cloud which reduces the amount of solar energy reaching the sea surface and so further reduces the sea temperature.

If, on the other hand, one starts with excessive cloud amounts, these will reduce the incoming solar radiation reaching the sea surface and so reduce both sea surface temperature and evaporation leading in time to reduced cloud amounts. In this way the oceans act as a brake and prevent the development of extreme conditions.

Bibliography

Among the many sources studied in the course of preparing this edition, the following were found to be of special interest and are recommended for further reading:

ATKINSON, Major G. D. 1971 Forecaster's Guide to Tropical Meteorology. US Air Force, Air Weather Service, Technical Report No. 240.

CALLAGHAN, W. G. 1975 Ocean waves. Dublin, Meteorological Service, Technical Report No. 39.

COLE, F. W. 1975 Introduction to Meteorology. New York (Wiley).

DARBYSHIRE, M. and DRAPER, L. 1963 Forecasting wind-generated sea waves. *Engineering*, **195**, pp. 482–484.

DRAPER, L. 1971 Waves how high? An analysis of sea conditions. *Motor Boat and Yachting*, **114**, pp. 49–56.

DONN, W. L. 1975 Meteorology. New York (McGraw-Hill).

GENTRY, R. C. 1973 Origin, structure and effects of tropical cyclones. Regional Tropical Cyclone Seminar, Brisbane, May, pp. 53–67.

KOREVAAR, C. G. 1974 Methods employed in wave analysis. World Meteorological Organization Regional Training Seminar, Rome, 1–12 April, pp. 73–91.

NEUMANN, G. 1968 Ocean currents. Amsterdam (Elsevier).

WOLF-HODECK, F. 1972 Windabschätzungen aus Satellitenfotos. Abhandlungen, **127**.

INDEX

Printed in the United Kingdom by HMSO at Edinburgh Press
Dd 294904 C10 5/92 19593 3390/2 (200381)